Nonlinear Evolution and Difference Equations of Monotone Type in Hilbert Spaces

Behzad Djafari-Rouhani
Department of Mathematical Sciences
University of Texas at El Paso, Texas, USA

Hadi Khatibzadeh
Department of Mathematics
University of Zanjan, Zanjan, Iran

CRC Press
Taylor & Francis Group
Boca Raton London New York

CRC Press is an imprint of the
Taylor & Francis Group, an **informa** business
A SCIENCE PUBLISHERS BOOK

CRC Press
Taylor & Francis Group
6000 Broken Sound Parkway NW, Suite 300
Boca Raton, FL 33487-2742

Printed on acid-free paper
Version Date: 20190114

International Standard Book Number-13: 978-1-4822-2818-2 (Hardback)

Library of Congress Cataloging-in-Publication Data

Names: Djafari-Rouhani, Behzad, author. | Khatibzadeh, Hadi, author.
Title: Nonlinear evolution and difference equations of monotone type in Hilbert spaces / Behzad Djafari-Rouhani (Department of Mathematical Sciences, University of Texas at El Paso, Texas, USA), Hadi Khatibzadeh (Department of Mathematics, Zanjan University, Zanjan, Iran).
Description: Boca Raton, FL : CRC Press, 2019. | "A science publishers book." | Includes bibliographical references and index.
Identifiers: LCCN 2018060366 | ISBN 9781482228182 (hardback)
Subjects: LCSH: Evolution equations, Nonlinear. | Differential equations, Nonlinear. | Differential equations. | Hilbert space.
Classification: LCC QA377.3 .D53 2019 | DDC 51/.35--dc23
LC record available at https://lccn.loc.gov/2018060366

Visit the Taylor & Francis Web site at
http://www.taylorandfrancis.com

and the CRC Press Web site at
http://www.crcpress.com

Preface

This book deals with first and second order evolution equations, as well as difference equations of monotone type. There are already many comprehensive books and papers on the subject. Just to mention a few, the books by V. Barbu, H. Brézis, G. Morosanu and N. Apreutesei. We apologize in advance for any contributing authors missing in our references. The novelty of our book is the approach taken of emphasizing the qualitative and long time behavior of the solutions. This approach was first introduced by the first author in his Ph.D. thesis and his subsequent papers, and later continued and extended with the second author in his Ph.D. thesis and subsequent papers to second order evolution equations and difference equations. Namely, we show that in most cases, the existence of solutions is actually equivalent to the zero set of the maximal monotone operator to be nonempty, and give a characterization of the limit as time goes to infinity, of the solution or its averages when either one exists. Most of the results related to this approach appear in book form for the first time.

We have tried to make the book self-contained as far as possible, so that in addition to researchers in the field, it can be also fruitfully used by advanced undergraduate and graduate students.

The first author would like to dedicate this book to his beloved teacher and mentor, the late Professor Shizuo Kakutani of Yale University.

Behzad Djafari-Rouhani
Hadi Khatibzadeh

Contents

Part I

Preliminaries

1 Preliminaries of Functional Analysis

In this chapter, we review the preliminaries and prerequisites that are needed in the sequel. Most proofs are omitted, and the reader is referred to the main references for this chapter that are [ADA, BRE2, DEB-MIK, YOS].

1.1 INTRODUCTION TO HILBERT SPACES

Unless otherwise stated, all vector spaces considered in this book are over the field of real numbers.

Definition 1.1.1 *Let X be a real vector space. A function $\|\cdot\| : X \to \mathbb{R}$ is called a norm on X if the following properties hold:*
1) $\|x\| \geq 0,\ \forall x \in X$ *and* $\|x\| = 0$ *iff* $x = 0$.
2) $\|\alpha x\| = |\alpha|\|x\|,\ \forall \alpha \in \mathbb{R},\ \forall x \in X.$
3) $\|x+y\| \leq \|x\| + \|y\|$ *(triangle inequality).*

A vector space X with a norm $\|\cdot\|$ is called a normed linear space. The normed linear space $(X, \|\cdot\|)$ is said to be a Banach space if it is complete as a metric space, for the metric $d(x,y) := \|x-y\|$, induced by the norm $\|\cdot\|$.

Definition 1.1.2 *Let X and Y be Banach spaces. The mapping $T : X \to Y$ is called a linear mapping if the following properties hold:*
1) $T(x+y) = Tx + Ty,\ \forall x, y \in X$
2) $T(\lambda x) = \lambda Tx,\ \forall x \in X$ *and* $\lambda \in \mathbb{R}$

Proposition 1.1.3 *Let X and Y be Banach spaces. For each linear mapping $T : X \to Y$, the following statements are equivalent:*
(i) T is uniformly continuous
(ii) T is continuous at 0
(iii) There is a positive constant M such that $\|Tx\| \leq M\|x\|,\ \forall x \in X$.

Proof. Obviously (i) implies (ii). Suppose T is continuous at 0, then there exists $r > 0$ such that if $\|x\| \leq r$, then $\|Tx\| < 1$. Given $0 \neq y \in X$, since $\|\frac{ry}{\|y\|}\| = r$, we have $\|T(\frac{ry}{\|y\|})\| < 1$. Then $\|Ty\| \leq \frac{1}{r}\|y\|$. Hence (ii) implies (iii). Finally, (iii) implies that T is Lipschitz continuous, and therefore (i) is satisfied. □

Definition 1.1.4 *For each linear mapping $T : X \to Y$, we define*

$$\|T\| = \sup_{x \neq 0} \frac{\|Tx\|}{\|x\|}.$$

The linear mapping T is said to be bounded if $\|T\| < +\infty$. The space of all bounded linear mappings from X to Y with the above norm is denoted by $L(X,Y)$, and is a normed linear space. $L(X,Y)$ is a Banach space if Y is a Banach space. From Proposition 1.1.3, we get

$$\|Tx\| \leq \|T\|\|x\|, \quad \forall x \in X.$$

In Definition 1.1.4, if $Y = \mathbb{R}$, the space $L(X,\mathbb{R})$ is denoted by X^* and is called the dual or conjugate space of X. The elements of the space X^* are called bounded linear functionals on X.

In this book, we concentrate on some special types of Banach spaces, called Hilbert spaces, that are the immediate generalization of the Euclidean spaces $\mathbb{R}^n$ to the infinite dimensional case with very similar properties, and are defined as follows.

Definition 1.1.5 *Let H be a real vector space. A function $(\cdot,\cdot) : H \times H \to \mathbb{R}$ is called an inner product on H if the following conditions hold:*
1) $(\alpha x + y, z) = \alpha(x,z) + (y,z), \quad \forall x,y,z \in H$, *and* $\alpha \in \mathbb{R}$
2) $(x,y) = (y,x), \quad \forall x,y \in H$
3) $(x,x) \geq 0, \forall x \in H$, *and* $(x,x) = 0$ *if and only if* $x = 0$.

The vector space H with inner product $(\cdot,\cdot)$ is called an inner product space.

Example 1.1.6 *The simplest example of an inner product space is $\mathbb{R}$ with multiplication. More generally, $\mathbb{R}^n$ with $(x,y) = \sum_{i=1}^n x_i y_i$ is an inner product space.*

Example 1.1.7 *The space l^2 of all real sequences $(x_1, x_2, x_3, \ldots)$ such that $\sum_{i=1}^{+\infty} |x_i|^2 < +\infty$, with inner product defined by*

$$(x,y) := \sum_{i=1}^{+\infty} x_i y_i,$$

for each $x = (x_1, x_2, \ldots)$ and $y = (y_1, y_2, \ldots)$ is an infinite dimensional inner product space.

Example 1.1.8 *The space $L^2(\mathbb{R})$ of all square integrable functions on $\mathbb{R}$, with inner product defined by*

$$(f,g) = \int_{-\infty}^{+\infty} f(x)g(x)dx$$

is an inner product space. More generally, $L^2(\mathbb{R}^n)$ with inner product defined by

$$(f,g) = \int_{\mathbb{R}^n} f(x)g(x)dx$$

is an inner product space. We often use $L^2((a,b))$ or $L^2(\Omega)$, where Ω is a subset of $\mathbb{R}^n$.

Example 1.1.9 *The space $C([a,b])$ of all real-valued continuous functions on $[a,b]$ with inner product defined by*

$$(f,g) = \int_a^b f(x)g(x)dx$$

is an inner product space.

We recall below without proof, some of the important identities and inequalities in an inner product space.
1) (Cauchy-Schwarz inequality): Let $x, y \in H$, then

$$|(x,y)| \leq \sqrt{(x,x)}\,\sqrt{(y,y)}.$$

Moreover, in the above inequality, equality holds if and only if x and y are linearly dependent.
2) $\|x+y\|^2 = \|x\|^2 + 2(x,y) + \|y\|^2$.
3) (Parallelogram identity): $\|x+y\|^2 + \|x-y\|^2 = 2\|x\|^2 + 2\|y\|^2$.
4) (Polarization identity): $4(x,y) = \|x+y\|^2 - \|x-y\|^2$.

In $(H,(\cdot,\cdot))$, defining $\|x\| := \sqrt{(x,x)}$, then $\|\cdot\|$ is a norm on H that is induced by the inner product $(\cdot,\cdot)$. Properties 1 and 2 of the norm are easy to show. The triangle inequality follows from Cauchy-Schwarz inequality. The inner product space $(H,(\cdot,\cdot))$ is said to be a Hilbert space, if it is complete for the norm induced by its inner product. The space $C([a,b])$ in Example 1.1.9 is not a Hilbert space, because it is not complete. The spaces in the other Examples 1.1.6–1.1.8 are Hilbert spaces. Other important examples of Hilbert spaces include Sobolev spaces. They will be studied in Section 3 of this chapter.

Theorem 1.1.10 (*Riesz representation theorem*) *Let H be a Hilbert space. For every $y \in H$, the mapping $f : H \to \mathbb{R}$ defined by $f(x) = (x,y), \forall x \in H$, is a bounded linear functional on H with $\|f\| = \|y\|$. Conversely, let $f : H \to \mathbb{R}$ be a bounded linear functional on H. Then there is a unique element y in H such that $f(x) = (x,y), \forall x \in H$; in addition $\|f\| = \|y\|$.*

Theorem 1.1.11 (*Best approximation theorem*) *Let H be a Hilbert space and C be a nonempty closed and convex subset of H. Then for each $x \in H$, there is a unique element $u \in C$ which is the nearest point to x. That is $\|x-u\| = \mathrm{Min}\{\|x-v\|; v \in C\}$. In addition, u is characterized by the following property:*

$$\begin{cases} u \in C \\ (x-u, v-u) \leq 0, \quad \forall v \in C. \end{cases}$$

This unique point u is denoted by $P_C x$, and is called the metric projection of x onto C.

Proposition 1.1.12 *Let C be a nonempty closed and convex subset of a Hilbert space H. Then the mapping $P_C : H \to C$ has the following properties:*
1) $(P_Cx - P_Cy, x-y) \geq \|P_Cx - P_Cy\|^2$,
2) $\|P_Cx - P_Cy\| \leq \|x-y\|$,
for each $x, y \in H$.

Corollary 1.1.13 (*Best approximation on subspaces*) *Suppose $M \subset H$ is a closed linear subspace of H, and $x \in H$. Then $u = P_M x$ is characterized by*

$$\begin{cases} u \in M \\ (x-u, y) = 0, \quad \forall y \in M. \end{cases}$$

That is $P_M x$ is the orthogonal projection of x onto M.

Definition 1.1.14 *Let $a : H \times H \to \mathbb{R}$ be a bilinear form (that is a linear mapping with respect to each argument separately). a is said to be*
(i) continuous if there is a constant C such that

$$|a(u,v)| \leq C\|u\|\|v\|, \ \forall u, v \in H$$

(ii) coercive if there is a constant $\alpha > 0$ such that

$$|a(u,u)| \geq \alpha\|u\|^2, \ \forall u \in H$$

(iii) symmetric if

$$a(u,v) = a(v,u), \ \forall u, v \in H$$

Now we recall the following theorem which is very useful for solving linear partial differential equations, and we refer to [BRE2], page 140, for its proof.

Theorem 1.1.15 (*Lax-Milgram*) *Assume that $a(u,v)$ is a continuous coercive bilinear form on H. Then, given any $\phi \in H$, there exists a unique element $u \in H$ such that*

$$a(u,v) = (\phi, v), \ \forall v \in H.$$

Moreover, if a is symmetric, then u is characterized by the following property:

$$u \in H \text{ and } \frac{1}{2}a(u,u) - (\phi,u) = \min_{v\in H}\{\frac{1}{2}a(v,v) - (\phi,v)\}$$

1.2 WEAK TOPOLOGY AND WEAK CONVERGENCE

Let $(X, \|\cdot\|)$ be a Banach space with dual X^*. For each $x \in X$ and $f \in X^*$, we denote $f(x)$ by $\langle f, x\rangle$.

Definition 1.2.1 *The weakest topology on the space X such that all elements of X^* are continuous is called the weak topology on X.*

The topology generated by the norm of X is called the strong topology. Obviously every open set in the weak topology is also open in the strong topology and similarly for closed sets, but the converse is not true. The following theorem shows that the converse holds for convex sets.

Theorem 1.2.2 (*Mazur*) *Let C be a convex subset of a Banach space X; then C is closed in the strong topology if and only if it is closed in the weak topology.*

Definition 1.2.3 *A sequence $\{x_n\}$ in X is said to be weakly convergent to $x \in X$ if*

$$\langle f, x_n \rangle \to \langle f, x \rangle, \quad \forall f \in X^*$$

as $n \to +\infty$.

In this text, we denote the weak convergence by $\rightharpoonup$. The following are some important properties of the weak convergence which are also valid in Banach spaces, and we refer the reader to [BRE2] for their proofs.

Theorem 1.2.4 1) *If $x_n \to x$ then $x_n \rightharpoonup x$.*
2) *If $x_n \rightharpoonup x$, then $\|x_n\|$ is bounded and $\|x\| \leq \liminf_{n\to+\infty} \|x_n\|$.*
3) *if $x_n \rightharpoonup x$ and $f_n \to f$ in X^*, then $\langle f_n, x_n \rangle \to \langle f, x \rangle$.*

In a Hilbert space H, by Theorem 1.1.10, the weak convergence of a sequence x_n to x is equivalent to

$$(x_n - x, y) \to 0, \ \forall y \in H$$

as $n \to +\infty$.

The following is from Opial [OPI] and is frequently used as a useful tool to prove weak convergence.

Lemma 1.2.5 (*Opial*) *Let $\{x_n\}$ be a sequence in a Hilbert space H and $F \subset H$ be nonempty, closed and convex such that the following conditions are satisfied:*
1) $\lim_{n\to+\infty} \|x_n - p\|$ *exists for each $p \in F$.*
2) *every weak cluster point of the sequence x_n is in F.*
Then $\{x_n\}$ converges weakly to an element of F.

The following criterion is very useful for the strong convergence.

Proposition 1.2.6 *[BAR1] If H is a Hilbert space and $(x_n)_n \subset H$ is a weakly convergent sequence to some $x \in H$, and if $\limsup_{n\to\infty} \|x_n\| \leq \|x\|$, then $x_n \to x$ strongly in H.*

1.3 REFLEXIVE BANACH SPACES

Definition 1.3.1 *Let X be a Banach space with dual X^* and bidual X^{**} (the dual of X^*). Let the mapping $J : X \to X^{**}$ be defined as follows: For each $x \in X$,*

$$\langle J(x), y^* \rangle = \langle y^*, x \rangle, \quad \forall y^* \in X^*.$$

Then J is called the canonical embedding of X into X^{**}. It is well known that J is an isometry. X is said to be reflexive if J is surjective.

Theorem 1.1.10 shows that every Hilbert space is reflexive. The following theorem is useful to prove weak convergence results as shown in future chapters.

Theorem 1.3.2 *[YOS] (Eberlein-Šmulyan's theorem) X is a reflexive Banach space if and only if every bounded sequence in X contains a weakly convergent subsequence in X.*

1.4 DISTRIBUTIONS AND SOBOLEV SPACES

1.4.1 VECTOR-VALUED FUNCTIONS

Let $[0,T]$ be a fixed finite interval and X be a real Banach space. The function $f:[0,T]\to H$ is said to be absolutely continuous if for each $\varepsilon>0$ there exists $\delta>0$ such that for each partition $\{0=t_0<t_1<\cdots<t_n=T\}$ of the interval $[0,T]$, $\sum_{i=1}^{n}|t_i-t_{i-1}|<\delta$ implies $\sum_{i=1}^{n}\|f(t_i)-f(t_{i-1}\|<\varepsilon$. It is well-known that if $X=\mathbb{R}$, every absolutely continuous function $f(t)$ on the real interval $[0,T]$ is almost everywhere differentiable on $(0,T)$, and can be recovered as the integral of its derivative.

Proposition 1.4.1 *[BAR1] Let H be a Hilbert space. Then every H-valued absolutely continuous function $f(t)$ on $[0,T]$ is a.e. differentiable on $(0,T)$ and*

$$f(t)=f(0)+\int_0^t \frac{df}{ds}(s)ds,\quad 0\le t\le T$$

1.4.2 L^p SPACES

Let $(X,\|\cdot\|)$ be a Banach space and $U\subset\mathbb{R}^n$ be Lebesgue measurable and $1\le p<+\infty$. The space of all equivalence classes of measurable functions $f:U\to X$ with respect to almost everywhere equality such that the function $x\mapsto\|f(x)\|^p$ is Lebesgue integrable on U is denoted by $L^p(U;X)$. It is a Banach space with the norm

$$\|f\|_{L^p(U;X)}=(\int_U\|f(x)\|^p dx)^{\frac{1}{p}}.$$

Similarly $L^\infty(U;X)$ is the space of all equivalence classes of measurable functions $f:U\to X$ such that $x\mapsto\|f(x)\|$ is essentially bounded in U. This is a Banach space with the norm

$$\|f\|_{L^\infty(U;X)}=\operatorname*{ess\,sup}_{x\in U}\|f(x)\|.$$

When $X=\mathbb{R}$, we denote $L^p(U;X)$ by $L^p(U)$, and if $U=(a,b)$, we denote it by $L^p(a,b;X)$. We denote $L^p_{\text{loc}}(0,+\infty;X)$ for $1\le p\le+\infty$, the space of all equivalence classes with respect to almost everywhere equality of measurable functions $f:(0,+\infty)\to X$ such that the restriction of f on each finite interval $(0,T)$ belongs to $L^p(0,T;X)$.

Theorem 1.4.2 *Assume that* vol $\Omega = \int_\Omega 1\,dx < \infty$ *and* $1 \le p \le q \le \infty$. *Then the following embedding holds:* $L^q(\Omega) \hookrightarrow L^p(\Omega)$.

Theorem 1.4.3 *Let* H *be a real Hilbert space with the scalar product* $(\cdot,\cdot)$, *then* $L^2(\Omega; H)$ *is a real Hilbert space with respect to the scalar product*

$$\langle u, v\rangle = \int_\Omega (u(x), v(x))\,dx,\ \forall u, v \in L^2(\Omega; H).$$

Theorem 1.4.4 (*Fatou's lemma*) *If* $\Omega \subseteq \mathbb{R}^N$ *is a measurable set and* $(f_n)_n$ *is a sequence of nonnegative and measurable real functions defined on* Ω, *then*

$$\int_\Omega \liminf_{n\to\infty} f_n(x)\,dx \le \liminf_{n\to\infty} \int_\Omega f_n(x)\,dx.$$

Theorem 1.4.5 (*Lebesgue's theorem*) *If* $\Omega \subseteq \mathbb{R}^N$ *is a measurable set,* $\{f_n\}_n$ *is a sequence of measurable real functions which is almost everywhere convergent on* Ω, *and if there exists a nonnegative function* $g \in L^1(\Omega)$ *such that* $|f_n(x)| \le g(x)$, $\forall n \in \mathbb{N}, and\ for\ a.e.\ x \in \Omega$, *then*

$$\lim_{n\to\infty} \int_\Omega f_n(x)\,dx = \int_\Omega \left(\lim_{n\to\infty} f_n(x)\right) dx.$$

1.4.3 SCALAR DISTRIBUTIONS AND SOBOLEV SPACES

Suppose that U is an open subset of $\mathbb{R}^n$. We denote by $C_0^\infty(U)$, the space of all infinitely differentiable functions $f : U \to \mathbb{R}$ whose $\text{supp}(f) = \overline{\{x \in U;\ f(x) \neq 0\}}$ is a compact subset of U. The elements of $C_0^\infty(U)$ are called test functions. Now suppose that K is a compact subset of U. We denote by $\mathcal{D}_K(U)$ the set of all $f \in C_0^\infty(U)$ with $\text{supp}(f) \subset K$. The space $\mathcal{D}_K(U)$ endowed with the family of seminorms

$$p_{K,j}(f) = \sup_{x\in X,\ |\alpha|\le j} |D^\alpha f(x)|,\ \ j \in \mathbb{N}$$

is a Fréchet space (i.e. a metrizable locally convex and complete space), where $|\alpha| = \alpha_1 + \cdots + \alpha_n$ and $D^\alpha = \frac{\partial^{|\alpha|}}{\partial x_1^{\alpha_1}\cdots\partial x_n^{\alpha_n}}$ for $\alpha = (\alpha_1, \ldots, \alpha_n) \in \mathbb{N}^n$. If K_1 and K_2 are two compact subsets of U such that $K_1 \subset K_2$, the topology of $\mathcal{D}_{K_1}(U)$ coincides with the inductive topology from $\mathcal{D}_{K_2}(U)$ over $\mathcal{D}_{K_1}(U)$. It is obvious that $C_0^\infty(U) = \cup_{K\subset U} \mathcal{D}_K(U)$ where the union is taken over all compact subsets of U. The space $C_0^\infty(U)$ endowed with the inductive limit topology is denoted by $\mathcal{D}(U)$.

Definition 1.4.6 *A linear and continuous functional on* $\mathcal{D}(U)$ *is called a distribution on* U.

The space of all distributions on U (that is the dual space of $\mathscr{D}(U)$) is denoted by $\mathscr{D}'(U)$. If $u : U \to \mathbb{R}$ is measurable and Lebesgue integrable on every compact subset of U, then the functional

$$\mathscr{D}(U) \ni f \mapsto \int_U u(x)f(x)dx$$

is a distribution on U that we denote again by u. The function u is identified with the corresponding distribution.

Definition 1.4.7 *The derivative of order α of a distribution $u \in \mathscr{D}'(U)$ is a distribution $\mathscr{D}^\alpha u$ defined by*

$$D^\alpha u(f) = (-1)^{|\alpha|} u(D^\alpha f), \quad \forall f \in \mathscr{D}(U)$$

Now suppose $1 \leq p \leq +\infty$ and $k \in \mathbb{N}$.

Definition 1.4.8 *The space*

$$W^{k,p}(U) = \{u;\ D^\alpha u \in L^p(U),\ \forall \alpha \in \mathbb{N}^n,\ \ |\alpha| \leq k\}$$

(where $D^\alpha u$ is the distributional derivative of u) is called the Sobolev space of order k.

Theorem 1.4.9 *$W^{k,p}(U)$ where $1 \leq p < +\infty$ and $k \in \mathbb{N}$ is a Banach space with the norm*

$$\|u\|_{W^{k,p}(U)} = \Big(\sum_{|\alpha| \leq k} \|D^\alpha u\|^p_{L^p(U)}\Big)^{\frac{1}{p}}$$

and $W^{k,\infty}(U)$ for $k \in \mathbb{N}$ is a Banach space with the norm

$$\|u\|_{W^{k,\infty}(U)} = \max_{|\alpha| \leq k} \|D^\alpha u\|_{L^\infty(U)}$$

Proof. See [ADA] □

The completion of $\mathscr{D}(U)$ with respect to the topology of $W^{k,p}(U)$ is denoted by $W_0^{k,p}(U)$. When $p = 2$, we denote $H^k(U) = W^{k,2}(U)$ and $H_0^k(U) = W_0^{k,2}(U)$.

Theorem 1.4.10 *Both $H^k(U)$ and $H_0^k(U)$ are Hilbert spaces with the inner product*

$$(u,v)_{H^k(U)} = \sum_{|\alpha| \leq k} \int_U D^\alpha u . D^\alpha v dx.$$

We denote the dual space of $H_0^k(U)$ by $H^{-k}(U)$. Since $\mathscr{D}(U)$ is dense in $H_0^k(U)$, and the embedding $i : \mathscr{D}(U) \to H_0^k(U)$ is continuous, then $H^{-k}(U)$ is embedded in $\mathscr{D}'(U)$ densely and continuously (by means of the adjoint operator $i^* : H^{-k}(U) \to \mathscr{D}'(U)$). If U is a bounded domain in $\mathbb{R}^n$ with boundary Γ sufficiently smooth, then

$$H_0^1(U) = \{u \in H^1(U);\ \ u|_\Gamma \equiv 0\}$$

where $u|_\Gamma$ is the trace of u on Γ.

1.4.4 VECTOR DISTRIBUTIONS AND SOBOLEV SPACES

Let $U=(a,b)$ with $-\infty \leq a < b \leq +\infty$ and $\mathscr{D}'(a,b;X)$ be the space of all continuous linear operators from $\mathscr{D}(a,b)$ to X. The elements of $\mathscr{D}'(a,b;X)$ are called vector distributions on (a,b) with values in X. If $u:(a,b)\to X$ is Bochner integrable on every bounded interval $I\subset(a,b)$, then as in the scalar case, we can define a vector distribution which we can denote again by u as follows

$$u(f)=\int_a^b f(t)u(t)dt, \ \forall f\in\mathscr{D}(a,b).$$

Now for $u\in\mathscr{D}'(a,b;X)$, we denote $u^{(j)}$ by the distributional derivative of order $j\in\mathbb{N}$ i.e.

$$u^{(j)}(f)=(-1)^j u(\frac{d^j f}{dt^j}), \ \forall f\in\mathscr{D}(a,b)$$

where by convention $u^{(0)}=u$. Let $k\in\mathbb{N}$ and $1\leq p\leq+\infty$

$$W^{k,p}(a,b;X)=\{u\in\mathscr{D}'(a,b;X);\ u^{(j)}\in L^p(a,b;X),\ j=0,1,\dots,k\}$$

Theorem 1.4.11 *For every $k\in\mathbb{N}$ and $1\leq p<+\infty$ the space $W^{k,p}(a,b;X)$ with norm*

$$\|u\|_{W^{k,p}(a,b;X)}=(\sum_{j=0}^{k}\|u^{(j)}\|^p_{L^p(a,b;X)})^{\frac{1}{p}}$$

is a Banach space. Moreover for each $k\in\mathbb{N}$, $W^{k,\infty}(a,b;X)$ is a Banach space with norm

$$\|u\|_{W^{k,\infty}(a,b;X)}=\max_{0\leq j\leq k}\|u^{(j)}\|_{L^\infty(a,b;X)}$$

As in the scalar case, for $p=2$ we denote $H^k(a,b;X)$ for $W^{k,2}(a,b;X)$.

Theorem 1.4.12 *Let X be a Hilbert space with inner product $(\cdot,\cdot)$. Then for every $k\in\mathbb{N}$, $H^k(a,b;X)$ is also a Hilbert space with the inner product*

$$(u,v)_{H^k(a,b;X)}=\sum_{j=0}^{k}\int_a^b (u^{(j)}(t),v^{(j)}(t))_X dt$$

We define $W^{k,p}_{\text{loc}}(a,b;X)$ for $1\leq p\leq+\infty$ and $k\in\mathbb{N}$ to be the space of all distributions $u\in\mathscr{D}'(a,b;X)$ such that $u\in W^{k,p}(t_1,t_2;X)$ for each bounded interval $(t_1,t_2)\subset(a,b)$. For the bounded interval (a,b), $1\leq p\leq+\infty$ and $k\in\mathbb{N}$ we denote by $A^{k,p}(a,b;X)$ the space of all absolutely continuous functions $u:(a,b)\to X$ such that $\frac{d^j u}{dt^j}$ exists and is defined almost everywhere for $j=1,2,\dots,k$, and is absolutely continuous and $\frac{d^k u}{dt^k}\in L^p(a,b;X)$.

Theorem 1.4.13 *Let $v\in L^p(a,b;X)$, $1\leq p\leq+\infty$, then the following conditions are equivalent:*
(1) $v\in W^{k,p}(a,b;X)$
(2) *There exists $v_1\in A^{k,p}(a,b;X)$ such that $v_1(t)=v(t)$ almost everywhere on (a,b)*

Definition 1.4.14 *If $\Omega = \mathbb{R}_+ = [0,\infty)$ and $1 \leq p \leq \infty$, then $L^p_{\mathrm{loc}}(0,\infty;X)$ is the locally convex space of all measurable functions $u:[0,\infty) \to X$ with the property that the restriction of u to every interval $[0,T]$ $(T \in (0,\infty))$ is in $L^p(0,T;X)$.*

We also define the spaces

$$W^{2,2}_{\mathrm{loc}}(0,T;X) = \left\{u:[0,T] \to X;\ u \in W^{2,2}(\eta, T-\eta;X),\ \forall\, \eta > 0\right\}$$

and

$$W^{2,2}_{\mathrm{loc}}(0,\infty;X) = \{u:[0,\infty) \to X;\ u \in W^{2,2}_{\mathrm{loc}}(0,T;X),\ \forall\, T > 0\}.$$

Theorem 1.4.15 *If X is a real Banach space and $u \in W^{1,p}(0,\infty;X)$ (with $1 \leq p < \infty$), then $\|u(t)\| \to 0$ as $t \to \infty$.*

REFERENCES

ADA. R. A. Adams, Sobolev Spaces, Academic Press, New York, 1975.

BAR. V. Barbu, Nonlinear Semigroups and Differential Equations in Banach Spaces, Noordhoff, Leyden, 1976.

BRE. H. Brézis, Functional Analysis, Sobolev Spaces and Partial Differential Equations, Universitext. Springer, New York, 2011.

DEB-MIK. L. Debnath and P. Mikusinski, Introduction to Hilbert Spaces with Applications, Academic Press, Inc., San Diego, CA, 1999.

MOR. G. Morosanu, Nonlinear Evolution Equations and Applications, Editua Academiei (and D. Reidel Publishing Company), Bucuresti, 1988.

OPI. Z. Opial, Weak convergence of the sequence of successive approximations for nonexpasive mappings, Bull. Amer. Math. Soc. 73 (1967), 591–597.

YOS. K. Yosida, Functional Analysis, third ed., Springer Verlag, Berlin, 1971.

2 Convex Analysis and Subdifferential Operators

2.1 INTRODUCTION

Convexity has an essential role in optimization, variational inequalities, evolution equations and many other branches of nonlinear analysis. This chapter is a quick review of convex analysis. Our aim is to provide the reader with the essential facts on convex sets and convex functions that will be needed in the subsequent chapters of the book.

2.2 CONVEX SETS AND CONVEX FUNCTIONS

This section is devoted to some definitions and elementary properties of convex sets and functions. Let X be a real Banach space with the dual X^*. For each $x^* \in X^*$ and $x \in X$, we denote $x^*(x)$ by $\langle x^*, x\rangle$.

Definition 2.2.1 *A subset C of a Banach space X is called convex if for all $x, y \in C$ and each $\lambda \in [0,1]$, $\lambda x + (1-\lambda)y \in C$.*

A ball in a normed linear space, an affine subspace of a linear space and a segment joining two points in a linear space are some examples of convex sets. Also it is clear that the intersection of a family of convex sets is convex.

Definition 2.2.2 *Let C be a subset of a Banach space X. The "convex hull" of C is the intersection of all convex subsets of X containing C. In other words, it is the smallest convex subset of X containing C. It is denoted by* $\mathrm{conv}(C)$. *The closed convex hull of C is the smallest closed convex subset of X containing C. It is denoted by* $\overline{\mathrm{conv}}(C)$

A simplex in a Banach space X is the closed convex hull of a finite number of points in X.

Definition 2.2.3 *Let C be a convex subset of X. Then $f : X \to (-\infty, +\infty]$ is called:*
i) convex if

$$f(\lambda x + (1-\lambda)y) \le \lambda f(x) + (1-\lambda) f(y), \quad \forall x, y \in C, \ \forall \lambda \in [0,1]$$

ii) strictly convex if

$$f(\lambda x + (1-\lambda)y) < \lambda f(x) + (1-\lambda) f(y), \quad \forall x \neq y \in C, \ \forall \lambda \in (0,1)$$

iii) strongly convex if

$$f(\lambda x + (1-\lambda)y) \le \lambda f(x) + (1-\lambda) f(y) - \lambda(1-\lambda)\|x-y\|^2, \ \forall x, y \in C, \ \forall \lambda \in (0,1)$$

Obviously

$$\text{strongly convex} \Rightarrow \text{strictly convex} \Rightarrow \text{convex}$$

The norm $\|\cdot\|$ of every Hilbert space H is a strictly convex function on H, but it is not strongly convex. However, $\|\cdot\|^2$ is a strongly convex function on H, and therefore also strictly convex, and hence convex.

Proposition 2.2.4 *Let $f : X \to (-\infty, +\infty]$ be a convex function. Then* $\text{dom}\, f := \{x \in X : f(x) < +\infty\}$, *which is called the effective domain of f is a convex set.*

Definition 2.2.5 *The convex function $f : X \to (-\infty, +\infty]$ is called proper if* $\text{dom}\, f \neq \emptyset$

Proposition 2.2.6 *If $f : X \to (-\infty, +\infty]$ is convex, then the sub-level set $L_r^f := \{x \in X : f(x) \leq r\}$ is a convex set.*

Definition 2.2.7 *Let $f : X \to (-\infty, +\infty]$. The epi-graph of f is defined as* $\text{epi} f := \{(x,r) \in X \times \mathbb{R} : f(x) \leq r\}$.

Proposition 2.2.8 *$f : X \to (-\infty, +\infty]$ is a convex function if and only if* $\text{epi} f$ *is a convex subset of $X \times \mathbb{R}$.*

The following proposition is called Jensen's inequality.

Proposition 2.2.9 *Let $f : X \to (-\infty, +\infty]$. Then f is convex if and only if for all finite families $(\lambda_i)_{i\in I}$ in $]0,1[$ such that $\sum_{i\in I} \lambda_i = 1$, and $(x_i)_{i\in I}$ in X, we have*

$$f(\sum_{i\in I} \lambda_i x_i) \leq \sum_{i\in I} \lambda_i f(x_i).$$

2.3 CONTINUITY OF CONVEX FUNCTIONS

This section contains some continuity properties of convex functions.

Theorem 2.3.1 *Let $f : X \to (-\infty, +\infty]$ be proper and convex and let $x_0 \in$* $\text{dom} f$. *Then the following are equivalent:*
i) f is locally Lipschitz near x_0.
ii) f is continuous at x_0.
iii) f is bounded on a neighborhood of x_0.
iv) f is bounded above on a neighborhood of x_0.
Moreover, if anyone of these conditions holds, then f is continuous and locally Lipschitz on $\text{int}(\text{dom} f)$.

Proof. (i)$\Rightarrow$(ii)$\Rightarrow$(iii)$\Rightarrow$(iv): Are clear.
(iv)$\Rightarrow$(ii): Take $\rho > 0$ such that $\eta = \sup f(B(x_0,\rho)) < +\infty$. Given $\varepsilon > 0$, choose $\alpha \in (0,1)$ such that $\alpha(\eta - f(x_0)) < \varepsilon$, and let $x \in B(x_0, \alpha\rho)$. The convexity of f yields

$$f(x) - f(x_0) = f\big((1-\alpha)x_0 + \frac{\alpha(x-(1-\alpha)x_0)}{\alpha}\big) - f(x_0)$$

$$\begin{aligned}&\leq \alpha\big(f(x_0+\frac{(x-x_0)}{\alpha})-f(x_0)\big)\\&\leq \alpha\big(\eta-f(x_0)\big).\end{aligned}$$

Similarly

$$\begin{aligned}f(x_0)-f(x)&=f(\frac{x}{1+\alpha}+\frac{\alpha}{1+\alpha}\frac{(1+\alpha)x_0-x}{\alpha})-f(x)\\&\leq\frac{\alpha}{1+\alpha}\big(f(x_0+\frac{(x_0-x)}{\alpha})-f(x)\big)\\&\leq\frac{\alpha}{1+\alpha}\big((\eta-f(x_0))+(f(x_0)-f(x))\big),\end{aligned}$$

which after rearranging implies that $f(x_0)-f(x)\leq\alpha(\eta-f(x_0))$. Altogether, we showed that $|f(x)-f(x_0)|\leq\alpha(\eta-f(x_0))<\varepsilon,\forall x\in B(x_0,\alpha\rho)$. Therefore f is continuous at x_0.
(iii)⇒(i): Choose $\rho>0$ such that $\delta:=2\sup|f(B(x_0,2\rho))|<+\infty$. Take distinct points x and y in $B(x_0,\rho)$ and set

$$z=x+(\frac{1}{\alpha}-1)(x-y),\ \text{ where } \alpha=\frac{\|x-y\|}{\|x-y\|+\rho}<\frac{\|x-y\|}{\rho}.$$

Then $x=\alpha z+(1-\alpha)y$ and $\|z-x_0\|\leq\|z-x\|+\|x-x_0\|=\rho+\|x-x_o\|\leq 2\rho$. Therefore, y and z belong to $B(x_0,2\rho)$ and hence, by the convexity of f, we have:

$$\begin{aligned}f(x)&=f(\alpha z+(1-\alpha)y)\\&\leq f(y)+\alpha(f(z)-f(y))\\&\leq f(y)+\alpha\delta\\&\leq f(y)+(\frac{\delta}{\rho})\|x-y\|.\end{aligned}$$

Thus $f(x)-f(y)\leq(\frac{\delta}{\rho})\|x-y\|$. Interchanging the roles of x and y, we conclude that $|f(x)-f(y)|\leq(\frac{\delta}{\rho})\|x-y\|$, which implies (i).
So far, we have shown that the statements (i)–(iv) are equivalent to each other. Now assume that (iv) holds, and say $\eta=\sup f(B(x_0,\rho))<+\infty$ for some $\rho>0$. Then f is continuous and locally Lipschitz near x_0. Take $x\in\text{int}(\text{dom}f)$ and $\gamma>0$ such that $B(x,\gamma)\subset\text{dom}f$. We may take $x\neq x_0$, otherwise there is nothing to prove. Now set

$$y=x_0+\frac{1}{1-\alpha}(x-x_0),\ \text{ where } \alpha=\frac{\gamma}{\gamma+2\|x-x_0\|}\in(0,1)$$

Then $y\in B(x,\gamma)$. Now take $z\in B(x,\alpha\rho)$ and set $w=x_0+\frac{(z-x)}{\alpha}=\frac{(z-(1-\alpha)y)}{\alpha}$. Then $w\in B(x_0,\rho)$ and $z=\alpha w+(1-\alpha)y$. Consequently,

$$f(z)\leq\alpha f(w)+(1-\alpha)f(y)\leq\alpha\eta+(1-\alpha)f(y)$$

Therefore f is bounded above on $B(x,\alpha\rho)$. It follows now from the first part of the proof that f is continuous and locally Lipschitz near x. □

In a finite dimensional Banach space, the local boundedness from above of a convex function follows from Proposition 2.2.9. Therefore, Theorem 2.3.1 implies the following corollary.

Corollary 2.3.2 *Suppose that X is finite-dimensional and let $f : X \to \mathbb{R}$ be convex. Then f is continuous on X.*

Proof. Let $x \in X$. Since X is finite-dimensional, there exist a finite family $\{y_1, \ldots, y_n\}$ in X and $\rho > 0$ such that $B(x,\rho) \subset \operatorname{conv}\{y_1, \ldots, y_n\}$. Consequently, by Proposition 2.2.9, $\sup f(B(x,\rho)) \leq \sup f(\operatorname{conv}\{y_1, \ldots, y_n\}) \leq \operatorname{Max}_{i=1,\ldots,n} f(y_i) < +\infty$. The conclusion follows now from Theorem 2.3.1. □

Remark 2.3.3 *Let $f : X \to (-\infty, +\infty]$, $x \in X$ and denote by $\mathcal{N}_s$ (resp. $\mathcal{N}_w$) the family of all neighborhoods of x in the strong (resp. weak) topology. We denote*

$$\liminf_{y \to x} f(y) = \sup_{W \in \mathcal{N}_s} \inf_{y \in W - \{x\}} f(y)$$

and

$$\liminf_{y \rightharpoonup x} f(y) = \sup_{W \in \mathcal{N}_w} \inf_{y \in W - \{x\}} f(y)$$

Definition 2.3.4 *$f : X \to (-\infty, +\infty]$ is called (weakly) lower semi-continuous at $x \in X$ if ($\liminf_{y \rightharpoonup x} f(y) \geq f(x)$) $\liminf_{y \to x} f(y) \geq f(x)$. f is called (weakly) lower semi-continuous if it is (weakly) lower semi-continuous at each point of X.*

Proposition 2.3.5 *$f : X \to (-\infty, +\infty]$ is (weakly) lower semi-continuous if L_r^f is (weakly) closed for each $r \in \mathbb{R}$.*

The following theorem is an immediate consequence of Mazur's theorem.

Theorem 2.3.6 *For any convex function $f : X \to (-\infty, +\infty]$ the following are equivalent:*
i) f is lower semi-continuous.
ii) f is weakly lower semi-continuous.

The following theorem provides some additional information on the continuity of convex functions.

Theorem 2.3.7 *Let $f : X \to (-\infty, +\infty]$ be a proper, convex and lower semi-continuous function. Then f is continuous, and in fact locally Lipschitz, at the points in the interior of* dom$f$.

Proof. Assume that $\operatorname{int}(\operatorname{dom} f) \neq \emptyset$, and let $x \in \operatorname{int}(\operatorname{dom} f)$. We have: $\operatorname{dom} f = \cup_{n=1}^{\infty} L_n^f$. Since f is lower semi-continuous, $\operatorname{dom} f$ and L_n^f are closed subsets of H, it then follows from Baire category theorem, that there is n such that $\operatorname{int}(L_n^f) \neq \emptyset$. Suppose that $B(x_0, \rho) \subset L_n^f$, for some $x_0 \in L_n^f$ and some $\rho > 0$. Then we have: $f(B(x_0, \rho)) \leq n$, and hence condition (iv) of Theorem 2.3.1 is satisfied. Now the result follows from the last assertion in Theorem 2.3.1. □

2.4 MINIMIZATION PROPERTIES

Convex functions play an essential role in optimization theory because of their nice minimization properties. It is important to know whether a local minimum is also a global minimum. Another important question is whether a critical point is a local minimum. The convexity guarantees these properties. In this short section we mention some minimization properties of convex functions.

Theorem 2.4.1 *Let C be a convex subset of X and f be a convex function on C. Then*
i) A local minimum point of f is a global minimum.
ii) The set of all minimum points of f is convex.
iii) The set of all minimum points of f is at most a singleton if f is strictly convex.

Proof. (i): Suppose that x_0 is a local minimum point of f; i.e. there exists $r > 0$ such that $f(x_0) \leq f(x)$ for all $x \in B(x_0, r)$. Choose $y \in X$ arbitrary, and $0 < \lambda < 1$ such that $\lambda \|y - x_0\| < r$. Let $x = \lambda y + (1-\lambda)x_0$. Then

$$\|x - x_0\| = \lambda \|y - x_0\| < r$$

Therefore $x \in B(x_0, r)$. By hypothesis,

$$f(x_0) \leq f(x) = f(\lambda y + (1-\lambda)x_0) \leq \lambda f(y) + (1-\lambda) f(x_0)$$

Hence $f(x_0) \leq f(y)$.
(ii): Suppose that x_0 is a minimum point of f. The set of all minimum points of f is $\{x \in X;\ f(x) = f(x_0)\}$. But we have:

$$\{x \in X;\ f(x) = f(x_0)\} = \{x \in X;\ f(x) \leq f(x_0)\}$$

and the last set is convex because f is convex and by using Proposition 2.2.6.
(iii): If a strictly convex function f has two distinct minimum points, then the value of f on the segment joining these two points will be strictly less than the minimum value of the function, and this is a contradiction. □

We know that for a differentiable function, a minimum point is a critical point, but the converse may not be true. In the following theorem, we show that the converse is also true for convex functions.

Theorem 2.4.2 *Let C be a convex subset of X and f be a convex differentiable function on C. Then a critical point $x_0 \in C$ is a global minimum for f.*

Proof. Let x_0 be a critical point of f and let x be an arbitrary point in C. By the convexity of f, we have

$$f(\lambda x + (1-\lambda)x_0) \leq \lambda f(x) + (1-\lambda) f(x_0), \forall \lambda \in (0,1)$$

Then

$$\frac{f(\lambda x + (1-\lambda)x_0) - f(x_0)}{\lambda} \leq f(x) - f(x_0)$$

By letting $\lambda \to 0$, we get

$$0 = \langle \nabla f(x_0), x - x_0 \rangle \leq f(x) - f(x_0)$$

which proves the theorem. □

The following theorem explores sufficient conditions for the existence of a minimum point for a convex function on a Hilbert space H. It is applied in optimization for instance in Tikhonov regularization methods and the proximal point algorithm for convex functions. Although it holds in the more general setting of reflexive Banach spaces, but for the sake of our future use in the sequel, we recall it in Hilbert spaces. For the proof, see page 71, Corollary 3.23 of [BRE2] or page 34, Theorem 1.10 of [MOR].

Theorem 2.4.3 *If $f : H \to (-\infty, +\infty]$ is a proper, convex and lower semi-continuous function on H, satisfying*

$$\lim_{\|x\| \to +\infty} f(x) = +\infty,$$

then there exists an $x_0 \in H$, such that $f(x_0) = \inf\{f(x) : \ x \in H\}$.

2.5 FENCHEL SUBDIFFERENTIAL

Proposition 2.5.1 *[BAR1, BAR-PRE] If $f : X \to (-\infty, +\infty]$ is a proper, convex and lower semi-continuous function, then f is bounded from below by an affine function, that means there exists $y \in X^*$ and $\mu \in \mathbb{R}$ such that*

$$f(x) \geq \langle y, x \rangle + \mu, \quad \forall x \in X.$$

We now introduce the notion of subdifferential mapping.

Definition 2.5.2 *Let $f : X \to (-\infty, +\infty]$, and $x \in \operatorname{dom} f$, the subdifferential of f at x, $\partial f(x)$, is defined by*

$$\partial f(x) = \{w \in X^* : \ f(y) - f(x) \geq \langle w, y - x \rangle, \ \forall y \in X\} \tag{2.1}$$

The elements $v \in \partial f(x)$ are called the subgradients of f at x, while the (possibly multivalued) map $x \mapsto \partial f(x)$ is called the subdifferential of f.

The following proposition is straightforward and easy to see.

Proposition 2.5.3 *f takes a minimum at x, if and only if $0 \in \partial f(x)$.*

The next proposition shows that ∂f is in fact an extension of the usual notion of derivative for nonsmooth functions. For a proof the reader can consult [FER].

Proposition 2.5.4 *If $x \in \operatorname{int}(\operatorname{dom} f)$ and f is differentiable at x, then $\partial f(x) = \{\nabla f(x)\}$.*

Subdifferential operators are defined even for nonconvex functions. The following proposition provides sufficient conditions for their domain to be nonempty.

Proposition 2.5.5 *[FER] Let $f : X \to (-\infty, +\infty]$ be a proper, convex and lsc function, then $\partial f(x)$ is nonempty for each $x \in \operatorname{int}(\operatorname{dom} f)$.*

2.6 THE FENCHEL CONJUGATE

Definition 2.6.1 *Let $f : X \to (-\infty, +\infty]$ be an arbitrary function (not necessarily convex). The Fenchel conjugate of f is the function $f^* : X^* \to [-\infty, +\infty]$ defined as*

$$f^*(x^*) := \sup_{x \in X}\{\langle x^*, x\rangle - f(x)\}$$

The conjugate of f is always a convex function. We can also define the biconjugate of f, $f^{**} : X \to [-\infty, +\infty]$, as follows:

$$f^{**}(x) = \sup_{x^* \in X^*}\{\langle x^*, x\rangle - f^*(x^*)\}.$$

Obviously, from the definitions we have $f^{**} \leq f$. However, equality holds with the conditions of the following theorem.

Theorem 2.6.2 *(Fenchel-Moreau) [BRE2] Let $f : X \to (-\infty, +\infty]$ be a proper, convex and lsc function, then $f^{**} = f$.*

In the following propositions, we state the relations between the conjugate of a function $f : X \to (-\infty, +\infty]$ and its subdifferential.

Proposition 2.6.3 *[LUC] Let $f : X \to (-\infty, +\infty]$. Then $x^* \in \partial f(x)$ if and only if $f(x) + f^*(x^*) = \langle x^*, x\rangle$.*

Proposition 2.6.4 *[LUC] Let $f : X \to (-\infty, +\infty]$. If $\partial f(x) \neq \emptyset$, then $f(x) = f^{**}(x)$. If $f(x) = f^{**}(x)$, then $\partial f(x) = \partial f^{**}(x)$.*

Corollary 2.6.5 *[LUC] Let $f : X \to (-\infty, +\infty]$. Then*

$$x^* \in \partial f(x) \Rightarrow x \in \partial f^*(x^*)$$

If $f(x) = f^{**}(x)$, then

$$x^* \in \partial f(x) \Leftrightarrow x \in \partial f^*(x^*).$$

REFERENCES

BAR. V. Barbu, Nonlinear Semigroups and Differential Equations in Banach Spaces, Noordhoff, Leyden, 1976.

BAR-PRE. V. Barbu and Th. Precupanu, Convexity and Optimization in Banach Spaces, Editors Academiei, Buchrest, 1986 (and D. Reidel Publishing Company).

BRE. H. Brézis, Functional Analysis, Sobolev Spaces and Partial Differential Equations, Universitext. Springer, New York, 2011.

FER. J. Ferrera, An Introduction to Nonsmooth Analysis, Elsevier/Academic Press, Amsterdam, 2014.

LUC. R. Lucchetti, Convexity and Well-posed Problems. CMS Books in Mathematics/Ouvrages de Mathmatiques de la SMC, 22. Springer, New York, 2006.

MOR. G. Morosanu, Nonlinear Evolution Equations and Applications, Editua Academiei (and D. Reidel Publishing Company), Bucuresti, 1988.

3 Maximal Monotone Operators

3.1 INTRODUCTION

In this chapter we review some essential properties of maximal monotone operators that we will need in this book. These nonlinear and possibly set-valued operators were first introduced by Minty [MIN], and have important applications in partial differential equations, evolution equations, nonlinear semigroups, optimization and variational inequalities. Maximal monotone operators can be defined in Banach spaces, but for the purpose of this book, we will study them only in Hilbert spaces. The main references for this chapter are [BRE1, BRE2, MOR, BAR1].

3.2 MONOTONE OPERATORS

Definition 3.2.1 *A nonlinear possibly set-valued mapping $A : H \to 2^H$ is said to be*
(i) *monotone if*

$$(x^* - y^*, x - y) \geq 0,$$

(ii) *α-strongly monotone for $\alpha > 0$, if*

$$(x^* - y^*, x - y) \geq \alpha \|x - y\|^2,$$

for each $x, y \in H$ and $x^* \in Ax$ and $y^* \in Ay$. A is said to be
(iii) *strictly monotone if*

$$(x^* - y^*, x - y) > 0,$$

for each $x, y \in H$ such that $x \neq y$. The domain of A is defined as

$$D(A) := \{x \in H : \ A(x) \neq \varnothing\}.$$

From now on, we denote a monotone operator by $A : D(A) \subset H \to H$, which assigns to each $x \in D(A)$, a subset $A(x)$ of H. The range of A is defined as

$$\mathscr{R}(A) := \{x^* \in H : \ \exists x \in D(A) \ \text{ such that } x^* \in A(x)\}$$

The graph of a monotone operator A is defined as the following subset of $H \times H$:

$$G(A) := \{[x, y] : \ x \in D(A) \ \text{ and } \ y \in A(x)\}.$$

Each monotone operator is usually identified by its graph. For each monotone operator A, its inverse A^{-1} whose graph is defined as $G(A^{-1}) := \{[y, x] : \ [x, y] \in A\}$ is also a monotone operator.

3.3 MAXIMAL MONOTONICITY

Definition 3.3.1 *A monotone operator $A : D(A) \subset H \to H$ is said to be maximal monotone if $G(A)$ (the graph of A) is not properly contained in the graph of any other monotone operator in H.*

Theorem 3.3.2 *Let $A : D(A) \subset H \to H$ be a monotone operator. Then A is maximal monotone if and only if for every $\lambda > 0$ (equivalently, for some $\lambda > 0$), $R(I + \lambda A) = H$.*

Proof. We first prove the necessity. It suffices to prove this for $\lambda = 1$. Given $z_0 \in H$, we should find $x_0 \in H$ such that $(y - (z_0 - x_0),\ x - x_0) \geq 0$ for all $[x, y] \in A$. Then the maximality of A implies that $z_0 - x_0 \in Ax_0$. For $[x, y] \in A$, define the weakly compact set $C_{x,y}$ by

$$C_{x,y} = \{x_0 \in H :\ (y + x_0 - z_0, x - x_0) \geq 0\}$$

It suffices to show that the family $\{C_{x,y}\}_{[x,y]\in A}$ has the finite intersection property. To this end take $[x_i, y_i] \in A$ for $i = 1, 2, \cdots, n$. Let $\Delta = \{(\lambda_1, \cdots, \lambda_n) :\ \lambda_i \geq 0; \sum_{i=1}^n \lambda_i = 1\}$ denote the n-dimensional simplex and consider the function $f : \Delta \times \Delta \to \mathbb{R}$ defined by

$$f(\lambda, \mu) = \sum_{i=1}^{n} \mu_i (y_i + x(\lambda) - z_0, x(\lambda) - x_i)$$

with $x(\lambda) = \sum_{i=1}^n \lambda_i x_i$. Clearly f is convex and continuous with respect to the first argument and linear with respect to the second one. The Von Neumann Minimax Theorem (see Theorem 1.1 of [BRE1]) implies the existence of $\lambda_0 \in \Delta$ such that

$$\max_{\mu\in\Delta} f(\lambda_0, \mu) = \max_{\mu\in\Delta} \min_{\lambda\in\Delta} f(\lambda, \mu) \leq \max_{\mu\in\Delta} f(\mu, \mu).$$

Now the monotonicity of A implies

$$\begin{aligned} f(\mu, \mu) &= \sum_{i=1}^{n} \mu_i (y_i, x(\mu) - x_i) + (x(\mu) - z_0, x(\mu) - x(\mu)) \\ &= \sum_{i,j=1}^{n} \mu_i \mu_j (y_i, x_j - x_i) \\ &= \frac{1}{2} \sum_{i,j=1}^{n} \mu_i \mu_j (y_i - y_j, x_j - x_i) \leq 0 \end{aligned}$$

which shows that $f(\lambda_0, \mu) \leq 0$ for all $\mu \in \Delta$. Choosing for μ the extreme points of Δ, we get $(y_i + x(\lambda_0) - z_0,\ x(\lambda_0) - x_i) \leq 0$ for all i, which implies that $x(\lambda_0) \in \cap_{i=1}^n C_{x_i, y_i}$. Conversely, take $[u, u^*] \in H \times H$ such that $(u^* - v^*, u - v) \geq 0$ for all $[v, v^*] \in A$. We shall prove that $[u, u^*] \in A$. Since $I + A$ is surjective, there is $[w, w^*] \in A$ such that $w + w^* = u + u^*$. Then $(u^* - w^*, u - w) = -\|u - w\|^2 \geq 0$ which implies that $u = w$ and $u^* = w^*$, and therefore $[u, u^*] \in A$. □

Remark 3.3.3 *If A is maximal monotone, then for each $x \in D(A)$*

$$A(x) = \{y \in H;\ (y-v, x-u) \geq 0,\ \forall [u,v] \in A\}$$

Therefore for each $x \in D(A)$, the set $A(x)$ is closed and convex in H. By Theorem 1.1.11, it has an element with minimum norm that we denote by $A^0(x)$. Therefore, for each $x \in D(A)$, $A^0(x)$ is identified by the following two properties:
(1) $A^0(x) \in A(x)$
(2) $\|A^0(x)\| = \inf\{\|y\| :\ y \in A(x)\}$.
$A^0 : D(A) \subset H \to H$ is a single-valued monotone operator which is called the "minimal section" of A.

Proposition 3.3.4 *Assume that $A : D(A) \subset H \to H$ is maximal monotone. Then A is demiclosed, i.e. if $x_n \to x \in H$, $y_n \rightharpoonup y \in H$ and $[x_n, y_n] \in A$, then $[x,y] \in A$.*

Proof. Let $[x_n, y_n] \in A$, then we have

$$(x_n - u, y_n - v) \geq 0, \qquad \forall [u,v] \in A.$$

Passing to the limit in this inequality, we get:

$$(x-u, y-v) \geq 0, \quad \forall [u,v] \in A.$$

Then the maximality of A implies that $[x,y] \in A$. This completes the proof. □

If A and B are two monotone operators, then $A+B$ with the domain $D(A) \cap D(B)$ is defined by

$$(A+B)(x) = \{x^* + y^*;\ x^* \in A(x) \text{ and } y^* \in B(x)\}.$$

It is clear that $A+B$ is also a monotone operator, but it is not necessarily maximal monotone. The following theorem gives a sufficient condition for the maximal monotonicity of $A+B$. For the proof see [MOR].

Theorem 3.3.5 *Let $A : D(A) \subset H \to H$ and $B : D(B) \subset H \to H$ be two maximal monotone operators such that $0 \in \operatorname{int}(D(A) - D(B))$, then $A+B$ is also maximal monotone.*

3.4 RESOLVENT AND YOSIDA APPROXIMATION

Let A be a maximal monotone operator and $\lambda > 0$. We define $J_\lambda := (I + \lambda A)^{-1}$ which is called the resolvent of A, and $A_\lambda := \frac{1}{\lambda}(I - J_\lambda)$ which is called the Yosida approximation of A. By Theorem 3.3.2, $D(J_\lambda) = D(A_\lambda) = H$, $\forall \lambda > 0$. In the following theorem, we collect some properties of the resolvent and the Yosida approximation of maximal monotone operators that we will need in this book.

Theorem 3.4.1 *Let $A : D(A) \subset H \to H$ be a maximal monotone operator. Then, for every $\lambda > 0$,*
(1) J_λ is nonexpansive (i.e. Lipschitz with constant 1).
(2) $A_\lambda(x) \in A(J_\lambda(x)), \ \forall x \in H$.
(3) A_λ is monotone and Lipschitz with constant $\frac{1}{\lambda}$.
(4) $\|A_\lambda(x)\| \leq \|A^0(x)\|, \ \forall x \in D(A)$.
(5) $\lim_{\lambda \to 0} A_\lambda(x) = A^0(x), \ \forall x \in D(A)$.
(6) $\overline{D(A)}$ is convex.
(7) $\lim_{\lambda \to 0} J_\lambda(x) = \mathrm{Proj}_{\overline{D(A)}} x, \ \forall x \in H$.

Proof. (1) From the definition of resolvent, we have

$$J_\lambda(x) - J_\lambda(y) + \lambda[A(J_\lambda(x)) - A(J_\lambda(y))] \ni x - y$$

Multiplying the above inclusion by $J_\lambda(x) - J_\lambda(y)$, and using the monotonicity of A, we get:

$$\|J_\lambda(x) - J_\lambda(y)\|^2 \leq (x - y, J_\lambda(x) - J_\lambda(y)).$$

This implies (1).
(2) Let $y = A_\lambda(x)$. By the definition of the Yosida approximation, we have: $x - \lambda y = J_\lambda(x)$. Now the definition of J_λ implies that $y \in A(x - \lambda y)$ which yields the result.
(3) Since J_λ is nonexpansive, then $I - J_\lambda$ is monotone and therefore A_λ is monotone too. By the definition of J_λ and A_λ we have:

$$x - y = J_\lambda(x) - J_\lambda(y) + \lambda[A_\lambda(x) - A_\lambda(y)].$$

Multiplying both sides by $A_\lambda(x) - A_\lambda(y)$, and using the monotonicity of A, we get:

$$\lambda\|A_\lambda(x) - A_\lambda(y)\|^2 \leq (x - y, A_\lambda(x) - A_\lambda(y))$$

which gives the desired result.
(4) By the definitions of J_λ and A_λ, we have

$$A_\lambda(x) = \lambda^{-1}(J_\lambda(x + \lambda A^0(x)) - J_\lambda(x)), \ \forall x \in D(A), \ \lambda > 0.$$

Now (1) yields the result.
(5) Let $x \in D(A)$. By Part (4), and Theorem 1.3.2, there exists a sequence $\lambda_n \to 0$ such that $A_{\lambda_n}(x)$ converges weakly to a point p. The monotonicity of A implies that

$$(A_{\lambda_n}(x) - v, J_{\lambda_n}(x) - u) \geq 0, \quad \forall [u, v] \in A.$$

By (4), $J_{\lambda_n}(x) \to x$, therefore the above inequality implies that:

$$(p - v, x - u) \geq 0, \ \forall [u, v] \in A.$$

Since A is maximal, $p \in A(x)$. Again (4) implies that

$$\|p\| \leq \limsup_{\lambda \to 0} \|A_\lambda(x)\| \leq \|A^0(x)\|.$$

Therefore $p = A^0(x)$. Now Proposition 1.2.6 of Chapter 1 implies that $A_\lambda(x) \to A^0(x)$ as $\lambda \to 0$.
(6) and (7) Let $C = \overline{\text{conv}}(D(A))$ and $x \in H$. The monotonicity of A implies that:

$$(A_\lambda(x) - v, J_\lambda(x) - u) \geq 0, \quad \forall [u,v] \in A.$$

Now by the definition A_λ we get:

$$(x - J_\lambda(x) - \lambda v, J_\lambda(x) - u) \geq 0, \ \forall [u,v] \in A$$

This yields

$$\|J_\lambda(x)\|^2 \leq (x - \lambda v, J_\lambda(x) - u) + (J_\lambda(x), u), \ \forall [u,v] \in A \tag{3.1}$$

Now, let us choose a sequence $\lambda_n \to 0$ such that $J_{\lambda_n}(x)$ converges weakly to q. Taking the limit in the above inequality we get:

$$\|q\|^2 \leq \limsup_{n \to +\infty} \|J_{\lambda_n}(x)\|^2 \leq (x, q - u) + (q, u), \ \forall u \in D(A)$$

i.e.

$$(x - q, u - q) \leq 0, \ \forall u \in D(A).$$

This inequality obviously holds for all $u \in C$. Theorem 1.2.2 shows that C is weakly closed and since $J_\lambda(x) \in C$, we deduce that $q \in C$. Therefore by Theorem 1.1.11, the inequality

$$(x - q, u - q) \leq 0, \ \forall u \in C$$

implies that $q = \text{Proj}_C x$. Since the sequence λ_n is arbitrary, we see that $J_\lambda(x) \rightharpoonup \text{Proj}_C x$ as $\lambda \to 0$. Now (3.1) yields

$$\limsup_{\lambda \to 0} \|J_\lambda(x)\|^2 \leq (x, q - u) + (q, u), \ \forall u \in C.$$

By taking $u = q$, we get

$$\limsup_{\lambda \to 0} \|J_\lambda(x)\| \leq \|q\|.$$

Now Theorem 1.2.6 implies that $J_\lambda(x) \to \text{Proj}_C x$ as $\lambda \to 0$. Finally, we prove that $C = \overline{D(A)}$. Since $J_\lambda(x) \in D(A)$, $\forall x \in H$, and $J_\lambda(z) \to z$ for all $z \in C$, we see that $C \subset \overline{D(A)}$. Therefore $\overline{D(A)} = C$. □

An important class of maximal monotone operators consists of the subdifferential of proper, convex and lower semicontinuous functions. This important result was proved by Rockafellar in [ROC].

Theorem 3.4.2 *Let $\varphi : H \to (-\infty, +\infty]$ be a proper, convex and lsc function. Then*
(1) *$\partial\varphi$ is a maximal monotone operator.*
(2) $\overline{D(\varphi)} = \overline{D(\partial\varphi)}$.

Proof. (1) Monotonicity is easy. We prove the maximality. By Theorem 3.3.2, it suffices to show that $\mathscr{R}(I+\partial\varphi)=H$. Let $x^*\in H$ and consider the convex function

$$\tilde{\varphi}(x)=\frac{1}{2}\|x\|^2+\varphi(x)-(x,x^*)$$

which is also proper and lsc from H to $(-\infty,+\infty]$. By Theorem 2.5.1,

$$\lim_{\|x\|\to\infty}\tilde{\varphi}(x)=+\infty.$$

Hence by Theorem 2.4.3, there exists $x_0\in D(\varphi)$ such that $\tilde{\varphi}(x_0)\le\tilde{\varphi}(x),\ \ \forall x\in H$. Then

$$\begin{aligned}\varphi(x_0)-\varphi(x)&\le(x^*,x_0-x)+\frac{1}{2}\|x\|^2-\frac{1}{2}\|x_0\|^2\\&\le(x^*-x,x_0-x),\ \ \forall x\in H.\end{aligned}$$

Now taking $x=tx_0+(1-t)v,\ \ 0<t<1,\ v\in H$ in this inequality, and using the convexity of φ, we obtain

$$\varphi(x_0)-\varphi(v)\le(x^*-x_0,x_0-v)+(1-t)\|x_0-v\|^2.$$

Taking the limit as $t\to 1$, in this inequality, we get

$$\varphi(x_0)-\varphi(v)\le(x^*-x_0,x_0-v),\ \ \ \forall v\in H,$$

which means that $x^*\in x_0+\partial\varphi(x_0)$, and proves the surjectivity of $I+\partial\varphi$, completing the proof.
(2) Let $x\in D(\varphi)$. We know that $J_\lambda x\in D(\partial\varphi)$. On the other hand, by Part (7) of Theorem 3.4.1, $J_\lambda(x)\to x$ as $\lambda\to 0$. Therefore $x\in\overline{D(\partial\varphi)}$ which shows that $\overline{D(\varphi)}\subset\overline{D(\partial\varphi)}$. The reverse inclusion is trivial because $D(\partial\varphi)\subset D(\varphi)$. □

Now we define the resolvent and Yosida approximation of a proper, convex and lsc function φ.

Definition 3.4.3 *Let $\varphi:H\to(-\infty,+\infty]$ be a proper, convex and lsc function. The Moreau-Yosida approximation of the function φ is defined by:*

$$\varphi_\lambda(x)=\inf\{\frac{1}{2\lambda}\|x-y\|^2+\varphi(y);\ y\in H\},\ x\in H,\ \lambda>0.$$

Theorem 3.4.4 *Let $\varphi:H\to(-\infty,+\infty]$ be a proper, convex and lsc function, and $A=\partial\varphi$. Then for each $\lambda>0$, φ_λ is convex and*
(1) $\varphi_\lambda(x)=\frac{1}{2\lambda}\|x-J_\lambda(x)\|^2+\varphi(J_\lambda(x)),\ \ \forall x\in H,\ \ \forall\lambda>0$, where J_λ is the resolvent of A.
(2) $\varphi(J_\lambda(x))\le\varphi_\lambda(x)\le\varphi(x),\ \ \forall x\in H,\ \ \forall\lambda>0$.
(3) $\lim_{\lambda\to 0}\varphi_\lambda(x)=\varphi(x),\ \ \forall x\in H$.
(4) $A_\lambda=\partial\varphi_\lambda,\ \ \forall\lambda>0$, where A_λ is the Yosida approximation of A.
(5) φ_λ is differentiable on H.

Proof. Let $\lambda > 0$ and $x \in H$. Consider the function

$$\psi(y) = \frac{1}{2\lambda}\|x-y\|^2 + \varphi(y), \quad y \in H.$$

Since by Theorem 2.5.1, φ has an affine minorant, we have

$$\lim_{\|y\|\to\infty} \psi(y) = +\infty$$

Then ψ achieves its minimum and therefore φ_λ is finite at each point of H. Convexity of φ_λ is easy to check.
(1) We have

$$\frac{1}{\lambda}(y-x) + \partial\varphi(y) \subset \partial\psi(y), \quad \forall y \in D(\partial\psi)$$

On the other hand $y = J_\lambda(x)$ is the unique solution of

$$0 \in \frac{1}{\lambda}(y-x) + \partial\varphi(y).$$

Therefore

$$0 \in \partial\psi(J_\lambda(x)).$$

Then by Proposition 2.5.3, $J_\lambda(x)$ is the minimizer of ψ.
(2) is a consequence of (1) and the definition of φ_λ.
(3) We first prove this for $x \in D(\varphi)$. By Part (7) of Theorem 3.4.1, for each $x \in D(\varphi)$ $\lim_{\lambda\to 0} J_\lambda x = x$. Now the result follows from the lower semicontinuity of φ, and by taking the limit as $\lambda \to 0$ in (2). To complete the proof, we should show that for every $x \notin D(\varphi)$, $\lim_{\lambda\to 0}\varphi_\lambda(x) = +\infty$. Suppose to the contrary there is a sequence $\lambda_n \to 0$ such that

$$\varphi_{\lambda_n}(x) \le C < +\infty. \tag{3.2}$$

Then by Part (2),

$$\varphi(J_{\lambda_n}(x)) \le C. \tag{3.3}$$

On the other hand, since by Theorem 2.5.1, φ has an affine minorant and $\{J_{\lambda_n}(x)\}$ is a bounded sequence, then $\{\varphi(J_{\lambda_n}(x))\}$ is also a bounded sequence. This fact, together with (3.2), (3.3) and Part (1) imply that $J_{\lambda_n}(x) \to x$. Letting $n \to +\infty$ in (3.3), the lower semicontinuity of φ implies that $x \in D(\varphi)$, which is a contradiction.
(4) By Part (1)

$$\begin{aligned}
\varphi_\lambda(x) - \varphi_\lambda(y) &= \frac{1}{2\lambda}\|x - J_\lambda(x)\|^2 - \frac{1}{2\lambda}\|y - J_\lambda(y)\|^2 + \varphi(J_\lambda(x)) - \varphi(J_\lambda(y)) \\
&\le \frac{1}{2\lambda}\|x - J_\lambda(x)\|^2 - \frac{1}{2\lambda}\|y - J_\lambda(y)\|^2 \\
&\quad + \frac{1}{\lambda}(x - J_\lambda(x), J_\lambda(x) - J_\lambda(y)) \\
&= \frac{1}{\lambda}(x - J_\lambda(x), x - y) - \|x-y\|^2
\end{aligned}$$

$$+2(x-y,J_\lambda(x)-J_\lambda(y))-\|J_\lambda(x)-J_\lambda(y)\|^2$$
$$\leq \frac{1}{\lambda}(x-J_\lambda(x),x-y).$$

Therefore we showed that:

$$\varphi_\lambda(x)-\varphi_\lambda(y)\leq \frac{1}{\lambda}(x-J_\lambda x,x-y),\quad \forall y\in H$$

which is the desired result by the definition of A_λ.
(5) By (4),

$$\varphi_\lambda(y)-\varphi_\lambda(x)\leq (A_\lambda(y),y-x),\quad \forall x,y\in H,\ \lambda>0$$

By Part (3) of Theorem 3.4.1, we have:

$$0\leq \varphi_\lambda(y)-\varphi_\lambda(x)-(A_\lambda(x),y-x)\leq (A_\lambda(y)-A_\lambda(x),y-x)$$
$$\leq \frac{1}{\lambda}\|x-y\|^2,\quad \forall x,y\in H,\quad \lambda>0.$$

Consequently, for every $\lambda>0$, φ_λ is Fréchet differentiable on H and

$$\partial\varphi_\lambda(x)=\nabla\varphi_\lambda(x)=A_\lambda(x),\quad \forall x\in H.$$

□

Remark 3.4.5 *By Part 1 of Theorem 3.4.4,*

$$J_\lambda(x)=\text{Argmin}\{\frac{1}{2\lambda}\|x-y\|^2+\varphi(y);\ y\in H\},$$

and it is called the resolvent of the convex function φ.

The following result is important in the theory of nonlinear operators [ROC], and its proof can be found in [BAR1].

Proposition 3.4.6 *Let $A=\partial f$, where $f:H\to(-\infty,+\infty]$ is a proper, convex and lower semi-continuous function. Then the following assertions are equivalent:*

$$(a)\ \lim_{\substack{\|x\|\to\infty\\ x\in D(f)}} f(x)/\|x\|=+\infty;$$

$$(b)\, R(A)=H \text{ and } A^{-1} \text{ is bounded.}$$

3.5 CANONICAL EXTENSION

Let Ω be a Lebesgue measurable set in $\mathbb{R}^n$, $n\geq 1$ and $A:D(A)\subset H\to H$ be a maximal monotone operator. We define $\bar{A}:D(\bar{A})\subset L^2(\Omega;H)\to L^2(\Omega;H)$ by

$$D(\bar{A})=\{u\in L^2(\Omega;H);\ \exists v\in L^2(\Omega;H)\text{ such that } v(\zeta)\in A(u(\zeta)),\ \text{a.e. }\zeta\in\Omega\},$$

$$\bar{A}(u) = \{v \in L^2(\Omega;H);\ v(\zeta) \in A(u(\zeta)),\ \text{ a.e. } \zeta \in \Omega\},\ \ \forall u \in D(\bar{A}).$$

If either Ω has finite measure or $0 \in A(0)$, then $\bar{A}$ is maximal monotone. Monotonicity is easy to see, and for maximality, we note that under either assumption above, for each $f \in L^2(\Omega;H)$, we have $u(t) = (I+A)^{-1}(f(t)) \in L^2(\Omega;H)$ and $u+\bar{A}(u) \ni f$. The operator $\bar{A}$ is called the canonical extension of A to $L^2(\Omega;H)$ or the realization of A on $L^2(\Omega;H)$. The canonical extensions of $(I+\lambda A)^{-1}$ and A_λ to $L^2(\Omega;H)$ are $(I+\lambda\bar{A})^{-1}$ and $(\bar{A})_\lambda$.

Theorem 3.5.1 *Let $U \subset \mathbb{R}^n$ with $\mu(U) < +\infty$ and let $\varphi : H \to (-\infty,+\infty]$ be a proper, convex and lsc function. Let the function $\Phi : L^2(U;H) \to (-\infty,+\infty]$ be defined by*

$$\Phi(u) = \begin{cases} \int_U \varphi(u(s))ds, & \varphi(u) \in L^1(U) \\ +\infty, & \textit{otherwise} \end{cases}$$

(Since φ is bounded from below by an affine function, it follows that $\int_U \varphi(u(s))ds$ cannot assume the value $-\infty$). Then Φ is proper, convex and lsc and $\partial\Phi = \bar{A}$, where $\bar{A}$ denotes the canonical extension of $A = \partial\varphi$ to $L^2(U;H)$. Moreover, we have

$$\Phi_\lambda(u) = \int_U \varphi_\lambda(u(s))ds,\ \ \forall\lambda > 0,\ \ \forall u \in L^2(U;H) \tag{3.4}$$

and

$$\overline{D(\Phi)}^{L^2(U;H)} = \{u \in L^2(U;H);\ u(s) \in \overline{D(\varphi)}^H,\ \ a.e.\ s \in U\} \tag{3.5}$$

Proof. Obviously Φ is convex, and since $D(\Phi)$ contains at least the constant function $u \equiv h$, with $h \in D(\varphi)$, Φ is proper. To show that Φ is lsc on $L^2(U;H)$, we prove that for every $r \in \mathbb{R}$, the sub-level set $\{u \in L^2(U;H) :\ \Phi(u) \le r\}$ is closed in $L^2(U;H)$. To this end, let $\{u_m\}$ be a sequence such that $\Phi(u_m) \le r$ and $u_m \to u$ in $L^2(U;H)$. We may assume that $u_m(t) \to u(t)$, a.e. $t \in U$. Now take $[x_0,y_0] \in \partial\varphi$ and define the function

$$\tilde{\varphi}(x) = \varphi(x) - \varphi(x_0) - (y_0, x - x_0).$$

Then $\tilde{\varphi}$ is convex, lsc and $\tilde{\varphi} \ge 0$ on H. Moreover for each $u \in L^2(U;H)$, $\tilde{\varphi}(u)$ is measurable. Hence, by making use of Fatou's lemma, we have

$$\liminf_{m\to+\infty} \int_U \tilde{\varphi}(u_m(t))dt \ge \int_U \tilde{\varphi}(u(t))dt.$$

This obviously implies that $\Phi(u) \le r$ and therefore Φ is lsc. To show that $\bar{A} = \partial\Phi$, it suffices to show that $\bar{A} \subset \partial\Phi$. Let $[u,v] \in \bar{A}$. Then $v(t) \in A(u(t))$, a.e. $t \in U$ and so

$$\varphi(u(t)) - \varphi(w(t)) \le (v(t), u(t) - w(t)),\ \text{ a.e. } t \in U,\ \ \forall w \in D(\Phi)$$

This shows that $\varphi(u) \in L^1(U)$ and

$$\Phi(u) - \Phi(w) \le \int_U (v(t), u(t) - w(t))dt,\ \ \forall w \in D(\Phi)$$

and hence $v \in \partial\Phi(u)$. Since $[u,v]$ is arbitrary, we have $\bar{A} \subset \partial\Phi$. Now we prove (3.4).

$$\begin{aligned}\Phi_\lambda(u) &= \frac{1}{2\lambda}\|u-(I+\lambda\bar{A})^{-1}(u)\|^2_{L^2(U;H)} + \Phi((I+\lambda\bar{A})^{-1}(u)) \\ &= \frac{1}{2\lambda}\int_U \|u(t)-(I+\lambda A)^{-1}(u(t))\|^2 dt + \int_U \varphi((I+\lambda A)^{-1}(u(t)))dt \\ &= \int_U \varphi_\lambda(u(t))dt, \quad \forall u \in L^2(U;H).\end{aligned}$$

Finally, we prove (3.5). We know that:

$$D(\Phi) \subset \{u \in L^2(U;H);\ u(t) \in D(\varphi),\ \text{ a.e. } t \in U\}.$$

Therefore it suffices to show that for any $u \in L^2(U;H)$ with $u(t) \in \overline{D(\varphi)}$, a.e. $t \in U$, we have $u \in \overline{D(\Phi)}$ (here the closure is with respect to $L^2(U;H)$). Let u be such a function and set

$$u_\lambda(t) = (I+\lambda A)^{-1}(u(t)),\ \ \lambda > 0,\ \text{ a.e. } t \in U$$

Then, by Theorems 3.4.1 and 3.4.2, we get

$$\lim_{\lambda\to 0}\|u_\lambda(t)-u(t)\| = 0,\ \text{ a.e. } t \in U \tag{3.6}$$

On the other hand, since $(I+\lambda A)^{-1}$ is nonexpansive,

$$\|u_\lambda(t)\| \le \|u(t)\| + C,\quad \text{a.e. } t \in U \tag{3.7}$$

where C is a constant. By Theorem 1.4.5 and (3.7) and (3.6), $u_\lambda \to u$ in $L^2(U;H)$. Since $u_\lambda = (I+\lambda\bar{A})^{-1}(u) \in D(\bar{A})$, this implies that $u \in \overline{D(\Phi)}$ in $L^2(U;H)$, which shows that (3.5) holds. □

REFERENCES

BAR. V. Barbu, Nonlinear semigroups and Differential Equations in Banach Spaces, Noordhoff, Leiden, 1976.

BAR-PRE. V. Barbu and Th. Precupanu, Convexity and Optimization in Banach Spaces, D. Reidel Publishing Co., Dordrecht; Editura Academiei Bucharest, 1986.

BRE1. H. Brézis, Opérateurs maximaux monotones et semi-groupes de contractions dans les espaces de Hilbert (French) North-Holland Mathematics Studies, No. 5. Notas de Matemática (50). North-Holland Publishing Co., Amsterdam-London; American Elsevier Publishing Co., Inc., New York, 1973.

BRE2. H. Brézis, Functional Analysis, Sobolev Spaces and Partial Differential Equations, Universitext. Springer, New York, 2011.

MOR. G. Morosanu, Nonlinear Evolution Equations and Applications, D. Reidel Publishing Co., Dordrecht; Editura Academiei Bucharest, 1988.

MIN. G. J. Minty, Monotone (nonlinear) operators in Hilbert space, Duke Math. J. 29 (1962) 341–346.

ROC. R. T. Rockafellar, On the maximal monotonicity of subdifferential mappings, Pacific J. Math. 33 (1970), 209–216.

Part II

Evolution Equations of Monotone Type

4 First Order Evolution Equations

4.1 INTRODUCTION

In this chapter we consider the nonhomogeneous Cauchy problem for a maximal monotone operator of the form

$$\begin{cases} -u'(t) \in A(u(t)) + f(t), & \text{a.e. on } (0,+\infty) \\ u(0) = u_0 \in \overline{D(A)} \end{cases}$$

and study the existence, uniqueness, periodicity and the asymptotic behavior of the solutions. In the next section we concentrate on the existence and uniqueness of solutions. After which study the existence of periodic solutions when $f(t)$ is a periodic function, and $A = \partial\varphi$, where φ is a proper, convex and lsc function on H. Here, we also study the asymptotic convergence of bounded solutions to periodic solutions. Solutions to the above Cauchy problem in the homogeneous case form a semigroup of nonexpansive mappings whose infinitesimal generator is the maximal monotone operator A. In the nonhomogeneous case, the trajectory of the solutions are almost nonexpansive curves (as will be defined later). Therefore, in order to study the asymptotic behavior of the solutions, we study the asymptotic behavior of nonexpansive and almost nonexpansive curves. The material in this chapter is generally adapted from [MOR, DJA1, DJA2, DJA3, DJA4, KHA-MOH].

4.2 EXISTENCE AND UNIQUENESS OF SOLUTIONS

Consider the following Cauchy problem:

$$\frac{du}{dt} + A(u) \ni f(t), \quad 0 < t < T \tag{4.1}$$

$$u(0) = u_0 \tag{4.2}$$

where $A : D(A) \subset H \to H$ is a (possibly multi-valued) operator, $u_0 \in H$ and f is a given function. First we introduce two concepts of solutions.

Definition 4.2.1 *Let $f \in L^1(0,T;H)$. The function $u \in C([0,T];H)$ is called a strong solution (or breifly, solution) of* (4.1) *and* (4.2) *if:*
(1) *u is absolutely continuous on each compact subinterval of* $(0,T)$.
(2) $u(t) \in D(A)$ *for almost every* $t \in (0,T)$.
(3) $u(0) = u_0$ *and u satisfies* (4.1) *for a.e.* $t \in (0,T)$.

Definition 4.2.2 *A function $u \in C([0,T];H)$ is a weak solution for* (4.1) *and* (4.2) *if there exist sequences $\{u_n\} \subset W^{1,\infty}(0,T;H)$ and $\{f_n\} \subset L^1(0,T;H)$ such that:*
(1) $\frac{du_n}{dt}(t) + A(u_n(t)) \ni f_n(t)$, *a.e.* $t \in [0,T)$, $n = 1,2,\cdots$
(2) $u_n \to u$ *in* $C([0,T];H)$
(3) $u(0) = u_0$, $f_n \to f$ *in* $L^1(0,T;H)$

Now we state a Gronwall type Lemma that we will need in the sequel.

Lemma 4.2.3 *Suppose that $\psi \in L^1(a,b)$ $(-\infty < a < b < +\infty)$ with $\psi(t) \geq 0$ for a.e. $t \in (a,b)$, and let C be a real constant. If $h \in C[a,b]$ satisfies the following inequality*

$$\frac{1}{2}h^2(t) \leq \frac{1}{2}C^2 + \int_a^t \psi(s)h(s)ds, \quad \forall t \in [a,b]$$

then

$$|h(t)| \leq |C| + \int_a^t \psi(s)ds, \quad \forall t \in [a,b].$$

Proof. Set $g(t) = \frac{1}{2}C^2 + \int_a^t \psi(s)h(s)ds$. Then obviously

$$g'(t) = \psi(t)h(t) \leq \psi(t)|h(t)|, \quad \text{for a.e. } t \in (a,b).$$

From the assumption we have $|h(t)| \leq \sqrt{2g(t)}$, therefore:

$$g'(t) \leq \psi(t)\sqrt{2g(t)}$$

Dividing both sides of the above inequality by $\sqrt{2g(t)}$ and then integrating from a to t, we get:

$$\int_a^t \frac{g'(s)}{\sqrt{2g(s)}}ds \leq \int_a^t \psi(s)ds$$

This shows that:

$$(2g(t))^{\frac{1}{2}} \leq (2g(a))^{\frac{1}{2}} + \int_a^t \psi(s)ds = |C| + \int_a^t \psi(s)ds$$

Since by assumption $|h(t)| \leq \sqrt{2g(t)}$, we get:

$$|h(t)| \leq |C| + \int_a^t \psi(s)ds,$$

which is the desired result. □

Theorem 4.2.4 *Suppose that $A : D(A) \subset H \to H$ is maximal monotone, $u_0 \in D(A)$ and $f \in W^{1,1}(0,T;H)$, then the Problem* (4.1) *and* (4.2) *has a unique strong solution $u \in W^{1,\infty}(0,T;H)$. Moreover, u is differentiable from the right at each point of $[0,T)$ and*

$$\frac{d^+u}{dt}(t) = (f(t) - A(u(t)))^o, \quad \forall t \in [0,T) \tag{4.3}$$

$$\left\|\frac{d^+u}{dt}(t)\right\| \leq \|(f(0)-A(u_0))^o\| + \int_0^t \left\|\frac{df}{ds}(s)\right\|ds, \ \forall t \in [0,T) \tag{4.4}$$

where $(f(t)-A(u(t)))^o$ is the least norm element of the set $f(t)-A(u(t))$. Moreover, if u,v are solutions corresponding to (u_0,f) and (v_0,g) in $D(A)\times W^{1,1}(0,T;H)$, then:

$$\|u(t)-v(t)\| \leq \|u_0-v_0\| + \int_0^t \|f(s)-g(s)\|ds, \ 0 \leq t \leq T \tag{4.5}$$

Proof. First we prove (4.5) which implies the uniqueness of the solution. It follows from the monotonicity of A that:

$$\frac{1}{2}\frac{d}{dt}\|u(t)-v(t)\|^2 \leq \|f(t)-g(t)\|\|u(t)-v(t)\|, \ \text{a.e. } t \in (0,T)$$

By integrating this inequality on $[0,t]$ we obtain:

$$\frac{1}{2}\|u(t)-v(t)\|^2 \leq \frac{1}{2}\|u_0-v_0\|^2 + \int_0^t \|f(s)-g(s)\|\|u(s)-v(s)\|ds, \ \forall t \in [0,T].$$

Now an application of Lemma 4.2.3 implies (4.5).

Next we show the existence of the solution to problem (4.1) and (4.2). A classical result from the theory of ordinary differential equations shows the existence of a unique solution $u_\lambda \in C^1([0,T];H)$ for the Cauchy problem

$$\frac{du_\lambda}{dt}(t) + A_\lambda(u_\lambda(t)) = f(t), \ 0 \leq t \leq T \tag{4.6}$$

$$u_\lambda(0) = u_0 \tag{4.7}$$

for each $\lambda > 0$, where A_λ is the Yosida approximation of A, because A_λ is Lipschitz continuous on H. Now since A_λ is monotone we get:

$$\frac{1}{2}\frac{d}{dt}\|u_\lambda(t+h)-u_\lambda(t)\|^2 \leq \|f(t+h)-f(t)\|\|u_\lambda(t+h)-u_\lambda(t)\|, \ \forall t,t+h \in [0,T]$$

and therefore integrating this inequality from 0 to t, we obtain:

$$\frac{1}{2}\|u_\lambda(t+h)-u_\lambda(t)\|^2 \leq \frac{1}{2}\|u_\lambda(h)-u_0\|^2 + \int_0^t \|f(s+h)-f(s)\|\|u_\lambda(s+h)-u_\lambda(s)\|ds,$$

for all $t,t+h \in [0,T]$. By Lemma 4.2.3, we get

$$\|u_\lambda(t+h)-u_\lambda(t)\| \leq \|u_\lambda(h)-u_0\| + \int_0^t \|f(s+h)-f(s)\|ds, \tag{4.8}$$

for all $t,t+h \in [0,T]$. Now if we divide (4.8) by $h>0$ and take the limit as $h \to 0$, we get:

$$\left\|\frac{du_\lambda}{dt}(t)\right\| \leq \|f(0)-A_\lambda(u_0)\| + \int_0^t \left\|\frac{df}{ds}(s)\right\|ds \leq \|f(0)\| + \|A^o(u_0)\| + \int_0^T \left\|\frac{df}{ds}(s)\right\|ds, \tag{4.9}$$

for $0 \le t \le T$. Therefore by (4.6)

$$\|A_\lambda(u_\lambda(t))\| \le \|f(t)\| + \|f(0)\| + \|A^o(u_0)\| + \int_0^T \|\frac{df}{ds}(s)\| \le \text{Const}, \quad 0 \le t \le T \tag{4.10}$$

Next we prove the convergence of u_λ in $C([0,T];H)$ as $\lambda \to 0$. To this aim, multiplying the equation

$$\frac{du_\lambda}{dt}(t) - \frac{du_\mu}{dt} + A_\lambda(u_\lambda(t)) - A_\mu(u_\mu(t)) = 0, \quad \lambda, \mu > 0, \, t \in [0,T]$$

by $u_\lambda(t) - u_\mu(t)$, we get:

$$\begin{aligned} \frac{1}{2}\frac{d}{dt}\|u_\lambda(t) - u_\mu(t)\|^2 = \\ &- \big(A_\lambda(u_\lambda(t)) - A_\mu(u_\mu(t)), \lambda A_\lambda(u_\lambda(t)) - \mu A_\mu(u_\mu(t))\big) \\ &- \big(A_\lambda(u_\lambda(t)) - A_\mu(u_\mu(t)), J_\lambda(u_\lambda(t)) - J_\mu(u_\mu(t))\big), \quad 0 \le t \le T \end{aligned} \tag{4.11}$$

Now it follows easily from (4.10) and (4.11) and the fact that

$$A_\lambda(u_\lambda(t)) \in A\big(J_\lambda(u_\lambda(t))\big),$$

that

$$\frac{d}{dt}\|u_\lambda(t) - u_\mu(t)\|^2 \le C(\lambda + \mu), \quad \forall \lambda, \mu > 0 \quad 0 \le t \le T$$

for some constant C. Therefore

$$\|u_\lambda(t) - u_\mu(t)\| \le c t^{\frac{1}{2}} (\lambda + \mu)^{\frac{1}{2}}, \quad 0 \le t \le T \tag{4.12}$$

where $c = C^{\frac{1}{2}}$. But (4.12) states that the net $\{u_\lambda\}$ is Cauchy and therefore there is $u \in C([0,T];H)$ such that

$$u_\lambda \to u \text{ as } \lambda \to 0 \text{ in } C([0,T];H) \tag{4.13}$$

It follows from (4.9) that $u \in W^{1,\infty}(0,T;H)$ and

$$\frac{du_\lambda}{dt} \to \frac{du}{dt} \tag{4.14}$$

in the weak star topology of $L^\infty(0,T;H)$ as $\lambda \to 0$. But then

$$A_\lambda(u_\lambda) \overset{*}{\rightharpoonup} f - \frac{du}{dt} \tag{4.15}$$

in $L^\infty(0,T;H)$ as $\lambda \to 0$. On the other hand from (4.10) we can see that

$$J_\lambda(u_\lambda) \to u \text{ in } C([0,T];H) \tag{4.16}$$

Denote the extension of A in $L^2(0,T;H)$ by $\bar{A}$ which is a maximal monotone operator; then $\bar{A}$ is demiclosed. This fact together with (4.15) and (4.16) implies that

$$f - \frac{du}{dt} \in \bar{A}(u)$$

i.e. u satisfies (4.1) almost everywhere. Moreover, (4.7) and (4.13) show that $u(0) = u_0$, and (4.16) implies that

$$u(t) \in D(A), \ \ \forall t \in (0,T]$$

because A is demiclosed. Now we prove that u has right derivatives and satisfies (4.3). Suppose that t_0 is an arbitrary point in $[0,T)$. Multiplying both sides of

$$\frac{d}{dh}(u(t_0+h) - u(t_0)) \in f(t_0+h) - A(u(t_0+h))$$

for almost every $h > 0$ with $t_0 + h < T$, by $u(t_0+h) - u(t_0)$, then by the monotonicity of A, we get the following inequality:

$$\frac{1}{2}\frac{d}{dh}\|u(t_0+h)-u(t_0)\|^2 \leq \{\|(f(t_0)-A(u(t_0)))^o\| + \|f(t_0+h)-f(t_0)\|\}\|u(t_0+h)-u(t_0)\|$$

Now intergrating this inequality on $[0,h]$ and applying Lemma 4.2.3, we get:

$$\|u(t_0+h)-u(t_0)\| \leq h\|(f(t_0)-A(u(t_0)))^o\| + \int_0^h \|f(t_0+s)-f(t_0)\|ds, \ \ \forall h > 0 \tag{4.17}$$

with $t_0 + h < T$. This gives

$$\limsup_{h\to+0} \frac{1}{h}\|u(t_0+h)-u(t_0)\| \leq \|(f(t_0)-A(u(t_0)))^o\| \tag{4.18}$$

On the other hand since u is a strong solution of (4.1) and (4.2), for each $[x,y] \in A$, we have

$$\frac{1}{2}\|u(t)-x\|^2 \leq \frac{1}{2}\|u(s)-x\|^2 + \int_0^t (f(s)-y, u(s)-x)ds \ \ 0 \leq s \leq t \leq T$$

This implies that

$$(u(t_0+h)-u(t_0), u(t_0)-x) \leq \frac{1}{2}\|u(t_0+h)-x\|^2 - \frac{1}{2}\|u(t_0)-x\|^2 \tag{4.19}$$

$$\leq \int_{t_0}^{t_0+h} (f(s)-y, u(s)-x)ds, \tag{4.20}$$

for all $[x,y] \in A$ and for all $h > 0$ such that $t_0 + h < T$. Now taking into acount (4.18), we can choose a subsequence $h_n \to 0$ such that

$$\frac{u(t_0+h_n)-u(t_0)}{h_n} \rightharpoonup p, \ \ \text{in } H.$$

Now taking $h = h_n$ in (4.20) and dividing the resulting inequality by h_n and taking the limit as $n \to +\infty$, we conclude that

$$(p - f(t_0) + y, x - u(t_0)) \geq 0, \quad \forall [x,y] \in A$$

Therefore by using (4.18) and the maximality of A we see that

$$p = (f(t_0) - A(u(t_0)))^o \tag{4.21}$$

Therefore p does not depend on the choice of the sequence h_n and so

$$\frac{u(t_0+h) - u(t_0)}{h} \rightharpoonup (f(t_0) - A(u(t_0)))^o \text{ as } h \to 0.$$

But this fact together with (4.18) shows that u is right differentiable at t_0 and

$$\frac{d^+u}{dt}(t_0) = (f(t_0) - A(u(t_0)))^o$$

i.e. u satisfies (4.3).

A similar argument gives

$$\|u(t+h) - u(t)\| \leq \|u(h) - u_0\| + \int_0^t \|f(s+h) - f(s)\|ds, \quad \forall t, t+h \in [0,T] \tag{4.22}$$

$$\|u(h) - u_0\| \leq \int_0^h \|(f(s) - A(u_0))^o\|ds, \quad \forall h \in [0,T] \tag{4.23}$$

Now from estimates (4.22) and (4.23) we conclude that:

$$\|u(t+h) - u(t)\| \leq \int_0^h \|(f(s) - A(u_0))^o\|ds + \int_0^t \|f(s+h) - f(s)\|ds, \quad \forall t, t+h \in [0,T]$$

and this obviously implies (4.4), which completes the proof of the theorem. □

Theorem 4.2.5 *Suppose that $A : D(A) \subset H \to H$ is maximal monotone and $u_0 \in \overline{D(A)}$ and $f \in L^1(0,T;H)$. Then there exists a unique weak solution of* (4.1) *and* (4.2) *satisfying*

$$\frac{1}{2}\|u(t) - x\|^2 \leq \frac{1}{2}\|u(s) - x\|^2 + \int_s^t (f(s) - y, u(s) - x)ds, \tag{4.24}$$

for all $0 \leq s \leq t \leq T$ and $[x,y] \in A$. Moreover, if u and v are weak solutions corresponding to

$$(u_0, f), (v_0, g) \in \overline{D(A)} \times L^1(0,T;H),$$

then the following inequality holds

$$\frac{1}{2}\|u(t) - v(t)\|^2 \leq \frac{1}{2}\|u(s) - v(s)\|^2 + \int_s^t (f(r) - g(r), u(r) - v(r))dr, \quad 0 \leq s \leq t \leq T \tag{4.25}$$

Proof. Suppose that $u_0 \in \overline{D(A)}$ and $f \in L^1(0,T;H)$. Then there exist sequences $\{u_0^n\} \subset D(A)$ and $\{f_n\} \subset W^{1,1}(0,T;H)$ such that $u_0^n \to u_0$ in H and $f_n \to f$ in $L^1(0,T;H)$. Now from Theorem 4.2.4, we know that for each $n \in \mathbb{N}$, $u_n \in W^{1,\infty}(0,T;H)$ exists such that

$$\begin{cases} \frac{du_n}{dt}(t) + A(u_n(t)) \ni f_n(t), \text{ a.e. } t \in (0,T) \\ u_n(0) = u_0^n \end{cases} \tag{4.26}$$

Moreover estimate (4.5) implies that:

$$\|u_n(t) - u_m(t)\| \leq \|u_0^n - u_0^m\| + \int_0^T \|f_n(s) - f_m(s)\| ds$$

Therefore $u \in C([0,T];H)$ exists such that $u_n \to u$ in $C([0,T];H)$ and in particular $u(0) = u_0$. Obviously the function u is only a weak solution of (4.1) and (4.2). It remains to prove (4.24) and (4.25). Obviously these hold for strong solutions, and therefore passing to the limit, they also hold for weak solutions. Note that uniqueness is an immediate consequence of (4.25), and the proof is now complete. □

4.3 PERIODIC FORCING

Lemma 4.3.1 *Suppose that $\varphi : H \to (-\infty, +\infty]$ is a proper, convex and lsc function and $u \in W^{1,2}(t_0,T;H)$ such that $u(t) \in D(\partial\varphi)$, a.e. $t \in (t_0,T)$ and $g \in L^2(t_0,T;H)$ exists such that*

$$g(t) \in \partial\varphi(u(t)), \text{ a.e. } t \in (0,T)$$

where $-\infty < t_0 < T < +\infty$. Then the function $t \mapsto \varphi(u(t))$ is absolutely continuous on $[t_0,T]$ and for almost every $t \in (t_0,T)$ we have

$$\frac{d}{dt}\varphi(u(t)) = (h, \frac{du}{dt}(t)), \ \forall h \in \partial\varphi(u(t)) \tag{4.27}$$

Proof. Obviously, for each $\lambda > 0$ the function $t \mapsto \varphi_\lambda(u(t))$ is differentiable a.e. on (t_0,T) (remember from Chapter 2, $\varphi_\lambda(x) := \inf_{y\in H}\{\varphi(y) + \frac{1}{2\lambda}\|x-y\|^2\}$, which is called the Yosida regularization of φ) and

$$\begin{aligned} \frac{d}{dt}\varphi_\lambda(u(t)) &= (\nabla\varphi_\lambda(u(t)), \frac{du}{dt}(t)) \\ &= ((\partial\varphi)_\lambda(u(t)), \frac{du}{dt}(t)), \quad \text{a.e. } t \in (t_0,T) \end{aligned}$$

Therefore

$$\varphi_\lambda(u(t)) - \varphi_\lambda(u(s)) = \int_s^t ((\partial\varphi)_\lambda(u(r)), \frac{du}{dr}(r))dr, \ \ s,t \in [t_0,T] \tag{4.28}$$

Note that for almost all $t \in (t_0,T)$

$$\lim_{\lambda\to 0}(\partial\varphi)_\lambda(u(t)) = (\partial\varphi)^o(u(t))$$

and

$$\|(\partial\varphi)_\lambda(u(t))\| \leq \|(\partial\varphi)^o(u(t))\| \leq \|g(t)\|$$

consequently

$$\partial\varphi_\lambda(u) \to (\partial\varphi)^o(u) \quad \text{in } L^2(t_0,T;H),$$

therefore by taking the limit in (4.28) as $\lambda \to 0$, we get

$$\varphi(u(t)) - \varphi(u(s)) = \int_s^t ((\partial\varphi)^o(u(r)), \frac{du}{dr}(r))dr, \quad s,t \in [t_0,T]$$

This implies that the function $t \mapsto \varphi(u(t))$ is absolutely continuous on $[t_0,T]$. Now choose $t_1 \in [t_0,T]$ such that both $u(t)$ and $\varphi(u(t))$ are differentiable at $t = t_1$ and $u(t_1) \in D(\partial\varphi)$, then for each $h \in \partial\varphi(u(t_1))$, we have

$$\varphi(u(t_1)) - \varphi(v) \leq (h, u(t_1) - v), \quad \forall v \in H.$$

Now taking $v = u(t_1 \pm \varepsilon)$ in the above inequality where $\varepsilon > 0$, then dividing the resulting inequality by ε and taking the limit as $\varepsilon \to 0$, we get

$$\frac{d}{dt}\varphi(u(t)) \mid_{t=t_1} = (h, \frac{du}{dt}(t_1)),$$

which completes the proof. □

Lemma 4.3.2 *Suppose that u is a weak solution of* (4.1) *with* $A = \partial\varphi$ *and* $f \in L^2(0,T;H)$, *then*

$$\int_0^T t\|\frac{du}{dt}\|^2 dt \leq \int_0^T t\|f(t)\|^2 dt + 2(\|u_0 - x_0\| + \int_0^T \|f(t)\|dt)^2,$$

where $x_0 \in (\partial\varphi)^{-1}(0)$

Proof. Take $[x_0,y_0] \in \partial\varphi$. We define $\tilde{\varphi}(x) = \varphi(x) - \varphi(x_0) - (y_0, x - x_0)$. Then problem (4.1) is equivalent to:

$$\frac{du}{dt} + \partial\tilde{\varphi}(u) \ni f(t) - y_0, \quad 0 < t < T$$

Hence without loss of generality we may assume that

$$\min_{x\in H}\varphi(x) = \varphi(x_0) = 0$$

and so $\varphi(x) \geq 0$. Multiplying both sides of (4.1) by $t\frac{du}{dt}$ and using Lemma 4.3.1, we get

$$t\|\frac{du}{dt}(t)\|^2 + t\frac{d}{dt}\varphi(u(t)) = t(f(t), \frac{du}{dt}(t)), \quad \text{a.e. } t \in (0,T)$$

Then after integration from 0 to T, we obtain

$$\int_0^T t\|\frac{du}{dt}\|^2 dt + T\varphi(u(T)) = \int_0^T t(f(t), \frac{du}{dt}(t))dt + \int_0^T \varphi(u(t))dt$$

Since $\varphi \geq 0$, a straightforward computation shows that

$$\int_0^T t\|\frac{du}{dt}(t)\|^2 dt \leq \int_0^T t\|f(t)\|^2 dt + 2\int_0^T \varphi(u(t))dt \tag{4.29}$$

On the other hand by (4.1), we have

$$\varphi(u(t)) \leq (f(t) - \frac{du}{dt}(t), u(t) - x_0), \quad \text{a.e. } t \in (0,T)$$

Therefore

$$\int_0^T \varphi(u(t))dt \leq \frac{1}{2}\|u_0 - x_0\|^2 + \int_0^T \|f(t)\|\|u(t) - x_0\|dt \tag{4.30}$$

Now multiplying (4.1) by $u(t) - x_0$ and integrating on $[0,t]$, by using Lemma 4.2.3, we get

$$\|u(t) - x_0\| \leq \|u_0 - x_0\| + \int_0^T \|f(t)\|dt, \quad 0 \leq t \leq T \tag{4.31}$$

From (4.30) and (4.31), we get

$$\int_0^T \varphi(u(t))dt \leq (\|u_0 - x_0\| + \int_0^T \|f(t)\ |dt)^2 \tag{4.32}$$

Now (4.29) and (4.32) imply that

$$\int_0^T t\|\frac{du}{dt}\|^2 dt \leq \int_0^T t\|f(t)\|^2 dt + 2(\|u_0 - x_0\| + \int_0^T \|f(t)\|dt)^2, \tag{4.33}$$

which is the desired inequality. □

Theorem 4.3.3 *Suppose that $A = \partial\varphi$ where $\varphi : H \to (-\infty, +\infty]$ is a proper, convex and lsc function and $f \in L^2_{\text{loc}}(\mathbb{R}^+; H)$ is a $T-$periodic function ($T > 0$). Then the Equation* (4.1) *has a bounded solution if and only if it has at least a periodic solution with period $T > 0$. In this case, all solutions of* (4.1) *are bounded on $\mathbb{R}^+$, and for each solution $z(t)$, $t \geq 0$ there is a T-periodic solution p of* (4.1) *such that*

$$z(t) - p(t) \rightharpoonup 0, \quad \text{as } t \to +\infty. \tag{4.34}$$

Moreover, any two periodic solutions differ by a constant and

$$\frac{dz_n}{dt} \to \frac{dp}{dt}, \quad \text{as } n \to +\infty \tag{4.35}$$

in $L^2(0,T;H)$, where $z_n(t) = z(t+nT)$, $n = 1, 2, \cdots$.

Proof. First note that if $z(t)$, $t \geq 0$ is a bounded solution of (4.1) on $\mathbb{R}^+$, then every other solution $v(t)$, $t \geq 0$ is also bounded because

$$\|z(t) - v(t)\| \leq \|z(0) - v(0)\|.$$

Now we intend to prove that the boundedness of solutions on $\mathbb{R}^+$ implies the existence of a periodic solution. To this aim, define $Q : \overline{D(A)} \to \overline{D(A)}$ as $Q(x) = z(T;x)$ where $z(t;x)$, $t \geq 0$ is the solution of (4.1) with initial condition $x \in \overline{D(A)}$. It is easy to see that Q is nonexpansive. For each $x \in \overline{D(A)}$ the sequence $\{Q^n(x)\}$ is bounded because $Q^n(x) = z(nT;x)$. By Browder-Petryshyn fixed point theorem, Q has at least a fixed point. But this means that (4.1) has at least a T-periodic solution. To prove the relations (4.34) and (4.35) we use Opial's lemma. Suppose that $z(t)$, $t \geq 0$ is a bounded solution of (4.1). Denote the set of all T-periodic solutions of (4.1) by F. Define $z_n : [0,2T] \to H$ by $z_n(t) = z(t+nT)$, $n = 1,2,\cdots$. Extending each $p \in F$ by periodicity on the interval $[0,2T]$, we deduce that F is a subset of $L^2(0,2T;H)$. We have:

$$\begin{aligned}
\|z_n(t) - p(t)\| &\leq \|z_n(0) - p(0)\| \\
&= \|z_{n-1}(T) - p(T)\| \\
&\leq \|z_{n-1}(t) - p(t)\| \\
&\leq \|z_{n-1}(0) - p(0)\|, \quad 0 \leq t \leq T,\ p \in F
\end{aligned} \tag{4.36}$$

Therefore

$$\lim_{n \to +\infty} \|z_n(t) - p(t)\| = l(p) \tag{4.37}$$

exists uniformly on $[0,T]$, and consequently

$$\|z_n - p\|_{\mathscr{H}} \to \rho(p) := (2T)^{\frac{1}{2}} l(p), \quad \forall p \in F \tag{4.38}$$

where $\mathscr{H} = L^2(0,2T;H)$. Therefore the first condition in Opial's lemma is satisfied. Now without loss of generality suppose that $\text{Min}\varphi = \varphi(x_0) = 0$ where $x_0 \in D(A)$. By Lemma 4.3.2, we have

$$\int_0^{2T} t\|\frac{dz_n}{dt}(t)\|^2 dt \leq \int_0^{2T} t\|f(t)\|^2 dt + 2(\|z_n(0) - x_0\| + \int_0^{2T} \|f(t)\| dt)^2 \leq C \tag{4.39}$$

This implies that

$$\int_T^{2T} t\|\frac{dz_n}{dt}(t)\|^2 dt \leq C$$

i.e.

$$\{\frac{dz_n}{dt}\} \text{ is bounded in } \mathscr{H} \tag{4.40}$$

Suppose that $q \in \mathscr{H}$ is a weak cluster point of $\{z_n\}$ i.e. $z_{n_k} \rightharpoonup q \in \mathscr{H}$. By (4.40), $\frac{dq}{dt} \in \mathscr{H}$ and

$$\frac{dz_{n_k}}{dt} \rightharpoonup \frac{dq}{dt} \text{ in } \mathscr{H} \tag{4.41}$$

Let $\bar{A}$ be the canonical extension of A to $\mathscr{H}$. By Theorem 3.5.1, $\bar{A} = \partial\Phi$ where $\Phi : \mathscr{H} \to (-\infty, +\infty]$ is defined by

$$\Phi(v) = \begin{cases} \int_0^{2T} \varphi(v) dt & \text{if } \varphi(v) \in L^1(0,2T) \\ +\infty, & \text{otherwise} \end{cases}$$

It is obvious that

$$\Phi(z_n) - \Phi(p) \le (f - \frac{dz_n}{dt}, z_n - p)_{\mathscr{H}}, \quad \forall p \in F$$

$$\Phi(p) - \Phi(v) \le (f - \frac{dp}{dt}, p - v)_{\mathscr{H}}, \quad \forall v \in D(\Phi),\ \forall p \in F$$

Consequently,

$$\Phi(z_n) - \Phi(v) \le -(\frac{dz_n}{dt} - \frac{dp}{dt}, z_n - p)_{\mathscr{H}} + (f - \frac{dp}{dt}, z_n - v)_{\mathscr{H}}, \quad \forall v \in D(\Phi),\ \forall p \in F \tag{4.42}$$

On the other hand from (4.37) we have

$$(\frac{dz_n}{dt} - \frac{dp}{dt}, z_n - p)_{\mathscr{H}} = (\frac{1}{2})(\|z_n(2T) - p(2T)\|^2 - \|z_n(0) - p(0)\|^2) \to 0, \quad \forall p \in F \tag{4.43}$$

Now by using (4.42) and (4.43) we can see that

$$\Phi(q) - \Phi(v) \le (f - \frac{dp}{dt}, q - v)_{\mathscr{H}}, \quad \forall v \in D(\Phi)$$

which implies that

$$f - \frac{dp}{dt} \in \bar{A}(q), \quad \forall p \in F \tag{4.44}$$

By Corollary 2.6.5, since $\partial\Phi^* = \bar{A}^{-1}$, where Φ^* is the conjugate function of Φ, we have

$$\Phi^*(f - z_n') - \Phi^*(f - p') \le (z_n, -z_n' + p')_{\mathscr{H}}$$

and

$$\Phi^*(f - p') - \Phi^*(f - v) \le (p, -p' + v)_{\mathscr{H}}, \quad v \in \mathscr{H},$$

where $z_n' = \frac{dz_n}{dt}$ and $p' = \frac{dp}{dt}$. Summing up the two inequalities above, we get:

$$\begin{aligned} \Phi^*(f - z_n') - \Phi^*(f - v) &\le -(z_n - p, z_n' - p')_{\mathscr{H}} - (p, z_n' - v)_{\mathscr{H}} \\ &= \frac{1}{2}(\|z_n(0) - p(0)\|^2 - \|z_n(2T) - p(2T)\|^2) \\ &\quad - (p, z_n' - v)_{\mathscr{H}} \end{aligned}$$

Now letting $n \to +\infty$, by (4.37), and since Φ^* is lsc, we obtain:

$$\Phi^*(f - q') - \Phi^*(f - v) \le -(p, q' - v)_{\mathscr{H}}$$

Again by Corollary 2.6.5, this shows that

$$f - \frac{dq}{dt} \in \bar{A}(p), \quad \forall p \in F \tag{4.45}$$

Therefore by Lemma 4.3.1 we have

$$\frac{d}{dt}\Phi(q) = (f - \frac{dp}{dt}, \frac{dq}{dt}) \tag{4.46}$$

and

$$\frac{d}{dt}\Phi(p) = (f - \frac{dq}{dt}, \frac{dp}{dt}) = (f - \frac{dp}{dt}, \frac{dp}{dt}) \tag{4.47}$$

for almost every $t \in (0,2T)$ and $p \in F$. Therefore

$$\begin{aligned} 2T\Phi(q(2T)) - \int_0^{2T} \Phi(q)dt &= \int_0^{2T} (f - \frac{dp}{dt}, \frac{dq}{dt})tdt \\ &= \int_0^{2T} ((f, \frac{dq}{dt}) + \|\frac{dp}{dt}\|^2 - 2(\frac{dq}{dt}, \frac{dp}{dt}))tdt \end{aligned} \tag{4.48}$$

On the other hand, by taking $v = q$ in (4.42) we obtain

$$\limsup_{k\to+\infty} \Phi(z_{n_k}) \le \Phi(q)$$

and therefore since Φ is lsc, we get:

$$\lim_{k\to+\infty} \Phi(z_{n_k}) = \Phi(q) \tag{4.49}$$

Similarly, we also have:

$$2T\Phi(z_n(2T)) - \int_0^{2T} \Phi(z_n)dt = \int_0^{2T} (f - \frac{dz_n}{dt}, \frac{dz_n}{dt})tdt \tag{4.50}$$

Now in (4.50), replacing n by n_k, substracting it from (4.48) and taking the limit as $k \to +\infty$, we get

$$\limsup_{k\to+\infty} \int_0^{2T} t\|\frac{dz_{n_k}}{dt} - \frac{dp}{dt}\|^2 dt = \limsup_{k\to+\infty} 2T\left(\Phi(q(2T)) - \Phi(z_{n_k}(2T))\right) \tag{4.51}$$

On the other hand, since $z_{n_k} \rightharpoonup q$ in $\mathscr{H}$, then by (4.41) we get

$$z_{n_k}(t) \rightharpoonup q(t), \text{ as } k \to +\infty, \ \forall t \in [0,2T] \tag{4.52}$$

Now it follows from (4.51), (4.52) and the fact that Φ is lsc, that

$$\lim_{k\to+\infty} \int_0^{2T} t\|\frac{dz_{n_k}}{dt} - \frac{dp}{dt}\|^2 dt = 0, \ \forall p \in F \tag{4.53}$$

and therefore

$$\frac{dq}{dt} = \frac{dp}{dt}, \ \forall p \in F$$

Therefore by (4.44), $q \in F$. Moreover this shows that every two periodic solutions differ by a constant. Thus all conditions in Opial's lemma are satisfied, therefore $p \in F$ exists such that $z_n \rightharpoonup p$ in $\mathscr{H}$. In fact this means that $q = p$. Everything that was proved for z_{n_k} is valid for z_n. Then (4.53) with z_n instead of z_{n_k} implies (4.35) because

$$T\int_0^T \|\frac{dz_n}{dt} - \frac{dp}{dt}\|^2 dt \le \int_T^{2T} t\|\frac{dz_n}{dt} - \frac{dp}{dt}\|dt$$

Finally (4.34) follows from the fact that $z_n(t) \rightharpoonup p(t)$ as $n \to +\infty$ uniformly with respect to $t \in [0,T]$. The proof is now complete. □

4.4 NONEXPANSIVE SEMIGROUP GENERATED BY A MAXIMAL MONOTONE OPERATOR

Definition 4.4.1 *Let C be a nonempty closed subset of H. A continuous semigroup of contractions (or breifly, a semigroup of contractions) on C is a family of operators $\{S(t) : C \to C;\ t \geq 0\}$ such that:*
(1) $S(t+s)x = S(t)S(s)x, \quad \forall x \in C, \ \forall t \geq 0$
(2) $S(0)x = x, \quad \forall x \in C$
(3) *for each $x \in C$ the mapping $t \to S(t)x$ is continuous on $[0, +\infty)$*
(4) $\|S(t)x - S(t)y\| \leq \|x - y\|, \quad \forall x, y \in C, \ \forall t \geq 0$

If besides 4 all of the other above conditions are satisfied, then $S(t)$ is called a continuous semigroup or briefly, a semigroup.

The infinitesimal generator of a semigroup $\{S(t) : \ C \to C;\ t \geq 0\}$, say G, is defined by

$$G(x) = \lim_{h \to 0^+} \frac{S(h)x - x}{h} \tag{4.54}$$

where $D(G)$ is the set of all $x \in C$ for which the strong limit in (4.54) exists in H.

Suppose that $A : D(A) \subset H \to H$ is a maximal monotone operator. Consider the following Cauchy problem

$$\frac{du}{dt} + A(u) \ni 0, \quad t \geq 0 \tag{4.55}$$

$$u(0) = u_0 \tag{4.56}$$

From Theorem 4.2.4, we know that for every $u_0 \in D(A)$, there exists a strong solution $u(t)$, $t \geq 0$ of (4.55) and (4.56). Set

$$S(t)u_0 = u(t), \quad t \geq 0$$

It is easy to check that for each $t \geq 0$, $S(t)$ is nonexpansive on $D(A)$ and therefore it can be uniquely extended to $\overline{D(A)}$. Moreover it is obvious that $S(t) : \overline{D(A)} \to \overline{D(A)}$; $t \geq 0$ is a continuous semigroup of nonexpansive mappings as was defined in Definition 4.4.1. Taking into account (4.3) we can also see that the infinitesimal generator of this semigroup is $-A^o$, where A^o is the minimal section of A. The given semigroup is called the semigroup generated by $-A$. On the other hand, by a result of Crandall and Pazy (see [BRE] p. 114), we know that if C is a closed convex subset of H, and $\{S(t) : C \to C;\ t \geq 0\}$ is a semigroup of nonexpansive mappings, then there exists a unique maximal monotone operator A with $\overline{D(A)} = C$, such that $(S(t))_{t \geq 0}$ coincides with the semigroup generated by $-A$.

4.5 ERGODIC THEOREMS FOR NONEXPANSIVE SEQUENCES AND CURVES

Definition 4.5.1 *A mapping $T : D \subset H \to H$ is called nonexpansive if*

$$\|Tx - Ty\| \leq \|x - y\|, \quad \forall x, y \in D,$$

where D is a nonempty subset of H.

The set of all fixed points of T is denoted by $F(T)$, therefore $F(T) = \{x \in D;\ \ Tx = x\}$. The first fixed point theorem for nonexpansive self-mappings was proved in 1965, simultaneously by Browder [BRO], Göhde [GOH] and Kirk [KIR], extending Banach's contraction principle.

Theorem 4.5.2 *Suppose that $T : D \to D$ is a nonexpansive mapping, where D is a nonempty, closed and convex subset of a Hilbert space H. Then T has a fixed point, and $F(T)$ is a closed and convex subset of H.*

Although in the Banach fixed point theorem, all orbits converge to the unique fixed point of T, this fact does not hold for a nonexpansive mapping, and orbits may not converge at all. As a simple example, consider $T : [-1,1] \to [-1,1]$ defined by $Tx = -x$. Clearly, for each non-zero point in the interval $[-1,1]$, the Picard iterates do not converge. Baillon [BAI] proved that the Cesaro means of the Picard iterates of any nonexpansive mapping T, always converge weakly to a fixed point of T, provided that $F(T) \neq \emptyset$. In the following, we state Baillon's mean ergodic theorem.

Theorem 4.5.3 *Let C be a closed convex subset of a Hilbert space H, and T be a nonexpansive mapping from C into itself. If the set $F(T)$ of fixed points of T is nonempty, then for each $x \in C$, the Cesaro means*

$$S_n(x) := \frac{1}{n} \sum_{k=0}^{n-1} T^k x$$

converge weakly to some $y \in F(T)$. If we define $P : C \to C$ by $Px = y$, then P is a nonexpansive retraction of C onto $F(T)$ such that $PT = TP = P$ and Px is contained in the closed convex hull of $\{T^n x : \ n = 1,2,ldots\}$ for each $x \in C$. This retraction is called an "ergodic retraction".

If C is not convex, then $F(T)$ may be empty, and then Baillon's proof is not applicable. To avoid the convexity assumption on C, we are going to introduce the notion of nonexpansive sequences.

4.5.1 ALMOST NONEXPANSIVE SEQUENCES

Definition 4.5.4 *A sequence $\{x_i\}$ in a Hilbert space H is called a nonexpansive sequence if*

$$\|x_{i+1} - x_{j+1}\| \leq \|x_i - x_j\|,$$

for all $i, j \geq 0$.

Similarly a nonexpansive curve is defined as follows:

Definition 4.5.5 *A curve $u : \mathbb{R}^+ \to H$ is called nonexpansive if*

$$\|u(t+h) - u(s+h)\| \leq \|u(t) - u(s)\|, \quad \forall t,s \geq 0, \ \forall h \geq 0.$$

Let $\{x_n\}$ be a sequence in H; let $s_n = \frac{1}{n}\sum_{i=0}^{n-1} x_i$.

Definition 4.5.6 *The sequence $\{x_n\}$ is said to be almost nonexpansive (in short ANES) if*

$$\|x_{i+k} - x_{j+k}\|^2 \leq \|x_i - x_j\|^2 + \varepsilon(i,j), \quad \forall i,j,k \geq 0,$$

where

$$\lim_{i,j \to +\infty} \varepsilon(i,j) = 0.$$

Obviously every nonexpansive sequence is an ANES.

Remark 4.5.7 *A bounded sequence $\{x_n\}$ in H that satisfies*

$$\|x_{i+k} - x_{j+k}\| \leq \|x_i - x_j\| + \varepsilon_1(i,j), \quad \forall i,j,k \geq 0$$

where

$$\lim_{i,j \to +\infty} \varepsilon_1(i,j) = 0$$

is an ANES.

Definition 4.5.8 *Given a bounded sequence $\{x_n\}$ in H, the asymptotic center c of $\{x_n\}$ is defined as follows (see [EDE]): for every $q \in H$, let $\varphi(q) = \lim_{n\to+\infty} \sup \|x_n - q\|^2$. Then φ is a continuous strictly convex function on H, satisfying $\varphi(q) \to +\infty$ as $\|q\| \to +\infty$. Thus φ achieves its minimum on H at a unique point c, called the asymptotic center of the sequence x_n.*

Remark 4.5.9 *a) $\omega_w(x_n)$ denotes the weak ω-limit set of x_n.*
b) $F(x_n)$ (in short F, when there is no confusion) denotes the following (possibly empty) subset of H:

$$F = \{q \in H : \lim_{n\to+\infty} \|x_n - q\| \text{exists}\}.$$

Note that if $F \neq \emptyset$, then the sequence $\{x_n\}$ is bounded.

Lemma 4.5.10 *F is a closed convex (possibly empty) subset of H.*

Proof. The closedness follows from the inequality

$$\begin{aligned}
\big|\|x_n - q\| - \|x_{n+k} - q\|\big| &= \big|\|x_n - q\| - \|x_n - q_m\| + \|x_n - q_m\| \\
&\quad - \|x_{n+k} - q_m\| + \|x_{n+k} - q_m\| - \|x_{n+k} - q\|\big| \\
&\leq \big|\|x_n - q\| - \|x_n - q_m\|\big| + \big|\|x_n - q_m\| \\
&\quad - \|x_{n+k} - q_m\|\big| + \big|\|x_{n+k} - q_m\| - \|x_{n+k} - q\|\big| \\
&\leq 2\|q_m - q\| + \big|\|x_n - q_m\| - \|x_{n+k} - q_m\|\big|.
\end{aligned}$$

The convexity follows from the equalities

$$\|x_n - (\lambda q_1 + (1-\lambda)q_2)\|^2 = \|\lambda(x_n - q_1) + (1-\lambda)(x_n - q_2)\|^2$$

$$= \lambda^2\|x_n - q_1\|^2 + (1-\lambda)^2\|x_n - q_2\|^2 + 2\lambda(1-\lambda)(x_n - q_1, x_n - q_2)$$

and

$$2(x_n - q_1, x_n - q_2) = \|x_n - q_1\|^2 + \|x_n - q_2\|^2 - \|q_1 - q_2\|^2.$$

□

Proposition 4.5.11 *For an ANES $\{x_n\}$ in H, the weak limit p of any weakly convergent subsequence $\{s_{m_l}\}$ of s_n (if any) belongs to F(i.e. $\lim_{n\to+\infty}\|x_n - p\|$ exists).*

Proof. For each $k, i \geq 0$ and each $m \geq 1$, by polarization identity, we have:

$$\begin{aligned}
2(x_k - x_{k+m}, x_i - p) &= 2(x_k - p, x_i - p) - 2(x_{k+m} - p, x_i - p) \\
&= \|x_k - p\|^2 + \|x_i - p\|^2 - \|x_k - x_i\|^2 - \|x_{k+m} - p\|^2 \\
&\quad - \|x_i - p\|^2 + \|x_{k+m} - x_i\|^2 \\
&= \|x_k - p\|^2 - \|x_{k+m} - p\|^2 + \|x_{k+m} - x_i\|^2 - \|x_k - x_i\|^2
\end{aligned}$$

Now summing both sides from $i = 0$ to $i = n-1$ and dividing by n, we get:

$$\begin{aligned}
2(x_k - x_{k+m}, s_n - p) &= \|x_k - p\|^2 - \|x_{k+m} - p\|^2 + \frac{1}{n}\sum_{i=0}^{n-1}\|x_{k+m} - x_i\|^2 \\
&\quad - \frac{1}{n}\sum_{i=0}^{n-1}\|x_k - x_i\|^2 \\
&= \|x_k - p\|^2 - \|x_{k+m} - p\|^2 + \frac{1}{n}\sum_{i=0}^{m-1}\|x_{k+m} - x_i\|^2 \\
&\quad + \frac{1}{n}\sum_{i=m}^{n-1}\|x_{k+m} - x_i\|^2 - \frac{1}{n}\sum_{i=0}^{n-1}\|x_k - x_i\|^2 \\
&= \|x_k - p\|^2 - \|x_{k+m} - p\|^2 + \frac{1}{n}\sum_{i=0}^{m-1}\|x_{k+m} - x_i\|^2 \\
&\quad + \frac{1}{n}\sum_{i=0}^{n-m-1}\|x_{k+m} - x_{i+m}\|^2 - \frac{1}{n}\sum_{i=0}^{n-m-1}\|x_k - x_i\|^2 \\
&\quad - \frac{1}{n}\sum_{i=n-m}^{n-1}\|x_k - x_i\|^2 \\
&\leq \|x_k - p\|^2 - \|x_{k+m} - p\|^2 + \frac{1}{n}\sum_{i=0}^{m-1}\|x_{k+m} - x_i\|^2 \\
&\quad + \frac{1}{n}\sum_{i=0}^{n-1}\varepsilon(i,k).
\end{aligned}$$

Let $\varepsilon > 0$ be given, and choose N_0 so that

$$\varepsilon(i,j) \leq \varepsilon, \quad \forall i,j \geq N_0.$$

Then taking $k \geq N_0$, and replacing n by m_l and letting $l \to +\infty$, we get

$$\|x_{k+m} - p\|^2 \leq \|x_k - p\|^2 + \varepsilon, \quad \forall k \geq N_0, \ \forall m \geq 1.$$

Hence

$$\limsup_{n\to+\infty} \|x_n - p\|^2 \leq \|x_k - p\|^2 + \varepsilon, \quad \forall k \geq N_0.$$

Thus

$$\limsup_{n\to+\infty} \|x_n - p\|^2 \leq \liminf_{n\to+\infty} \|x_n - p\|^2 + \varepsilon, \quad \forall \varepsilon > 0,$$

which implies that $\lim_{n\to+\infty} \|x_n - p\|$ exists, and therefore $p \in F$. □

Lemma 4.5.12 *If $p_1, p_2 \in F$ and $q_1, q_2 \in \omega_w(\{x_n\})$, then $(p_1 - p_2, q_1 - q_2) = 0$. In particular, $F \cap \overline{conv}\omega_w(\{x_n\})$contains at most one point.*

Proof. Assume $x_{n_k} \rightharpoonup q_1$ as $k \to +\infty$ and $x_{m_l} \rightharpoonup q_2$ as $l \to +\infty$, we have:

$$\lim_{n\to+\infty} \|x_n - p_i\|^2 = \varphi(p_i), \quad \text{for } i = 1,2,$$

and

$$\|x_n - p_2\|^2 = \|x_n - p_1\|^2 + \|p_1 - p_2\|^2 + 2(x_n - p_1, p_1 - p_2)$$

thus

$$\varphi(p_2) = \varphi(p_1) + \|p_1 - p_2\|^2 + 2(q_1 - p_1, p_1 - p_2)$$
$$\varphi(p_2) = \varphi(p_1) + \|p_1 - p_2\|^2 + 2(q_2 - p_1, p_1 - p_2).$$

Hence by subtraction, we get $(q_1 - q_2, p_1 - p_2) = 0$. □

Theorem 4.5.13 *Let $\{x_n\}$ be an ANES in H. Then the following are equivalent:*
i) $F \neq \emptyset$.
ii) $\liminf_{n\to+\infty} \|s_n\| < +\infty$.
iii) s_n converges weakly to $p \in H$.
Moreover under these conditions, we have:
a) $\overline{conv}\omega_w(\{x_n\}) \cap F = \{p\}$.
b) p is the asymptotic center of the sequence $\{x_n\}$.

Proof. (i)⇒(ii): If $F \neq \emptyset$, then $\{x_n\}$ is bounded hence $\|s_n\|$ is also bounded.

(ii)⇒(iii): Assume $\liminf_{n\to+\infty} \|s_n\| < +\infty$. Then by the reflexivity of H, s_n has a weakly convergent subsequence $\{s_{n_k}\}$ to some $p \in H$. Let us prove that $s_n \rightharpoonup p$ as $n \to +\infty$. In fact if there was another subsequence $s_{m_l} \rightharpoonup q \in H$ as $l \to +\infty$, then by Proposition 4.5.11, we have $p \in F$ and $q \in F$. (Thus $\|x_n\|$ is bounded.) Hence the sequence

$$\alpha_k = \|x_k - p\|^2 - \|x_k - q\|^2 = \|p\|^2 - \|q\|^2 + 2(x_k, q - p)$$

has a limit as $k \to +\infty$. Thus $\lim_{k\to+\infty}(x_k, q-p)$ exists and therefore $(p, q-p) = (q, q-p)$, which implies $\|q-p\|^2 = 0$ and hence $q = p$. Hence, every weakly convergent subsequence of s_n converges weakly to p, therefore (since $\|s_n\|$ is bounded), s_n converges weakly to p.

(iii)$\Rightarrow$(i): This follows from Proposition 4.5.11 since $p \in F$.

Let us now prove (a). We Obviously have $p \in \overline{conv}\,\omega_w(\{x_n\}) \cap F$, and the conclusion follows from Lemma 4.5.12.

Finally, let us prove (b). For every $u \in H$, we have:

$$\|x_n - p\|^2 = \|x_n - u\|^2 + \|u - p\|^2 + 2(x_n - u, u - p),$$

hence

$$\frac{1}{n}\sum_{i=0}^{n-1}\|x_i - p\|^2 = \frac{1}{n}\sum_{i=0}^{n-1}\|x_i - u\|^2 + \|u - p\|^2 + 2(s_n - u, u - p).$$

Since $p \in F$, we have

$$\lim_{n\to+\infty}\|x_n - p\|^2 = \varphi(p),$$

therefore letting $n \to +\infty$, we get that

$$\varphi(p) \le \limsup_{n\to+\infty}\|x_n - u\|^2 - \|p - u\|^2 = \varphi(u) - \|p - u\|^2, \quad \forall u \in H.$$

This implies that p is the asymptotic center of $\{x_n\}$. □

Proposition 4.5.14 *Assume $\{x_n\}$ is an ANES in H which is weakly asymptotically regular (i.e. $x_{n+1} - x_n \rightharpoonup 0$). Then $\omega_w(\{x_n\}) \subset F$ (in particular $\omega_w(\{x_n\}) \neq \emptyset \Rightarrow \|x_n\|$ is bounded).*

Proof. By a similar proof as in Proposition 4.5.11, we get

$$\begin{aligned}
2(x_k - x_{k+m}, \frac{1}{n}\sum_{i=0}^{n-1} x_{n_j+i} - p) &\le \|x_k - p\|^2 - \|x_{k+m} - p\|^2 \\
&\quad + \frac{1}{n}\sum_{i=0}^{m-1}\|x_{k+m} - x_{n+i}\| + \frac{1}{n}\sum_{i=m}^{n-1}\varepsilon(k, n_j + i - m) \\
&\le \|x_k - p\|^2 - \|x_{k+m} - p\|^2 + \frac{M(m,k)}{n} \\
&\quad + \frac{1}{n}\sum_{i=m}^{n-1}\varepsilon(k, n_j + i - m),
\end{aligned}$$

where $M(m,k)$ is a constant depending only on m and k. Let $\varepsilon > 0$ be given, and choose N_0 so that

$$\varepsilon(i,j) < \varepsilon, \quad \forall i \ge N_0,\ \forall j \ge N_0.$$

Then taking $k \geq N_0$, $m \geq 1$, $n \geq 1$ fixed, and letting $j \to +\infty$, we get

$$0 \leq \|x_k - p\|^2 - \|x_{k+m} - p\|^2 + \frac{M(m,k)}{n} + \varepsilon, \quad \forall k \geq N_0, \ \forall m,n \geq 1.$$

Now letting $n \to +\infty$, we get

$$\|x_{k+m} - p\|^2 \leq \|x_k - p\|^2 + \varepsilon, \quad \forall k \geq N_0, \quad \forall m \geq 1.$$

Therefore by the same argument as in Proposition 4.5.11, we conclude that $\lim_{n\to+\infty} \|x_n - p\|$ exists and thus $p \in F$. □

Theorem 4.5.15 *Let $\{x_n\}$ be a weakly asymptotically regular ANES. Then the following are equivalent:*
i) $F \neq \emptyset$.
ii) $\liminf_{n\to+\infty} \|x_n\| < +\infty$.
iii) x_n converges weakly to $p \in H$.

Proof. $(i) \Rightarrow (ii)$: If $F \neq \emptyset$, then $\|x_n\|$ is bounded.
$(ii) \Rightarrow (iii)$: Assume $\liminf_{n\to+\infty} \|x_n\| < +\infty$. Then since a Hilbert space is reflexive, x_n has a weakly convergent subsequence $\{x_{n_k}\}$ to some $p \in H$.

Now if $x_{n_k} \rightharpoonup p \in H$ as $k \to +\infty$ and $x_{m_l} \rightharpoonup q \in H$ as $l \to +\infty$, then by Proposition 4.5.11, we have $p \in F$ and $q \in F$. Hence we must have $q = p$. This follows by the same argument as in Theorem 4.5.15 since $\lim_{n\to+\infty}(x_n, q-p)$ exists. Thus every weakly convergence subsequence of x_n converges weakly to p, therefore (since $\|x_n\|$ is bounded), x_n converges weakly to p.
$(iii) \Rightarrow (i)$: This follows from Proposition 4.5.14 since $p \in F$. □

4.5.2 ALMOST NONEXPANSIVE CURVES

Let $u \in C([0,+\infty[;H)$; in the sequel we refer to such u as a curve in H. Let $\sigma_T := \frac{1}{T}\int_0^T u(t)dt$. In this section, we prove some theorems for almost non-expansive curves in H, similar to those for sequences proved in the previous subsection, by giving appropriate definitions for curves. Since the proofs are similar to those of the previous subsection, we will simply state the theorems and omit their proofs.

Definition 4.5.16 *The curve $\{u(t)\}$ is almost non-expansive (shortly, ANEC), if*

$$\|u(r+h) - u(s+h)\|^2 \leq \|u(r) - u(s)\|^2 + \varepsilon(r,s),$$

where

$$\lim_{r,s\to+\infty} \varepsilon(r,s) = 0.$$

Remark 4.5.17 *A bounded sequence $\{u(t)\}$ in H that satisfies*

$$\|u(r+h) - u(s+h)\| \leq \|u(r) - u(s)| + \varepsilon_1(r,s), \quad \forall r,s,h \geq 0,$$

where

$$\lim_{r,s\to+\infty} \varepsilon_1(r,s) = 0$$

is an ANEC.

Definition 4.5.18 *Given a bounded curve $\{u(t)\}$ in H, the asymptotic center c of $\{u(t)\}$ is defined as follows (see [EDE]): for every $q \in H$, let $\varphi(q) = \lim_{t\to+\infty}\sup\|u(t)-q\|^2$. Then φ is a continuous strictly convex function on H, satisfying $\varphi(q) \to +\infty$ as $\|q\| \to +\infty$. Thus φ achieves its minimum on H at a unique point c, called the asymptotic center of the curve $u(t)$.*

Remark 4.5.19 *a) $\omega_w(\{u(t)\})$ denotes the weak ω-limit set of $\{u(t)\}$.*
b) $F(\{u(t)\})$ (in short F, when there is no confusion) denotes the following (possibly empty) subset of H:

$$F = \{q \in H : \lim_{t\to+\infty} \|u(t)-q\| exists\}.$$

Note that if $F \neq \emptyset$, then the curve $\{u(t)\}$ is bounded.

Lemma 4.5.20 *F is a closed convex (possibly empty) subset of H.*

Proof. Similar to Lemma 4.5.10. □

Proposition 4.5.21 *For an ANEC $\{u(t)\}$ in H, the weak limit p of any weakly convergent subnet $\{\sigma_{T_l}\}$ of σ_T (if any) belongs to F(i.e. $\lim_{t\to+\infty}\|u(t)-p\|$ exists).*

Proof. Similar to Proposition 4.5.11. □

Lemma 4.5.22 *If $p_1, p_2 \in F$ and $q_1, q_2 \in \omega_w(\{u(t)\})$, then $(p_1-p_2, q_1-q_2) = 0$. In particular, $F \cap \overline{conv}\omega_w(\{u(t)\})$ contains at most one point.*

Proof. Similar to Lemma 4.5.12. □

Theorem 4.5.23 *Let $\{u(t)\}$ be an ANEC in H. Then the following are equivalent:*
i) $F \neq \emptyset$.
ii) $\liminf_{t\to+\infty}\|\sigma_T\| < +\infty$.
iii) σ_T converges weakly to $p \in H$.
Moreover under these conditions we have:
a) $\overline{conv}\omega_w(\{u(t)\}) \cap F = \{p\}$.
b) p is the asymptotic center of the curve $\{u(t)\}$

Proof. Similar to Theorem 4.5.13. □

Proposition 4.5.24 *Assume $\{u(t)\}$ is an ANEC in H which is weakly asymptotically regular (i.e. $u(t+h)-u(t) \rightharpoonup 0$). Then $\omega(\{u(t)\}) \subset F$ (in particular $\omega_w(\{u(t)\}) \neq \emptyset \Rightarrow \|u(t)\|$ is bounded).*

Proof. Similar to Proposition 4.5.14. □

Theorem 4.5.25 *Let $\{u(t)\}$ be a weakly asymptotically regular ANEC. Then the following are equivalent:*
i) $F \neq \emptyset$.
ii) $\liminf_{t\to+\infty}\|u(t)\| < +\infty$.
iii) $u(t)$ converges weakly to $p \in H$.

Proof. Similar to Theorem 4.5.15. □

4.6 WEAK CONVERGENCE OF SOLUTIONS AND MEANS

In this section, by using the results of the previous sections on almost nonexpansive sequences and curves, we study the ergodic convergence of solutions to the following quasi-autonomous dissipative system

$$\begin{cases} \frac{du}{dt} + Au \ni f \\ u(0) = u_0 \end{cases} \tag{4.57}$$

where A is a monotone operator in H, $u_0 \in H$, and $f \in L^1_{\text{loc}}\big((0,+\infty);H\big)$. We prove the weak ergodic convergence of the weak solution $u(t)$ of this equation when $f \in L^1\big((0,+\infty);H\big)$, or more generally $f - f_\infty \in L^1\big((0,+\infty);H\big)$, for some $f_\infty \in H$.

First we recall the following lemmas. See Brézis [BRE].

Lemma 4.6.1 *Let A be a monotone operator in H, and $f,g \in L^1\big((0,T);H\big)$; then if u and v are respectively weak solutions of the equations*

$$\frac{du}{dt} + Au \ni f \quad \textit{and} \quad \frac{dv}{dt} + Av \ni g \quad \textit{on} \quad [0,T]$$

we have

$$\|u(t) - v(t)\| \le \|u(s) - v(s)\| + \int_s^t \|f(\theta) - g(\theta)\| d\theta, \quad \forall 0 \le s \le t \le T$$

Corollary 4.6.2 *Let A be a monotone operator in H and $f \in L^1_{\text{loc}}\big((0,+\infty);H\big)$. Then if u is a weak solution of the equation $\frac{du}{dt} + Au \ni f$, we have*

$$\|u(r+h) - u(s+h)\| \le \|u(r) - u(s)\| + \int_s^{s+h} \|f(\theta + (r-s)) - f(\theta)\| d\theta, \ \forall h \ge 0, \ \forall r \ge s \ge 0$$

Proof. It is enough to apply Lemma 4.6.1 with $g(t) = f(t + (r-s))$ and $v(t) = u(t + (r-s))$. □

Proposition 4.6.3 *If for each $T > 0$, u is a weak solution of the system* (4.57) *on $[0,T]$, and if $\sup_{t \ge 0} \|u(t)\| < +\infty$ and*

$$\lim_{s,r \to +\infty_{r>s}} \int_s^{+\infty} \|f(\theta + (r-s)) - f(\theta)\| d\theta = 0,$$

then the curve $u(t)$ is ANEC in H.

Proof. This follows from Corollary 4.6.2 and Remark 4.5.17 by taking

$$\varepsilon_1(r,s) = \begin{cases} \int_s^{+\infty} \|f(\theta + (r-s)) - f(\theta)\| d\theta & \text{if} \quad r \ge s \\ \int_r^{+\infty} \|f(\theta + (s-r)) - f(\theta)\| d\theta & \text{if} \quad s \ge r \end{cases}$$

□

The following theorem which is the main result of this section, follows from Theorem 4.5.23 and Proposition 4.6.3.

Theorem 4.6.4 *If u is a weak solution of the system* (4.57) *on every interval* $[0,T]$, *and satisfies* $\sup_{t>0}\|u(t)\| < +\infty$, *and if $f - f_\infty \in L^1((0,+\infty);H)$ for some $f_\infty \in H$, then $\sigma_T = \frac{1}{T}\int_0^T u(t)dt$ converges weakly to the asymptotic center of the curve $u(t)$.*

Finally, taking into account Proposition 4.6.3 and Theorem 4.5.25, we have the following theorem for the weak convergence of solutions to (4.57).

Theorem 4.6.5 *If u is a weak solution of the system* (4.57) *on every interval* $[0,T]$, *and satisfies* $\sup_{t>0}\|u(t)\| < +\infty$, *and for each $h \geq 0$, $u(t+h) - u(t) \rightharpoonup 0$, and if $f - f_\infty \in L^1((0,+\infty);H)$ for some $f_\infty \in H$, then $u(t)$ converges weakly to the asymptotic center of the curve $u(t)$.*

4.7 ALMOST ORBITS

First we define the concept of an almost-orbit of a semi-group, first introduced by Miyadera [MIY]. Let C be a nonempty closed and convex subset of H, and $\{T(t) : C \to C;\ t \geq 0\}$ be a nonlinear semi-group (defined in Section 4). An almost orbit for $T(t)$ is a function $u : \mathbb{R}^+ \to C$ such that

$$\lim_{t\to+\infty}[\sup_{h\geq 0}\|u(t+h) - T(h)u(t)\|] = 0.$$

In the following, we show that every almost-orbit of a semi-group of non-expansive mappings is an ANEC.

Lemma 4.7.1 *Let C be a nonempty closed and convex subset of H, and $u(\cdot)$ be a bounded almost orbit of $\{T(t) : C \to C;\ t \geq 0\}$. Then $u(\cdot)$ is an ANEC in H.*

Proof. Let $\phi(t) = \sup_{h\geq 0}\|u(t+h) - T(h)u(t)\|$. Then $\lim_{t\to+\infty}\phi(t) = 0$. Now we have:

$$\|u(t+h) - u(s+h)\| \leq \phi(t) + \phi(s) + \|u(t) - u(s)\|, \quad \forall h,t,s \geq 0$$

The result follows therefore by letting $\varepsilon(t,s) = \phi(t) + \phi(s)$. □

Theorem 4.7.2 *Let $\{T(t) : C \to C;\ t \geq 0\}$ be a non-expansive semi-group. If every orbit of $T(t)$ converges strongly (resp. weakly) as t goes to ∞, so does every almost orbit of $T(t)$.*

Proof. Let τ be the corresponding topology (weak or strong topology), and suppose that $\tau - \lim_{t\to+\infty}T(t)x$ exists for all x. Let u be an almost-orbit of $T(t)$. Let $s \geq 0$ and set $\zeta(s) = \tau - \lim_{t\to+\infty}T(t-s)u(s)$. We have

$$\zeta(s+h) - \zeta(s) = \tau - \lim_{t\to+\infty}\{T(t-s-h)u(s+h) - T(t-s)u(s)\}$$

But for all $t \geq s+h$ the quantity $\|T(t-s-h)u(s+h)-T(t-s)u(s)\|$ is bounded above by $\|u(s+h)-T(h)u(s)\|$. By the $\tau-$lower semi-continuity of the norm, we get $\|\zeta(s+h)-\zeta(s)\| \leq \|u(s+h)-T(h)u(s)\|$. Since u is an almost-orbit of $T(t)$, the right hand side tends to 0 as $s \to +\infty$, uniformly with respect to $h \geq 0$. Therefore $\zeta(s)$ is a Cauchy net and converges strongly to a limit ζ_∞. Now

$$u(s+h)-\zeta_\infty = [u(s+h)-T(h)u(s)] + [T(h)u(s)-\zeta(s)] + [\zeta(s)-\zeta_\infty].$$

Given $\varepsilon > 0$, we can choose s large enough so that the first and the third terms on the right hand side are less than ε in norm, uniformly in h for the first term. Next, for such a fixed s, we let $h \to \infty$ so that the second term $\tau-$ converges to zero. Then $u(t)$ is $\tau-$ convergent to ζ_∞ as $t \to +\infty$. □

For a maximal monotone operator $A : D(A) \subset H \to H$, by Part (6) of Theorem 3.4.1, $\overline{D(A)}$ is closed and convex. Suppose $\{T(t) : \overline{D(A)} \to \overline{D(A)}; t \geq 0\}$ is a nonlinear semi-group generated by

$$\begin{cases} -u'(t) \in A(u(t)) \\ u(0) = x \in \overline{D(A)} \end{cases} \tag{4.58}$$

Proposition 4.7.3 *If $f \in L^1((0,+\infty);H)$, then every solution to* (4.57) *is an almost orbit of the semi-group $T(t)$ generated by* (4.58).

Proof. From (4.57) and (4.58), we have

$$\frac{1}{2}\frac{d}{d\tau}\|u(t+\tau)-T(\tau)u(t)\|^2 \leq \|f(t+\tau)\|\|u(t+\tau)-T(\tau)u(t)\|$$

Therefore

$$\frac{d}{d\tau}\|u(t+\tau)-T(\tau)u(t)\| \leq \|f(t+\tau)\|$$

Integrating the above inequality from $\tau = 0$ to $\tau = s$, we get

$$\|u(t+s)-T(s)u(t)\| \leq \int_0^s \|f(t+\tau)\|d\tau = \int_t^{t+s} \|f(\tau)\|d\tau \leq \int_t^\infty \|f(\tau)\|d\tau.$$

The result follows now by taking the supremum over $s \geq 0$, and then letting $t \to +\infty$ in the above inequality. □

4.8 SUB-DIFFERENTIAL AND NON-EXPANSIVE CASES

In this section, we study the weak convergence of solutions to (4.57) for two special cases. First, for a monotone operator that is the sub-differential of a proper convex and lower semi-continuous function φ, and then for operators of the form $I-T$ where I is the identity operator, and T is a non-expansive mapping on H. In both cases, we prove that if Argmin$\varphi \neq \emptyset$ (resp. $\mathrm{F}(T) \neq \emptyset$), then the trajectory of the solution to (4.57) converges weakly to a minimum point of φ (resp. to a fixed point of T).

Theorem 4.8.1 *Suppose that $u(t)$ is a solution to* (4.57)*, then we have:*
(i) If $A = \partial\varphi$, where φ is a proper, convex and lower semi-continuous function with $\mathrm{Argmin}\varphi \neq \emptyset$*, then $u(t)$ converges weakly to a minimum point of φ.*
(ii) If $A = I - T$, where I is the identity operator, and T is a non-expansive mapping on the Hilbert space H, with $\mathrm{F}(T) \neq \emptyset$*, then $u(t)$ converges weakly to a fixed point of T.*

Proof. By Theorem 4.7.2 and Proposition 4.7.3, it is enough to consider the homogeneous first order evolution Equation (4.58), and by Theorem 4.5.25 we only need to show the asymptotic regularity of the solutions to (4.58). For any monotone operator, since the trajectory of the solutions to (4.58) is non-expansive, i.e. the function $t \mapsto \|u(t+h) - u(t)\|$ is non-increasing, then $\|u'(t)\|$ is also non-increasing. (i) First, let's consider the case $A = \partial\varphi$. By Lemma 4.3.1, we have:

$$\frac{d}{dt}\varphi(u(t)) = (-u'(t), u'(t)) = -\|u'(t)\|^2$$

Therefore

$$\int_0^T \|u'(t)\|^2 dt \le \varphi(u(0)) - \varphi(u(T)) \le \varphi(u(0)) - \varphi(p)$$

where $p \in \mathrm{Argmin}\varphi$. By letting $T \to +\infty$, we get

$$\int_0^{+\infty} \|u'(t)\|^2 dt < +\infty$$

Since $\|u'(t)\|$ is non-increasing, it follows that $\|u'(t)\|$ goes to zero as $t \to +\infty$. Now

$$\|u(t+h) - u(t)\| = \|\int_t^{t+h} u'(s)ds\| \le \int_t^{t+h} \|u'(s)\|ds \le h^{\frac{1}{2}}(\int_t^{t+h} \|u'(s)\|^2 ds)^{\frac{1}{2}} \to 0$$

as $t \to +\infty$. This shows that $u(t)$ is asymptotically regular, and completes the proof of (i).
(ii) Now we consider the case $A = I - T$, where T is a non-expansive mapping on H. By the non-expansiveness of T

$$\|Tu(t) - p\| \le \|u(t) - p\|, \quad \forall p \in F(T)$$

Now by (4.58), we have

$$\|u'(t) + u(t) - p\|^2 \le \|u(t) - p\|^2$$

Hence

$$\|u'(t)\|^2 \le -\frac{d}{dt}\|u(t) - p\|^2$$

Integrating from $t = 0$ to T, we get

$$\int_0^T \|u'(t)\|^2 dt \le \|u(0) - p\|^2 - \|u(T) - p\|^2$$

Therefore

$$\int_0^\infty \|u'(t)\|^2 dt < +\infty$$

Since we know that $\|u'(t)\|$ is non-increasing, it follows that $\|u'(t)\| \to 0$ as $t \to +\infty$, and therefore with a similar proof as in case (i), $u(t)$ is asymptotically regular. □

Remark 4.8.2 *The weak convergence of solutions to* (4.57) *may be shown also with a more general condition on the monotone operator and the semi-group, which is the demi-positivity of the semi-group generated by* (4.58). *We refer the reader to [BRU] (see also [MOR]) for the definition and the result.*

4.9 STRONG ERGODIC CONVERGENCE

This section is devoted to the strong convergence of the sequence of means for solutions to (4.57), to the asymptotic center of the trajectory of solutions, as well as to a zero of the monotone operator A if A is maximal monotone. First we prove some strong ergodic convergence results for sequences. Recall from the introduction in Section 5, that the first weak ergodic convergence theorem for iterates of a non-expansive mapping was proved by Baillon [BAI]. He also proved that if the non-expansive mapping is odd, then the convergence of the means is strong. This result was extended by Brézis and Browder [BRE-BRO] to a more general summation method called strongly regular summation method. Set $y_n = \sum_{j=0}^{\infty} a_{n,j}x_j$, where $a_{n,j} \geq 0$, $\sum_{j=0}^{\infty} a_{n,j} = 1$, $a_{n,j} \to 0$ as $n \to +\infty$ for fixed j, and $\sigma_n = \sum_{j=0}^{\infty} |a_{n,j+1} - a_{n,j}| \to 0$ as $n \to +\infty$. In this section, we study the strong convergence of the sequence y_n, where x_n is a sequence satisfying:

$$\begin{cases} (x_j, x_{j+l}) \leq (x_k, x_{k+l}) + \varepsilon(k,l,j-k), \ \forall k,l \geq 0 \ \text{and} \ j \geq k, \\ \text{with } \varepsilon \text{ bounded and } \lim_{k,l,m \to +\infty} \varepsilon(k,l,m) = 0 \end{cases} \tag{4.59}$$

This condition which was introduced by Djafari Rouhani [DJA4] is more general than the conditions assumed by Brézis and Browder [BRE-BRO] and Wittmann [WIT].

Let $C = \{q \in H;\ \lim_{n\to+\infty}(x_n, q) \text{ exists}\}$, and $C_1 = \{q \in H;\ (x_n, q) \geq (x_{n+1}, q),\ \forall n \geq 0\}$. The proof of the following theorem is similar to that of Theorem 4.5.13.

Theorem 4.9.1 *Let x_n be a bounded sequence in H such that the weak limit of every weakly convergent subsequence of y_n belongs to C. Then y_n converges weakly to the unique element $p \in C \cap \overline{conv}\omega_w(\{x_n\})$. Moreover, if $\lim_{n\to+\infty}\|x_n\|$ exists, then p is the asymptotic center of the sequence x_n.*

Proposition 4.9.2 *If in Theorem 4.9.1 we make the stronger assumptions that $\|x_n\|$ is non-increasing and replacing C by C_1, then in addition, the sequence $z_n = P_{C_1}(-x_n)$ converges strongly to some $z \in C_1$ with $\|z\| \leq \|p\|$ and $\|p+z\|^2 \leq \|p\|^2 - \|z\|^2$.*

Proof. It is clear for the assumption that for any $q \in C_1$ the sequence $\|x_n + q\|$ is non-increasing. We have

$$\|x_{n+1} + z_{n+1}\|^2 \leq \|x_{n+1} + z_n\|^2 \leq \|x_n + z_n\|^2$$

where the first inequality holds since $z_{n+1} = P_{C_1}(-x_{n+1})$ and the second holds because $z_n \in C_1$. Hence the sequence $\|x_n + z_n\|^2$ is non-increasing. Now for all $q \in C_1$, we have:

$$(-x_n - z_n, q - z_n) = \frac{1}{2}(\|x_n + z_n\|^2 + \|q - z_n\|^2 - \|x_n + q\|^2) \leq 0$$

Replacing n by $n+k$ and q by $z_n \in C_1$, we get:

$$\|z_n - z_{n+k}\|^2 \leq \|x_{n+k} - z_n\|^2 - \|x_{n+k} - z_{n+k}\|^2 \leq \|x_n + z_n\|^2 - \|x_{n+k} - z_{n+k}\|^2 \to 0,$$

uniformly in $k \geq 0$. Therefore z_n is a Cauchy sequence in H, thus strongly convergent to some $z \in C_1$. Since the projection map P_{C_1} satisfies $(P_{C_1}x - P_{C_1}y, x - y) \geq \|P_{C_1} - P_{C_1}y\|^2$, $\forall x, y \in H$, putting $y = 0 \in C_1$ and $x = -x_n$ we get: $(z_n, x_n) \leq -\|z_n\|^2$, hence

$$\sum_{j=0}^{\infty} a_{n,j}(z_j, x_j) = \sum_{j=0}^{\infty} a_{n,j}(z_j - z, x_j) + (z, \sum_{j=0}^{\infty} a_{n,j}x_j) \leq -\|z_n\|^2$$

Letting $n \to +\infty$, we get: $\|z\|^2 \leq -(z, p) \leq \|z\|\|p\|$ from which it follows that $\|z\| \leq \|p\|$ and $\|p + z\|^2 = \|p\|^2 + \|z\|^2 + 2(z, p) \leq \|p\|^2 - \|z\|^2$ completing the proof of the proposition. □

Example 4.9.3 *By taking $x_n = 1 + \frac{1}{n}$ in the real numbers, we have $z_n = 0$ for all $n \geq 0$ whereas $x_n \to 1$, showing therefore that in Proposition 4.9.2 we may have $\|z\| < \|p\|$*

Corollary 4.9.4 *Let x_n be a sequence in H satisfying (4.59). Then the sequence $y_n = \sum_{j=0}^{\infty} a_{n,j}x_j$ converges weakly to the unique element $p \in C \cap \overline{conv}(\omega_w\{x_n\})$ and $(x_n, p) \to \|p\|^2$. Moreover if $\lim_{n\to+\infty} \|x_n\|$ exists, then p is the asymptotic center of the sequence x_n.*

Proof. We note that (4.59) implies that the sequence x_n is bounded. Hence y_n has a weakly convergent subsequence. Assume $y_{n_r} \rightharpoonup q$. Let $\varepsilon > 0$ given; choose n_0 so that $(x_j, x_{j+l}) \leq (x_k, x_{k+l}) + \varepsilon$ for all $k \geq n_0$, $l \geq n_0$ and $j \geq k + n_0$. Then, we have:

$$\begin{aligned}(x_j - q, x_k - x_{k+l}) &= (x_j, x_k) - (x_j, x_{k+l}) + (q, x_{k+l} - x_k)\\ &\geq (x_j, x_k) - (x_{j-l}, x_k) - \varepsilon + (q, x_{k+l} - x_k)\end{aligned}$$

for all $k \geq n_0$, $l \geq n_0$ and $j \geq l + k + n_0$. Thus multiplying by $a_{n,j}$, we get:

$$\begin{aligned}a_{n,j}(x_j - q, x_k - x_{k+l}) \geq{}& (a_{n,j}x_j, x_k) - (a_{n,j-l}x_{j-l}, x_k)\\ &+ (a_{n,j-l} - a_{n,j})(x_{j-l}, x_k) - a_{n,j}\varepsilon + (a_{n,j}q, x_{k+l} - x_k)\end{aligned}$$

for $k \geq n_0$, $l \geq n_0$ and $j \geq l+k+n_0$. But we have:

$$\begin{aligned}\sum_{j=l+k+n_0}^{\infty} |a_{n,j-l} - a_{n,j}| &\leq \sum_{j=l+k+n_0}^{\infty} \sum_{i=0}^{l-1} |a_{n,j-l+i} - a_{n,j-l+i+1}| \\ &= \sum_{i=0}^{l-1} \sum_{j=l+k+n_0}^{\infty} |a_{n,j-l+i} - a_{n,j-l+i+1}| \\ &\leq \sum_{i=0}^{l-1} \sum_{j=0}^{\infty} |a_{n,j} - a_{n,j+1}| \\ &= l\sigma_n \to 0\end{aligned}$$

as $n \to +\infty$. Hence for fixed $k \geq n_0$ and $l \geq n_0$, summing up over j and letting $n = n_r \to +\infty$, by using the strong regularity of $a_{n,j}$, we get:

$$0 \geq (q, x_k) - (q, x_k) - \varepsilon + (q, x_{k+l} - x_k) = (q, x_{k+l} - x_k) - \varepsilon$$

Now letting $l \to +\infty$ and then $k \to +\infty$, we get: $\limsup_{n \to +\infty}(q, x_n) \leq \liminf_{n \to +\infty}(q, x_n) + \varepsilon$, and since $\varepsilon > 0$ was arbitrary, we conclude that $\lim_{n \to +\infty}(x_n, q)$ exists, thus $q \in C$. Now the result follows by applying Theorem 4.9.1. Finally we have:

$$\begin{aligned}\lim_{n \to +\infty} (x_n, p) &= \lim_{n \to +\infty} (\sum_{j=0}^{\infty} a_{n,j} x_j, p) \\ &= \|p\|^2\end{aligned}$$

which completes the proof of the corollary. □

Before stating our main result, we need the following lemma from [BRE-BRO].

Lemma 4.9.5 *If $\{a_{n,j}\}$ is a strongly regular summation method, then $\{b_{n,j}\}$ defined by: $b_{n,0} = \sum_{j=0}^{\infty} a_{n,j}^2$ and $b_{n,r} = 2\sum_{j=0}^{\infty} a_{n,j} a_{n,j+r}$ for $r \geq 1$, is also a strongly regular summation method.*

Proof. We have

$$\sum_{r=0}^{\infty} b_{n,r} = (\sum_{j=0}^{\infty} a_{n,j})^2 = 1$$

Now, $a_{n,j} \leq a_{n,0} + \sum_{i=1}^{j} |a_{n,i} - a_{n,i-1}| \leq a_{n,0} + \sigma_n$ for all $j \geq 0$, hence $b_{n,0} = \sum_{j=0}^{\infty} a_{n,j}^2 \leq \sum_{j=0}^{\infty} a_{n,j}(\sigma_n + a_{n,0}) = \sigma_n + a_{n,0} \to 0$; and for $r \geq 1$, $b_{n,r} = 2\sum_{j=0}^{\infty} a_{n,j} a_{n,j+r} \leq 2(\sigma_n + a_{n,0}) \to 0$. Finally we have

$$\begin{aligned}\sum_{r=0}^{\infty} |b_{n,r+1} - b_{n,r}| &= |b_{n,1} - b_{n,0}| + 2\sum_{r=1}^{\infty} |\sum_{j=0}^{\infty} a_{n,j}(a_{n,j+r+1} - a_{n,j+r})| \\ &\leq |b_{n,1} - b_{n,0}| + 2\sum_{j=0}^{\infty} a_{n,j} \sum_{r=1}^{\infty} |a_{n,j+r+1} - a_{n,j+r}|\end{aligned}$$

$$\leq b_{n,1} - b_{n,0} + 2\sigma_n \to 0$$

This completes the proof. □

Now we state and prove our main result.

Theorem 4.9.6 *(i) Let x_n be a sequence in H satisfying* (4.59). *Then the sequence $y_n = \sum_{j=0}^{\infty} a_{n,j}x_j$ converges strongly in H to the unique element $p \in C \cap \overline{conv}(\omega_w\{x_n\})$. If, moreover, $\lim_{n\to+\infty} \|x_n\|$ exists, then p is the asymptotic center of x_n.*
(ii) If x_n satisfies

$$(x_j, x_{j+r}) \leq (x_i, x_{i+r}) + \varepsilon(i, j-i) \ \forall i, r \geq 0, \text{ and } j \geq i \text{ with } \lim_{i,m\to+\infty} \varepsilon(i,m) = 0, \tag{4.60}$$

then in addition p is the element of minimum norm in K, i.e. $p = P_K 0$.

Proof. (i) We already know from Corollary 4.9.4 that y_n converges weakly to $p \in C \cap \overline{conv}(\omega_w\{x_n\})$ and moreover $(x_n, p) \to \|p\|^2$. Therefore, to prove the strong convergence of y_n, all we need to show is that : $\limsup_{n\to+\infty} \|y_n\|^2 \leq \|p\|^2$. Let $M = \sup_{n\geq 0} \|x_n\|$and let $\{b_{n,r}\}$ denote the strongly regular summation method introduced in Lemma 4.9.5. Let $\varepsilon > 0$ given and choose n_0 so that $(x_j, x_{j+r}) \leq (x_k, x_{k+r}) + \varepsilon$ for all $k \geq n_0$, $r \geq n_0$, and $j \geq k + n_0$. Let $k \geq n_0$ fixed. We have:

$$\begin{aligned}
\|y_n\|^2 &= \sum_{i=0}^{\infty}\sum_{j=0}^{\infty} a_{n,i}a_{n,j}(x_i, x_j) \\
&= \sum_{j=0}^{\infty} a_{n,j}^2 \|x_j\|^2 + 2\sum_{j=0}^{\infty}\sum_{r=1}^{\infty} a_{n,j}a_{n,j+r}(x_j, x_{j+r}) \\
&= \sum_{j=0}^{\infty} a_{n,j}^2 \|x_k\|^2 + \sum_{j=0}^{\infty} a_{n,j}^2 (\|x_j\|^2 - \|x_k\|^2) \\
&\quad + 2\sum_{j=0}^{\infty}\sum_{r=1}^{\infty} a_{n,j}a_{n,j+r}(x_k, x_{k+r}) \\
&\quad + 2\sum_{j=0}^{\infty}\sum_{r=1}^{\infty} a_{n,j}a_{n,j+r}((x_j, x_{j+r}) - (x_k, x_{k+r})) \\
&= (x_k, \sum_{r=0}^{\infty} b_{n,r}x_{k+r}) + \sum_{j=0}^{\infty} a_{n,j}^2 (\|x_j\|^2 - \|x_k\|^2) \\
&\quad + 2\sum_{j=0}^{\infty}\sum_{r=1}^{\infty} a_{n,j}a_{n,j+r}((x_j, x_{j+r}) - (x_k, x_{k+r})) \\
&\leq (x_k, \sum_{r=0}^{\infty} b_{n,r}x_{k+r}) + M^2 b_{n,0} + \varepsilon \sum_{r=n_0}^{\infty} b_{n,r} \\
&\quad + 2\sum_{j=0}^{k+n_0-1}\sum_{r=1}^{n_0-1} a_{n,j}a_{n,j+r}((x_j, x_{j+r}) - (x_k, x_{k+r}))
\end{aligned}$$

$$\leq (x_k, \sum_{r=0}^{\infty} b_{n,r}x_{k+r}) + M^2 b_{n,0} + 4M^2 \sup_{0\leq j\leq 2n_0+k} a_{n,j}^2 + \varepsilon.$$

Now since by Lemma 4.9.5, $\{b_{n,r}\}$ is a strongly regular summation method, for $k \geq n_0$ fixed, we deduce from Corollary 4.9.4 that $\sum_{r=0}^{\infty} b_{n,r}x_{k+r}$ converges weakly to p. Hence letting $k \geq n_0$ fixed, and $n \to +\infty$ in the above inequality, we get: $\limsup_{n\to+\infty} \|y_n\|^2 \leq (x_k, p) + \varepsilon$ for all $k \geq n_0$. Now letting $k \to +\infty$, we get $\limsup_{n\to+\infty} \|y_n\|^2 \leq \|p\|^2 + \varepsilon$, which implies the desired result since $\varepsilon > 0$ was arbitrary.

(ii) Now assume that (4.60) holds and let us show that $p = P_K 0$. Let $z_n = P_{K_n} 0$; since $K_{n+1} \subset K_n$, $\|z_n\|$ is nondecreasing, hence convergent. We have $(0 - z_n, z_m - z_n) \leq 0$ for all $m \geq n$. Hence $\|z_n\|^2 \leq (z_n, z_m)$ for all $m \geq n$, which implies

$$\|z_n - z_m\|^2 = \|z_n\|^2 - 2(z_n, z_m) + \|z_m\|^2 \leq \|z_m\|^2 - \|z_n\|^2 \to 0,$$

as $m, n \to +\infty$. z_n converges strongly to some $z \in K$, and we have $(0 - z, y - z) \leq 0$ for all $y \in K$; hence $\|z\|^2 \leq (y, z) \leq \|y\|\|z\|$, which implies that $\|z\| \leq \|y\|$ for all $y \in K$, thus $z = P_K 0$. Since $p \in K$, we have $\|z\| \leq \|p\|$. Let $\varepsilon > 0$ given; we have:

$$\forall n > 0\ \exists k_n \geq n\ \exists \alpha_i \geq 0,\ \sum_{i=n}^{k_n} \alpha_i = 1 \text{ such that } \|\sum_{i=n}^{k_n} \alpha_i x_i\|^2 \leq \|z_n\|^2 + \varepsilon$$

choosing n_0 as in (i), we have:

$$\begin{aligned} \|\sum_{i=n}^{k_n} \alpha_i x_{i+l}\|^2 &= \sum_{i=n}^{k_n}\sum_{j=n}^{k_n} \alpha_i \alpha_j (x_{i+l}, x_{j+l}) \\ &\leq \sum_{i=n}^{k_n}\sum_{j=n}^{k_n} \alpha_i \alpha_j (x_i, x_j) + \varepsilon \\ &= \|\sum_{i=n}^{k_n} \alpha_i x_i\|^2 + \varepsilon \end{aligned}$$

for all $n \geq n_0$ and $l \geq n_0$. Therefore we get:

$$\begin{aligned} \|\sum_{i=n}^{k_n} \alpha_i \sum_{l=0}^{\infty} a_{N,l} x_{i+l}\|^2 &= \|\sum_{l=0}^{\infty} (a_{N,l} \sum_{i=n}^{k_n} \alpha_i x_{i+l})\|^2 \\ &\leq (\sum_{l=0}^{\infty} a_{N,l}) \sum_{l=0}^{\infty} a_{N,l} \|\sum_{i=n}^{k_n} \alpha_i x_{i+l}\|^2 \\ &= \sum_{l=0}^{\infty} a_{N,l} \|\sum_{i=n}^{k_n} \alpha_i x_{i+l}\|^2 \\ &\leq \sum_{l=0}^{\infty} a_{N,l} (\|\sum_{i=n}^{k_n} \alpha_i x_i\|^2 + \varepsilon) + \sum_{l=0}^{n_0-1} a_{N,l} \|\sum_{i=n}^{k_n} \alpha_i x_{i+l}\|^2 \end{aligned}$$

$$
\begin{aligned}
&\leq \|\sum_{i=n}^{k_n} \alpha_i x_i\|^2 + \varepsilon + \sum_{l=0}^{n_0-1} a_{N,l} \|\sum_{i=n}^{k_n} \alpha_i x_{i+l}\|^2 \\
&\leq \|z_n\|^2 + 2\varepsilon + M^2 \sum_{l=0}^{n_0-1} a_{N,l} \quad \text{for } n \geq n_0,
\end{aligned}
$$

where we have used Cauchy-Schwarz inequality for the first inequality above. Now for $n \geq n_0$ fixed, letting $N \to +\infty$, we have:

$$
\sum_{l=0}^{\infty} a_{N,l} x_{i+l} \to p \quad \text{for all } \; i = n, \cdots, k_n
$$

Hence we get $\|p\|^2 \leq \|z_n\|^2 + 2\varepsilon$ for all $n \geq n_0$. Letting $n \to +\infty$, we obtain $\|p\| \leq \|z\|$, since $\varepsilon > 0$ was arbitrary; now since $p \in K$, by uniqueness of $P_K 0$, we get $p = z = P_K 0$, completing the proof of the theorem. □

The following theorem is an immediate consequence of Theorem 4.9.6

Theorem 4.9.7 *Let $\{x_n\}$ be any sequence in H satisfying the following condition: $\lim_{i\to+\infty}(x_i, x_{i+m}) = \alpha_m$ exists uniformly in $m \geq 0$. Then $s_n = \frac{1}{n}\sum_{i=0}^{n-1} x_i$ converges strongly to the asymptotic center of $\{x_n\}$.*

Similarly, we have the following analogous result for curves.

Theorem 4.9.8 *Let $\{u(t)\}$ be any curve in H satisfying the following condition: $\lim_{t\to+\infty}(u(t), u(t+h)) = \alpha(h)$ exists uniformly in $h \geq 0$. Then $\sigma_T = \frac{1}{T}\int_0^T u(t)dt$ converges strongly to the asymptotic center of $\{u(t)\}$.*

The following result which follows from Theorem 4.9.8, is the main result of this section for the solutions to (4.57).

Theorem 4.9.9 *If u is a weak solution of the system* (4.57) *on every interval* $[0,T]$, *and satisfies $\lim_{t\to+\infty}(u(t), u(t+h)) = \alpha(h)$ exists uniformly in $h \geq 0$, then $\sigma_T = \frac{1}{T}\int_0^T u(t)dt$ converges strongly to the asymptotic center of the curve $u(t)$.*

4.10 STRONG CONVERGENCE OF SOLUTIONS

In this section, with additional assumptions on the monotone operator A or the convex function φ when $A = \partial\varphi$, we study the strong convergence of solutions to (4.57). By Theorem 4.7.2 and Proposition 4.7.3, we only need to study the strong convergence of solutions to (4.58).

In the following theorem, in addition to the strong convergence, we also determine the rate of convergence of the solutions.

Theorem 4.10.1 *Suppose that $u(t)$ is a solution to* (4.58), *where A is α-strongly monotone, and p is the unique element of $A^{-1}(0)$, then $\|u(t) - p\| = O(e^{-\alpha t})$.*

Proof. Multiplying both sides of (4.58) by $u(t)-p$, and using the α-strong monotonicity of A, we get:

$$\frac{1}{2}\frac{d}{dt}\|u(t)-p\|^2 \le -\alpha\|u(t)-p\|^2. \tag{4.61}$$

Since by (4.5), $\|u(t)-p\|$ is non-increasing (take $v \equiv p \in A^{-1}(0)$), if $\|u(t_0)-p\| = 0$ for some $t_0 > 0$, then $\|u(t)-p\| = 0$, for all $t \ge t_0$ and the result follows. Otherwise, dividing both sides of (4.61) by $\|u(t)-p\|^2$, we get:

$$\frac{d}{dt}\ln\|u(t)-p\|^2 \le -2\alpha.$$

Integrating from 0 to T, it follows that:

$$\ln\|u(T)-p\|^2 - \ln\|u(0)-p\|^2 \le -2\alpha T. \tag{4.62}$$

Therefore:

$$\|u(T)-p\| \le \|u(0)-p\|e^{-\alpha T},$$

which yields the theorem. □

Theorem 4.10.2 *Assume that $A = \partial\varphi$, where $\varphi : H \to (-\infty,+\infty]$ is a proper, convex and lower semi-continuous function such that $D(\varphi) = -D(\varphi)$ and*

$$\varphi(x) - \varphi(0) \ge \alpha(\varphi(-x) - \varphi(0)), \quad \forall x \in D(\varphi) \tag{4.63}$$

where α is some positive real number. Then there exists a minimum point p of φ such that $u(t)$ converges strongly to p.

Proof. Without loss of generality we may assume that $\varphi(0) = 0$. Also without loss of generality we may assume that $0 < \alpha \le 1$. In fact we have:

$$0 = \varphi(0) = \varphi(x + (-x)/2) \le 1/2\varphi(x) + 1/2\varphi(-x)$$

which implies that $\varphi(x) \ge -\varphi(-x)$. This inequality together with (4.63) implies that: $\varphi(x) \ge \mathrm{Max}\{\alpha\varphi(-x), -\varphi(-x)\}$. Therefore $\varphi(x) \ge 0$, for all $x \in D(\varphi)$. This shows that if (4.63) holds for $\alpha > 1$, it also holds for any $0 < \beta < \alpha$, and in particular for any $0 < \beta \le 1$. Therefore we may assume $0 < \alpha \le 1$. Now, for each $x \in D(\varphi)$, by (4.63) and the convexity of φ, we have:

$$\begin{aligned}
\varphi(0) &= \varphi(\frac{\alpha}{1+\alpha}x + \frac{1}{1+\alpha}(-\alpha x)) \\
&\le \frac{\alpha}{1+\alpha}\varphi(x) + \frac{1}{1+\alpha}\varphi(-\alpha x) \\
&\le \frac{\alpha}{1+\alpha}\varphi(x) + \frac{\alpha}{1+\alpha}\varphi(-x) \\
&\le \frac{\alpha}{1+\alpha}\varphi(x) + \frac{1}{1+\alpha}\varphi(x) \\
&= \varphi(x).
\end{aligned}$$

This implies that 0 is a minimum point of φ, or $0 \in A^{-1}(0)$. By Lemma 4.3.1, we have

$$\frac{d}{dt}\varphi(u(t)) = (u'(t), -u'(t)) = -\|u'(t)\|^2 \leq 0, \ \ a.e.\, t > 0.$$

Hence the function $\varphi(u(t))$ is non-increasing. Then by (4.63), the convexity of φ, and the sub-differential inequality, for $s \geq t$, we get:

$$\begin{aligned} \varphi(u(t)) &\geq \varphi(u(s)) \geq \alpha\varphi(-u(s)) \geq \varphi(-\alpha u(s)) \\ &\geq \varphi(u(t)) + \big(\frac{du}{dt}(t), \alpha u(s) + u(t)\big), \ \ a.e.\, t \in]0, s[\end{aligned} \tag{4.64}$$

For fixed $s > 0$, define $g : [0, s] \to \mathbb{R}$ by

$$g(t) = \frac{1}{2}(1+\alpha)(\|u(t)\|^2 - \|u(s)\|^2) - \frac{\alpha}{2}\|u(t) - u(s)\|^2$$

Then g is absolutely continuous and $g(s) = 0$. Moreover, it follows from (4.64) that for $0 < t \leq s$,

$$g'(t) = \big(u'(t), \alpha u(s) + u(t)\big) \leq 0, \ \ a.e.\, t \in]0, s[$$

Therefore $g(t) \geq g(s) = 0, \ \ \forall t \in]0, s[$. In other words,

$$(1+\alpha)(\|u(t)\|^2 - \|u(s)\|^2) \geq \alpha\|u(t) - u(s)\|^2, \ \ \forall\, t \in]0, s[,$$

which implies that $\|u(t)\|^2$ is non-increasing and also $u(t)$ converges strongly to some p, which by Theorem 4.8.1 belongs to Argminφ. □

4.11 QUASI-CONVEX CASE

Consider the following non-homogeneous evolution system

$$\begin{cases} -x'(t) &= \nabla\phi(x(t)) + f(t), \\ x(0) &= x_0 \in H. \end{cases} \tag{4.65}$$

where ϕ is a quasi-convex function as defined below.

When $f(t) \equiv 0$, Goudou and Munier [GOU-MUN] proved the weak convergence of solutions to (4.65) to a critical point of ϕ. They also proved the strong convergence of solutions to (4.65) with additional assumptions on ϕ. In this section, we study (4.65) with the condition $f \in L^1\big((0, +\infty); H\big)$. We prove the weak and strong convergence of solutions to (4.65) to a critical point of ϕ. These results extend the similar classical results on the asymptotic behavior of non-homogeneous gradient systems associated with convex functions which have been also extended to non-smooth convex functions (see Theorem 4.8.1 of this chapter as well as [BRU]). A function $\phi : H \to \mathbb{R}$ is said to be quasi-convex if

$$\phi(\lambda x + (1-\lambda)y) \leq \max\{\phi(x), \phi(y)\}, \ \ \forall x, y \in H, \ \ \forall \lambda \in [0, 1],$$

or equivalently every sub-level set of ϕ is convex. A differentiable function ϕ on H is quasi-convex if

$$\phi(x) \geq \phi(y) \Rightarrow (\nabla\phi(x), y-x) \leq 0.$$

A function ϕ is called pseudo-convex if the following condition is satisfied: If $\phi(y) > \phi(x)$, then there exists $\beta(x,y) > 0$ and $0 < \delta(x,y) \leq 1$, such that: $\phi(y) - \phi(tx+(1-t)y) \geq t\beta(x,y)$ for all $t \in (0, \delta(x,y))$. A differentiable function ϕ is pseudo-convex if and only if

$$\phi(x) > \phi(y) \Rightarrow (\nabla\phi(x), y-x) < 0.$$

Obviously convexity implies pseudo-convexity and pseudo-convexity implies quasi-convexity. We refer the reader to the interesting book by Cambini and Martein [CAM-MAR] for the definitions, properties and illustrative examples of convexity and its extensions. Throughout this section, we assume that $\phi : H \to \mathbb{R}$ is a continuously differentiable quasi-convex function with Argmin$\phi \neq \emptyset$, and $\nabla\phi$ Lipschitz continuous on bounded subsets of H.
The results are applicable even to one dimensional differential equations (where of course weak and strong convergence coincide). Consider the following nonlinear differential equation

$$\begin{cases} -x'(t) = \frac{2x(t)}{((x(t))^2+1)^2} + \frac{2}{(t+1)^3} - \frac{2(t+1)^6}{((t+1)^4+1)^2}, \\ x(0) = 1, \end{cases}$$

which is in the form (4.65) with $\phi(x) = \frac{x^2}{x^2+1}$ and $f(t) = \frac{2}{(t+1)^3} - \frac{2(t+1)^6}{((t+1)^4+1)^2}$. One can easily verify that $x(t) = \frac{1}{(t+1)^2}$ is a solution which converges to zero as $t \to +\infty$, as predicted by Theorem 4.11.3.

Lemma 4.11.1 *Suppose that $x(t)$ is a solution to (4.65). If* Argmin$\phi \neq \emptyset$, *then* $\lim_{t\to+\infty} \|x(t) - x\|$ *exists for each* $x \in$ Argminϕ.

Proof. Since $x \in$ Argminϕ, we have $\phi(x) \leq \phi(x(t))$ for all $t \geq 0$. By the quasi-convexity of ϕ, we have

$$\big(\nabla\phi(x(t)), x - x(t)\big) \leq 0.$$

Therefore

$$\begin{aligned} \frac{d}{dt}\|x(t)-x\|^2 &= 2\big(x'(t), x(t)-x\big) \\ &= 2\big(-\nabla\phi(x(t)) - f(t), x(t)-x\big) \\ &\leq 2\|f(t)\|\|x(t)-x\|. \end{aligned} \tag{4.66}$$

First we prove that $x(t)$ is bounded. By contradiction if $x(t)$ is unbounded, there is an increasing sequence $t_n \to +\infty$ such that $\|x(t_n) - x\| \to +\infty$, and $\|x(t_{n+1}) - x\| > 2\|x(t_n) - x\|$, and $\|x(s) - x\| < \|x(t_{n+1}) - x\|$, $\forall s \in (t_n, t_{n+1})$. To show the existence of such a sequence, let $f(t) = \|x(t) - x\|$. f is continuous, and by contradiction it is unbounded. Define a sequence t_n in this way: $f(t_n) = \max_{n \leq t \leq n+1} f(t)$. Then define the subsequence n_j as follows. Let $n_1 = 1$, and $n_{j+1} = \min\{k > n_j : f(t_k) > f(t_{n_j})\}$. Since f is unbounded, the sequence n_j exists. By the definition of t_{n_j}, we have

$$f(t_k) \leq f(t_{n_j}) < f(t_{n_{j+1}}), \quad \forall n_j < k < n_{j+1} \tag{4.67}$$

Therefore $f(t_{n_j})$ is unbounded, because $f(t_k)$ is unbounded and since $f(t_{n_j})$ is increasing, then $f(t_{n_j}) \to +\infty$ as $j \to +\infty$. Now, choose a subsequence of t_{n_j}, which we again denote it by t_{n_j} such that $f(t_{n_{j+1}}) \geq 2f(t_{n_j})$. Obviously this subsequence satisfies

$$f(t_k) \leq f(t_{n_{j+1}}), \quad \forall n_j \leq k \leq n_{j+1}$$

by (4.67). Now, integrating (4.66) from t_{n_j} to $t_{n_{j+1}}$, we get

$$\|x(t_{n_{j+1}}) - x\|^2 - \|x(t_{n_j}) - x\|^2 \leq 2\|x(t_{n_{j+1}}) - x\| \int_{t_{n_j}}^{t_{n_{j+1}}} \|f(t)\| dt.$$

Dividing both sides of the above inequality by $\|x(t_{n_{j+1}}) - x\|$, we get

$$\frac{3}{2}\|x(t_{n_j}) - x\| \leq 2\int_{t_{n_j}}^{t_{n_{j+1}}} \|f(t)\| dt.$$

We get a contradiction when $j \to +\infty$, because $\int_0^{+\infty} \|f(t)\| dt < +\infty$. Therefore $x(t)$ is bounded. Let $M := \sup_{t \geq 0} \|x(t) - x\|$. Now integrating (4.66) from s to $t > s$, we get

$$\|x(t) - x\|^2 - \|x(s) - x\|^2 \leq 2M \int_s^t \|f(\tau)\| d\tau.$$

Taking limsup as $t \to +\infty$ and liminf as $s \to +\infty$, we get that $\lim_{t \to +\infty} \|x(t) - x\|$ exists. □

Lemma 4.11.2 *Suppose that* $x(t)$ *is a solution to* (4.65) *and* $\text{Argmin}\phi \neq \emptyset$*, then* $\lim \phi(x(t))$ *exists.*

Proof. Since $\nabla\phi$ is bounded on bounded subsets of H, by Equation (4.65) and Lemma 4.11.1, we have

$$\begin{aligned}
\frac{d}{dt}\phi(x(t)) &= \big(\nabla\phi(x(t)), x'(t)\big) \\
&= \big(\nabla\phi(x(t)), -\nabla\phi(x(t)) - f(t)\big) \\
&= -\|\nabla\phi(x(t))\|^2 - \big(\nabla\phi(x(t)), f(t)\big) \\
&\leq \|\nabla\phi(x(t))\| \|f(t)\| \\
&= \|\nabla\phi(x(t)) - \nabla\phi(x)\| \|f(t)\| \\
&\leq L\|x(t) - x\| \|f(t)\| \\
&\leq LM\|f(t)\|,
\end{aligned}$$

where $M = \sup_{t \geq 0} \|x(t) - x\|$, L is the Lipschitz constant of $\nabla\phi$ and x is a critical point of ϕ. Now since $f \in L^1\big((0, +\infty); H\big)$, a similar proof as in Lemma 4.11.1 shows that $\lim_{t \to +\infty} \phi(x(t))$ exists. □

Theorem 4.11.3 *Suppose that $x(t)$ is a solution to* (4.65). *If* $\text{Argmin}\phi \neq \emptyset$*, then there is $x \in H$ such that $x(t) \rightharpoonup x$ as $t \to +\infty$ and $\nabla\phi(x) = 0$.*

Proof. We consider the following two cases:
1) $\lim_{t\to+\infty} \phi(x(t)) = \inf \phi$. Since $x(t)$ is bounded by Lemma 4.11.1, there is a sequence $\{t_n\}$ and $\tilde{x} \in H$ such that $x(t_n) \rightharpoonup \tilde{x}$, as $n \to +\infty$. By Lemma 4.11.2 and the weak lower semi-continuity of ϕ (by Mazur's lemma), we have

$$\phi(\tilde{x}) \leq \liminf_{n\to+\infty} \phi(x(t_n)) = \lim_{t\to+\infty} \phi(x(t)) = \inf \phi.$$

Therefore $\tilde{x} \in \text{Argmin}\phi$, which implies by Lemma 4.11.1 and Opial's lemma [OPI], that $x(t) \rightharpoonup x \in \text{Argmin}\phi$.
2) $\lim_{t\to+\infty} \phi(x(t)) > \inf \phi$. Then there exist $r > 0$, $t_0 > 0$ and $\tilde{x} \in \text{Argmin}\phi$ such that for all $t \geq t_0$ and every $y \in \bar{B}_r(\tilde{x})$, $\phi(y) \leq \phi(x(t))$. Now by the quasi-convexity of ϕ, we have $\big(y - x(t), \nabla\phi(x(t))\big) \leq 0$. If $\nabla\phi(x(t)) \neq 0$, then letting $y = \tilde{x} + r\frac{\nabla\phi(x(t))}{\|\nabla\phi(x(t))\|}$, we get:

$$\begin{aligned} r\|\nabla\phi(x(t))\| &\leq \big(x(t) - \tilde{x}, \nabla\phi(x(t))\big) \\ &\leq -\frac{1}{2}\frac{d}{dt}\|x(t) - \tilde{x}\|^2 + M\|f(t)\|, \end{aligned}$$

where $M = \sup_{t\geq 0} \|x(t) - \tilde{x}\|$. By (4.66), this inequality is obviously satisfied also if $\nabla\phi(x(t)) = 0$. Therefore $\nabla\phi(x(t)) \in L^1\big((0,+\infty);H\big)$. By (4.65), this implies that $x'(t) \in L^1\big((0,+\infty);H\big)$. Therefore by the same argument as in Lemma 4.11.1, there is $x \in H$ such that $x(t) \to x$. Now by the continuity of $\nabla\phi$, $\nabla\phi(x(t)) \to \nabla\phi(x)$. Since $\nabla\phi(x(t)) \in L^1\big((0,+\infty);H\big)$, there exists a sequence $t_n \to +\infty$ such that $\nabla\phi(x(t_n)) \to 0$. Therefore $\nabla\phi(x) = 0$. □

Remark 4.11.4 *Suppose that the assumptions of Theorem 4.11.3 are satisfied and ϕ is pseudo-convex. By the pseudo-convexity of ϕ, if $\nabla\phi(x) = 0$, then $x \in$* Argminϕ. *In fact, suppose to the contrary that $\nabla\phi(x) = 0$ but $x \notin$* Argminϕ. *Then there is an $x' \in H$ such that $\phi(x') < \phi(x)$. The pseudo-convexity of ϕ then implies that $\big(x' - x, \nabla\phi(x)\big) < 0$, which is a contradiction. Therefore if ϕ is pseudo-convex, then $x(t) \rightharpoonup x \in$* Argmin$\phi$.

Theorem 4.11.5 *Suppose that $x(t)$ is a solution to* (4.65). *If* $\text{Argmin}\phi \neq \emptyset$ *and either one of the following conditions is satisfied:*
a) $x \notin$ Argminϕ*, where x is a weak cluster point of $x(t)$,*
b) $\text{int}(\text{Argmin}\phi) \neq \emptyset$,
then $x(t) \to x$ and x is a critical point of ϕ.

Proof. a) Suppose that $x(t_n) \rightharpoonup x \notin \text{Argmin}\phi$. Then

$$\lim_{t\to+\infty} \phi(x(t)) = \liminf_{n\to+\infty} \phi(x(t_n)) \geq \phi(x) > \inf \phi$$

Now the result follows from Case (2) in the proof of Theorem 4.11.3.
b) If $\text{int}(\text{Argmin}\phi) \neq \emptyset$, then there exist $\tilde{x} \in \text{Argmin}\phi$, $r > 0$ and $t_0 > 0$ such that

for all $t \geq t_0$ and every $y \in \bar{B}_r(\tilde{x})$, $\phi(y) \leq \phi(x(t))$. Then by the quasi-convexity of ϕ, $\big(y - x(t), \nabla\phi(x(t))\big) \leq 0$. Now if $\nabla\phi(x(t)) \neq 0$, then letting $y = \tilde{x} + r\frac{\nabla\phi(x(t))}{\|\nabla\phi(x(t))\|}$, we have:

$$\begin{aligned} r\|\nabla\phi(x(t))\| &\leq \big(x(t) - \tilde{x}, \nabla\phi(x(t))\big) \\ &= \big(x(t) - \tilde{x}, -x'(t) - f(t)\big) \\ &\leq -\frac{1}{2}\frac{d}{dt}\|x(t) - \tilde{x}\|^2 + M\|f(t)\|, \end{aligned}$$

where $M = \sup_{t\geq 0}\|x(t) - \tilde{x}\|$. By (4.66), this inequality is obviously satisfied also if $\nabla\phi(x(t)) = 0$. Therefore $\nabla\phi(x(t)) \in L^1\big((0,+\infty);H\big)$. By (4.65), this implies that $x'(t) \in L^1\big((0,+\infty);H\big)$. Therefore by the same argument as in Lemma 4.11.1, there is $x \in H$ such that $x(t) \to x$ as $t \to +\infty$. On the other hand, $\nabla\phi(x(t_n)) \to 0$ for a sequence $t_n \to +\infty$ as $n \to +\infty$. Then the continuity of $\nabla\phi$ implies that $\nabla\phi(x) = 0$. □

Theorem 4.11.6 *Suppose that $x(t)$ is a solution to* (4.65) *and $f(t) \equiv 0$. If ϕ is even, then there is $x \in H$ such that $x(t) \to x$ as $t \to +\infty$, where $\nabla\phi(x) = 0$.*

Proof. It follows from the proof of Lemma 4.11.2 that $\phi(x(t))$ is non-increasing. Therefore for all $t \geq s$, $\phi(x(t)) \leq \phi(x(s))$. Then the quasi-convexity of ϕ implies that

$$\big(\nabla\phi(x(s)), x(t) - x(s)\big) \leq 0.$$

Since ϕ is even, $\phi(-x(t)) = \phi(x(t))$. Therefore again by the quasi-convexity of ϕ, for each $t \geq s$, we get:

$$\big(-x(t) - x(s), \nabla\phi(x(s))\big) \leq 0. \tag{4.68}$$

Summing up the last two inequalities, we get:

$$\big(x(s), \nabla\phi(x(s))\big) \geq 0 \Rightarrow \frac{d}{ds}\|x(s)\|^2 \leq 0.$$

Therefore $\|x(t)\|$ is non-increasing. By (4.68) for each $t \geq s$, we have

$$\big(x(t) + x(s), x'(s)\big) \leq 0 \Rightarrow \frac{d}{ds}\|x(s)\|^2 \leq -2\big(x(t), x'(s)\big).$$

Integrating this inequality from s to t, we get:

$$\|x(t)\|^2 \leq \|x(s)\|^2 - 2\|x(t)\|^2 + 2\big(x(t), x(s)\big), \quad \forall t > s$$

From the above inequality, we get:

$$\|x(t) - x(s)\|^2 = \|x(t)\|^2 + \|x(s)\|^2 - 2\big(x(t), x(s)\big) \leq 2(\|x(s)\|^2 - \|x(t)\|^2) \to 0,$$

as $t, s \to +\infty$. Therefore $x(t)$ is a Cauchy net. So $x(t)$ converges strongly to $x \in H$, and $\nabla\phi(x) = 0$, by Theorem 4.11.3. □

Definition 4.11.7 *Let $f : H \to (-\infty, +\infty]$ be proper, then f is said to be uniformly quasi-convex with modulus $\eta : [0, +\infty) \longrightarrow [0, +\infty)$ if η is increasing, vanishes only at 0, and $(\forall\, x, y \in dom\, f,\ \forall \alpha \in (0, 1))$*

$$f(\alpha x + (1-\alpha)y) + \alpha(1-\alpha)\eta(\|x-y\|) \leqslant \max\{f(x), f(y)\}$$

Example 4.11.8 *We define f and η as follows:*

$$f(x) = \begin{cases} x^2 & x \geqslant -1 \\ 4\sqrt{-x} - 3 & -4 \leqslant x \leqslant -1 \\ +\infty & x < -4 \end{cases}$$

and

$$\eta(x) = \frac{x^2}{16 + x^2}, \quad \forall x \in [0, +\infty).$$

f is not convex but it is uniformly quasi-convex with modulus η.

Theorem 4.11.9 *Suppose that $x(t)$ is a solution to* (4.65). *If* Argmin$\phi \neq \emptyset$ *and ϕ is uniformly quasi-convex with modulus η, then x(t) converges strongly to the unique element of* Argminϕ.

Proof. Let $\tilde{x}$ be the unique element of Argminϕ, then $\phi(\tilde{x}) \leq \phi(x(t))$. The uniform quasi-convexity of ϕ shows that

$$\left(\tilde{x} - x(t), \nabla\phi(x(t))\right) \ \leq -\eta(\|x(t) - \tilde{x}\|)$$

$$\Longrightarrow 0 \leq \eta(\|x(t) - \tilde{x}\|) \leq \left(x(t) - \tilde{x}, \nabla\phi(x(t))\right)$$

$$= \left(x(t) - \tilde{x}, -x'(t) - f(t)\right) \leq \frac{-1}{2}\frac{d}{dt}\|x(t) - \tilde{x}\|^2 + M\|f(t)\|,$$

where $M = \sup\limits_{t \geqslant 0}\|x(t) - \tilde{x}\|$. Now integrating both sides of the above inequality, we get:

$$\begin{aligned} 0 &\leq \int_0^{+\infty} \eta(\|x(t) - \tilde{x}\|)dt \\ &\leq \frac{1}{2}\|x(0) - \tilde{x}\| - \lim_{t\to\infty}\frac{1}{2}\|x(t) - \tilde{x}\| + M\int_0^{+\infty} \|f(t)\|dt < +\infty. \end{aligned}$$

Since $\lim\limits_{t\to\infty}\|x(t) - \tilde{x}\|$ exists and η is an increasing function which vanishes only at 0, it follows that $\lim\limits_{t\to\infty}\|x(t) - \tilde{x}\| = 0$. Hence, we conclude that $x(t) \to \tilde{x}$. □

REFERENCES

BAR. V. Barbu, Nonlinear semigroups and Differential Equations in Banach Spaces, Noordhoff, Leyden, 1976.

BAR-PRE. V. Barbu and Th. Precupanu, Convexity and Optimization in Banach Spaces, Editors Academiei, Buchrest, 1986 (and D. Reidel Publishing Company).

BAI. J. B. Baillon, Un théorème de type ergodique pour les contractions non linéaires dans un espace de Hilbert, C. R. Acad. Sci. Paris 280 (1975), A1511–A1514.

BRE. H. Brézis, "Opérateurs Maximaux Monotones et Semi-groupes de Contractions dans les Espaces de Hilbert", North-Holland Mathematics studies, Vol. 5, North-Holland Publishing Co., Amsterdam-London (1973).

BRE-BRO. H. Brézis and F. E. Browder, Nonlinear ergodic theorems, Bull. Amer. Math. Soc. 82 (1976), 959–961.

BRO. F.E. Browder, Nonexpansive nonlinear operators in a Banach space. Proc. Nat. Acad. Sci. USA 54, 1041–1044 (1965).

BRU. R. E. Bruck, Asymptotic convergence of nonlinear contraction semigroups in Hilbert space, J. Funct. Anal. 18 (1975), 15–26.

CAM-MAR. A. Cambini and L. Martein, Generalized Convexity and Optimization, Lecture Notes in Economics and Mathematicals Systems, 616. springer-Verlag, Berlin, 2009.

DJA1. B. Djafari Rouhani, Ergodic theorems for nonexpansive sequences in Hilbert spaces and related problems, Ph.D. Thesis, Yale University, Part I, pp. 1–76 (1981).

DJA2. B. Djafari Rouhani, Asymptotic behaviour of quasi-autonomous dissipative systems in Hilbert spaces, J. Math. Anal. Appl. 147 (1990), 465–476.

DJA3. B. Djafari Rouhani, Asymptotic behaviour of almost nonexpansive sequences in a Hilbert space, J. Math. Anal. Appl. 151 (1990), 226–235.

DJA4. B. Djafari Rouhani, An ergodic theorem for sequences in a Hilbert space, Nonlinear Anal. Forum, 4 (1999), 33–48.

EDE. M. Edelstein, The construction of an asymptotic center with a fixed-point property, Bull. Amer. Math. Soc. 78 (1972), 206–208.

GOH. D. Göhde, Zum Prinzip der kontraktiven Abbildung. Math. Nachr. 30 (1965), 251–258.

GOU-MUN. X. Goudou and J. Munier, The gradient and heavy ball with friction dynamical systems: the quasiconvex case, Math. Program., Ser. B 116 (2009), 173–191.

KHA-MOH. H. Khatibzadeh and V. Mohebbi, Non-homogeneous Continuous and Discrete Gradient Systems: The Quasi-convex Case, Bull. Iran. Math. Soc.

KIR. W.A. Kirk, A fixed point theorem for mappings which do not increase distances. Amer. Math. Monthly 72 (1965), 1004–1006.

LUC. R. Lucchetti, Convexity and well-posed problems. CMS Books in Mathematics/Ouvrages de Mathématiques de la SMC, 22. Springer, New York, 2006.

MIY. I. Miyadera, Nonlinear ergodic theorems for semigroups of Non-lipschitzian mappings in Banach Spaces II, Math J. Okayama, 43 (2001),123–135.

MOR. G. Morosanu, Nonlinear Evolution Equations and Applications, Editura Academiei (and D. Reidel Publishing Company), Bucharest, 1988.

OPI. Z. Opial, Weak convergence of the sequence of successive approximations for nonexpasive mappings, Bull. Amer. Math. Soc. 73 (1967), 591–597.

ROC. R. T. Rockafellar, On the maximal monotonicity of subdifferential mappings, Pacific Math. J. 33 (1970), 209–216.

WIT. R. Wittmann, Mean ergodic theorems for nonlinear operators, Proc. Amer. Math. Soc. 108 (1990), 781–788.

5 Second Order Evolution Equations

5.1 INTRODUCTION

In this chapter, the existence and asymptotic behavior of solutions to the second order evolution equation of monotone type is studied. Let $A : D(A) \subset H \to H$ be a maximal monotone operator. Consider the following second order evolution equation associated to A:

$$\begin{cases} p(t)u''(t) + q(t)u'(t) \in Au(t), & \text{a.e. on}(0, +\infty) \\ u(0) = u_0 \in D(A), & \sup_{t \geq 0} \|u(t)\| < +\infty \end{cases} \tag{5.1}$$

In the first section, we prove the existence of solutions to (5.1) with suitable assumptions on $p(t)$ and $q(t)$. The first results in this direction were proved by Barbu [BAR1, BAR2, BAR3] for the case $p(t) \equiv 1$ and $q(t) \equiv 0$, and by Véron [VER1, VER2] for more general cases. In Section 2, we state the existence theorem of Véron from [VER2]. Sections 3 and 4 are devoted to the nonhomogeneous case of the evolution equation with $p(t) \equiv 1$ and $q(t) \equiv 0$. In these sections, we consider the existence of solutions for a two point boundary value problem on an interval, as well as an existence result for a second order nonhomogeneous evolution equation on the positive axis. In Section 5, we consider the evolution equation

$$\begin{cases} u''(t) \in Au(t) + f(t), & \text{a.e. } t \in (0, +\infty) \\ u(0) = u_0 \in \overline{D(A)}, & \sup_{t \geq 0} \|u(t)\| < +\infty \end{cases} \tag{5.2}$$

where f is periodic with period $T > 0$. We prove the existence of a periodic solution to (5.2) and the weak convergence of solutions to a periodic solution. The results are due to Bruck [BRU1, BRU2]. In Section 6, the semigroup generated by (5.2) in the homogeneous case (i.e. when $f \equiv 0$) is considered and the square root of the maximal monotone operator generating this semigroup is defined. Sections 7 and 8 are devoted to the asymptotic behavior of solutions to (5.1) for the homogeneous and nonhomogeneous cases.

5.2 EXISTENCE AND UNIQUENESS OF SOLUTIONS

The main aim of this section is to study the existence of solutions to (5.1) with the following assumptions:

$$p \in W^{2,\infty}(0, +\infty), \quad q \in W^{1,\infty}(0, +\infty) \tag{5.3}$$

$$\exists \alpha > 0, \quad \text{such that} \quad \forall t \geq 0, \ p(t) \geq \alpha. \tag{5.4}$$

Also to show the uniqueness if moreover:

$$\int_0^{+\infty} e^{-\int_0^t \frac{q(s)}{p(s)} ds} dt = +\infty \tag{5.5}$$

Let $\rho(t) = \exp\left(\int_0^t \frac{q(s)-p'(s)}{p(s)} ds\right)$. Theorems 5.2.1 and 5.2.11 are the main results of this section.

5.2.1 THE STRONGLY MONOTONE CASE

In this subsection we consider the existence of solutions to (5.1), when the maximal monotone operator A is strongly monotone.

Theorem 5.2.1 *Let $A : D(A) \subset H \to H$ be a maximal monotone and β-strongly monotone operator such that $0 \in A(0)$. Then for each $u_0 \in D(A)$, there exists a unique continuously differentiable function $u \in H^2(0,+\infty;H)$ that satisfies*

$$\begin{cases} p(t)u''(t) + q(t)u'(t) \in Au(t), & a.e. \text{ on } [0,+\infty) \\ u(0) = u_0, \quad u(t) \in D(A), & a.e. \text{ on } [0,+\infty) \end{cases} \tag{5.6}$$

If u and v are solutions associated to the initial values u_0 and v_0 respectively, then

$$\|u(t) - v(t)\| \le \|u_0 - v_0\|. \tag{5.7}$$

The function $\|u\|$ on $[0,+\infty)$ is nonincreasing. Moreover $u \in H^2_\rho(0,+\infty;H)$.

Proof. First we prove the uniqueness. If ϕ is a differentiable function, then we have

$$\frac{1}{\rho}(p\rho\phi)' = p\phi' + q\phi$$

Let $w = u - v$. To show the inequality (5.7), by assumption, for almost all $t \in [0,+\infty)$ we have:

$$\big((p\rho w')'(t), w(t)\big) - \beta\rho(t)\|w(t)\|^2 \ge 0 \tag{5.8}$$

Since $\lim_{t\to+\infty} \|w(t)\| = 0$, then by contradiction, there is $t_1 > 0$ such that $\|w(t_1)\| > \|w(0)\|$ and $\|w(t_1)\| = \max_{0\le t<+\infty} \|w(t)\|$. Now integrating (5.8) by parts for $t > t_1$, since $(w,w')(t_1) = \frac{1}{2}\frac{d}{dt}\|w(t_1)\|^2 = 0$, we get:

$$\rho(t)p(t)(w,w')(t) \ge \int_{t_1}^t \rho(t)p(t)\|w'(t)\|^2 + \beta\int_{t_1}^t \rho(t)\|w(t)\|^2 dt > 0,$$

which is a contradiction with $\lim_{t\to+\infty} \|w(t)\| = \lim_{t\to+\infty} \|w'(t)\| = 0$. Since $0 \in A0$, the solution corresponding to the initial value 0 is 0. Therefore by (5.7) the function $\|u\|$ is nonincreasing. □

Now to prove the existence, we approximate u by a function u_T which is the solution to a similar problem on a finite interval $[0,T]$.
To prove the existence we first need the following lemmas.

Lemma 5.2.2 *The operator $\mathscr{B}$ defined by*

$$\begin{cases} D(\mathscr{B}) = \{u \in H^2(0,T;H) \ \ s.t. \ u(0) = u_0; \ u'(T) = 0\} \\ (\mathscr{B}u)(t) = -(\rho p u')'(t) \end{cases} \tag{5.9}$$

is maximal monotone on $L^2(0,T;H)$.

Proof. The monotonicity follows easily by integrating by parts. Let's prove the maximality. We need to show that for a given $f \in L^2(0,T;H)$ there is $u \in D(\mathscr{B})$ such that $u + \lambda\mathscr{B}u = f$ for fixed $\lambda > 0$. Let $\tilde{H}^1(0,T;H)$ be the subset of $H^1(0,T;H)$ consisting of the functions that are zero at 0. This is a Hilbert space with the same inner product.

In this space, let the bilinear form b_λ for $\lambda > 0$ be defined by

$$b_\lambda(\phi,\psi) = \int_0^T (\phi,\psi)(t)dt + \lambda\int_0^T \rho(t)p(t)(\phi'(t),\psi'(t))dt \tag{5.10}$$

Then b_λ is continuous, and $\mu > 0$ exists such that for each $\phi \in \tilde{H}(0,T;H)$

$$b_\lambda(\phi,\phi) \geq \mu\|\phi\|^2_{\tilde{H}(0,T;H)}$$

By Lax-Milgram Theorem, b_λ defines an isomorphism of $\tilde{H}(0,T;H)$ onto its dual in the following sense:

$$\forall L \in (\tilde{H}(0,T;H))', \ \exists\psi \in \tilde{H}(0,T;H)$$

such that

$$\forall\phi \in \tilde{H}(0,T;H), \ b_\lambda(\psi,\phi) = L(\phi).$$

For $f \in L^2(0,T;H)$, let $g = f - u_0$. Then g can be identified as an element of $(\tilde{H}(0,T;H))'$ because the mapping $\phi \mapsto \int_0^T (g,\phi)(t)dt$ is continuous on $\tilde{H}(0,T;H)$. Therefore $\tilde{u} \in \tilde{H}(0,T;H)$ exists such that for each $\phi \in \tilde{H}(0,T;H)$ we have

$$b_\lambda(\tilde{u},\phi) = \int_0^T (g,\phi)(t).$$

Therefore in the sense of vectorial distributions, we have:

$$\tilde{u} - \lambda(\rho p\tilde{u}')' = g \tag{5.11}$$

and since $g \in L^2(0,T;H)$, we have $\tilde{u}'' \in L^2(0,T;H)$. By choosing ϕ to be a function of class C^1 that is 0 at zero, since (5.11) is satisfied almost everywhere, we deduce that:

$$\int_0^T (g,\phi)(t)dt = b_\lambda(\tilde{u},\phi) = \int_0^T (\tilde{u},\phi)dt + \lambda\int_0^T \rho(t)p(t)(\tilde{u}',\phi')(t)dt \\ - \lambda\rho(T)p(T)(\tilde{u}',\phi)(T)$$

Since this relation holds for each ϕ, then comparing it to (5.10), we conclude that $\tilde{u}'(T) = 0$. So the function $u = \tilde{u} + u_0$ belonges to $D(\mathscr{B})$ and the relation $u + \lambda\mathscr{B}u = f$ is satisfied. □

Now let $\bar{A} = A - \beta I$. Then $\bar{A}$ is maximal monotone in H, $D(\bar{A}) = D(A)$, and $0 \in \bar{A}0$. Let $\bar{J}_\lambda$ and $\bar{A}_\lambda$ be respectively the resolvent and Yosida approximation of $\bar{A}$.

Lemma 5.2.3 *There is a unique function $u_{\lambda,T}$ that is twice continuously differentiable and belongs to $H^3(0,T;H)$ such that*

$$\begin{cases} p(t)u''_{\lambda,T}(t) + q(t)u'_{\lambda,T}(t) - \beta u_{\lambda,T}(t) = \bar{A}_\lambda(u_{\lambda,T}(t)), & 0 \le t \le T \\ u_{\lambda,T}(0) = u_0, \quad u'_{\lambda,T}(T) = 0 \end{cases} \tag{5.12}$$

Proof. For simplicity, we denote $u = u_{\lambda,T}$. Using the operator $\mathscr{B}$, (5.12) can be written as:

$$(\mathscr{B}u)(t) + \rho(t)\bar{A}_\lambda u(t) + \beta\rho u(t) = 0 \tag{5.13}$$

Define the operator $\mathscr{U}_\lambda$ on $L^2(0,T;H)$ by $(\mathscr{U}_\lambda u)(t) = (\rho\bar{A}_\lambda u)(t) + \beta(\rho u)(t)$. It is easy to see that $\mathscr{U}_\lambda$ is maximal monotone, strongly monotone and Lipschitzian on $L^2(0,T;H)$. By the previous lemma, $\mathscr{U}_\lambda + \mathscr{B}$ is surjective and u is the unique element in $(\mathscr{U}_\lambda + \mathscr{B})^{-1}(0)$. By the regularity of u and the Lipschitz property of $\bar{A}_\lambda$, we can get (5.12) and $u \in H^3(0,T;H)$. The uniqueness is proved in a similar way as in the previous theorem, by showing that $\|u\|$ is nonincreasing. □

In order to pass to the limit, we now need some estimates on the norm of $u_{\lambda,T}$ in

$$H^2(0,T;H) \cap H^2_\rho(0,T;H)$$

where $\|u\|^2_{L^2_\rho((0,+\infty);H)} := \int_0^{+\infty} \|u(t)\|^2 \rho(t)dt$.

Lemma 5.2.4 *There exists a constant M, independent of λ and T, such that if $u_{\lambda,T}$ is the solution to* (5.12)*, then*

$$\begin{cases} (i)\ \|u_{\lambda,T}\|_{H^2(0,T;H)} \le M \\ (ii)\ \|u_{\lambda,T}\|_{H^2_\rho(0,T;H)} \le M \end{cases} \tag{5.14}$$

Proof. Again for simplicity we denote $u_{\lambda,T} = u$. Using the monotonicity of $\bar{A}_\lambda$, for each $t \in [0,T]$ we have:

$$p(t)(u'',u)(t) + q(t)(u',u)(t) - \beta\|u(t)\|^2 \ge 0$$

Integrating the above inequality by parts on the interval $[0,T]$, we get:

$$-\int_0^T p(t)\|u'(t)\|^2 dt - \beta\int_0^T \|u(t)\|^2 dt + [p(u',u)]_0^T + \int_0^T (q-p')(t)(u',u)(t)dt \ge 0$$

Since $(u',u)(t) = \frac{1}{2}\frac{d}{dt}\|u(t)\|^2 \le 0$, and $u'(T) = 0$ and $p(t) \ge \alpha$, we get:

$$\alpha\int_0^T \|u'(t)\|^2 dt + \beta\int_0^T \|u(t)\|^2 \le -p(0)(u_0,u'(0)) + \|\frac{q-p'}{2}\|_{L^\infty}\|u_0\|^2$$

Set $w_h(t) = u(t+h) - u(t)$. Using the monotonicity of $\bar{A}_\lambda$ and Equation (5.12), we get:

$$(p(t+h)u''(t+h) - p(t)u''(t), w_h(t)) + (q(t+h)u'(t+h) - q(t)u'(t), w_h(t)) \\ - \beta\|w_h(t)\|^2 \geq 0$$

Dividing both sides of the above inequality by h^2, and taking the limit as $h \to 0$, since $u \in H^3(0,T;H)$, we get:

$$p(t)(u''',u')(t) + (q+p')(u'',u')(t) + (q'(t) - \beta)\|u'(t)\|^2 \geq 0 \tag{5.15}$$

Integrating by parts the above inequality on the interval $[0,T]$, we get:

$$-\int_0^T p(t)\|u''(t)\|^2 dt + [p(u'',u')]_0^T + \left[\frac{q}{2}\|u'\|^2\right]_0^T + \int_0^T \left(\frac{q'(t)}{2} - \beta\right)\|u'(t)\|^2 dt \geq 0$$

Therefore

$$\alpha \int_0^T \|u''(t)\|^2 dt \leq \left\|\frac{q'}{2} - \beta\right\|_{L^\infty} \int_0^T \|u'(t)\|^2 dt - \frac{q(0)}{2}\|u'(0)\|^2 - p(0)(u''(0), u'(0))$$

Clearly $u''(0)$ can be estimated directly from (5.12). Then

$$-\frac{q(0)}{2}\|u'(0)\|^2 - p(0)(u''(0),u'(0)) = -\beta(u_0,u'(0)) + \frac{q(0)}{2}\|u'(0)\|^2 - (\bar{A}_\lambda u_0, u'(0))$$

and since $u_0 \in D(\bar{A})$, by Theorem 3.4.1, we have:

$$\|\bar{A}_\lambda u_0\| \leq \|\bar{A}^0 u_0\|$$

Therefore:

$$\alpha \int_0^T \|u''(t)\|^2 dt \leq \left\|\frac{q'}{2} - \beta\right\|_{L^\infty} \int_0^T \|u'(t)\|^2 dt + M_1\|u'(0)\|^2 + N_1,$$

where M_1 and N_1 are independent of λ and T. We have therefore obtained the following estimates

$$\begin{cases} (i)\ \int_0^T \|u(t)\|^2 dt \leq A_0 + B_0\|u'(0)\|^2, \\ (ii)\ \int_0^T \|u'(t)\|^2 dt \leq A_1 + B_1\|u'(0)\|^2, \\ (iii)\ \int_0^T \|u''(t)\|^2 dt \leq A_2 + B_2\|u'(0)\|^2, \end{cases} \tag{5.16}$$

where A_i, B_i, $i = 0,1,2$, are positive constants independent of λ and T. Since

$$\|u'(0)\|^2 = -2\int_0^T (u'',u')(t)dt,$$

then by using the Cauchy-Shwarz inequality, and (5.16) (ii) and (iii), we get:

$$\|u'(0)\|^4 \leq 4(A_1 + B_1\|u'(0)\|^2)(A_2 + B_2\|u'(0)\|^2).$$

This inequality implies the boundedness of $\|u'(0)\|$ independently of λ and T, from which we get (5.14) (i). (5.14) (ii) is proved in a similar way. See [VER2] for details. □

Lemma 5.2.5 $\{u_\lambda\}_{\lambda>0}$ *is a Cauchy net in* $H^1(0,T;H)$.

Proof. We know that $\bar{A}_\lambda u_\lambda \in \bar{A}\bar{J}_\lambda u_\lambda$. Now replacing $\bar{J}_\lambda u_\lambda$ by its value obtained from the equality $\bar{A}_\lambda = \frac{I-\bar{J}_\lambda}{\lambda}$ replaced in (5.12), we obtain:

$$p(t)u_\lambda''(t)+q(t)u_\lambda' - \beta u_\lambda(t) \in \bar{A}[(1+\lambda\beta)u_\lambda(t) - \lambda p(t)u_\lambda''(t) - \lambda q(t)u_\lambda'(t)].$$

Now letting $x = \lambda u_\lambda - \mu u_\mu$ and $w = u_\lambda - u_\mu$, by the monotonicity of $\bar{A}$, we get:

$$((\rho p w')'(t) - \beta\rho(t)w(t), w(t)+\beta x(t) - p(t)x''(t) - q(t)x'(t)) \geq 0, \quad t\in[0,T]$$

Integrating the above inequality by parts on $[0,T]$, and using the given boundary conditions, we get:

$$\int_0^T \rho(t)p(t)\|w'(t)\|^2 dt + \beta\int_0^T \rho(t)\|w(t)\|^2 dt$$

$$\leq \int_0^T ((\rho p w')'(t) - \beta\rho(t)w(t), \beta x(t) - p(t)x''(t) - q(t)x'(t))dt.$$

By Lemma 5.2.4, w remains bounded in $H^2(0,T;H)$. Therefore there exists N independent of λ and μ such that

$$\|x\|_{H^2(0,T;H)} \leq (\lambda+\mu)N.$$

By developing the integral on the right hand side, we see that there exists K independent of λ and μ such that

$$\int_0^T \rho(t)p(t)\|w'(t)\|^2 dt + \int_0^T \beta\rho(t)\|w(t)\|^2 dt \leq (\lambda+\mu)K.$$

This completes the proof of the lemma, since the norms in $L^2(0,T;H)$ and $L^2_\rho(0,T;H)$ are equivalent. □

Proposition 5.2.6 *For each* $u_0 \in D(A)$, *there exists a unique continuously differentiable function* $u_T \in H^2(0,T;H)$ *that satisfies*

$$\begin{cases} p(t)u_T''(t)+q(t)u_T'(t) \in Au_T(t), & a.e. \text{ on } [0,T) \\ u_T(0) = u_0,\ u_T'(T) = 0,\ u_T(t)\in D(A), & a.e. \text{ on } [0,T) \end{cases} \tag{5.17}$$

If u_T *and* v_T *are the solutions to* (5.17) *associated respectively to* u_0 *and* v_0, *then*

$$\|u_T(t) - v_T(t)\| \leq \|u_0 - v_0\|, \quad \forall\, t\in[0,T] \tag{5.18}$$

The function $\|u_T\|$ *is nonincreasing and there exists a constant* M *independent of* T *such that*

$$\begin{cases} \|u_T\|_{H^2_\rho(0,T;H)} \leq M \\ \|u_T\|_{H^2(0,T;H)} \leq M \end{cases} \tag{5.19}$$

Proof. Unicity follows from (5.18) whose proof is similar to the one given in Theorem 5.2.1. Let u_T be the limit of $u_{\lambda,T}$ in $H^1(0,T;H)$, which exists by Lemma 5.2.5. Since $u''_{\lambda,T}$ converges to u''_T in $\mathscr{D}'(0,T;H)$, it follows from Lemma 5.2.4 that $u''_{\lambda,T}$ converges weakly to u''_T in $L^2(0,T;H)$, as λ tends to zero. Moreover, since

$$\|(u'_\lambda - u'_\mu)(t)\|^2 = -2\int_t^T (u'_\lambda - u'_\mu, u''_\lambda - u''_\mu)(t)dt$$

it follows that the net $(u_\lambda)_{\lambda>0}$ converges in $C^1([0,T];H)$. Let $\bar{\mathscr{A}}$ be the canonical extension of $\bar{A}$ to the space $L^2(0,T;H)$. $\bar{\mathscr{A}}$ is maximal monotone in $L^2(0,T;H)$(see [BRE]). By (5.12), in $L^2(0,T;H)$ we have:

$$pu''_\lambda + qu'_\lambda - \beta u_\lambda \in \bar{\mathscr{A}} \bar{J}_\lambda u_\lambda \tag{5.20}$$

and

$$\|\bar{J}_\lambda u_\lambda - u_T\| \leq \lambda \|\bar{A}_\lambda u_\lambda\| + \|u_\lambda - u_T\|.$$

By Lemma 5.2.4, u_λ is bounded in $H^2(0,T;H)$, hence $A_\lambda u_\lambda$ also is bounded in $L^2(0,T;H)$. Therefore, passing to the limit in (5.20) as $\lambda \to 0$, and since $\bar{\mathscr{A}}$ is demiclosed, we get:

$$\begin{cases} p(t)u''_T(t) + q(t)u'_T(t) - \beta u_T(t) \in \bar{A} u_T(t) \\ u_T(0) = u_0, \;\; u'_T(T) = 0, \;\; u_T(t) \in D(A), \;\; \text{a.e. on } [0,T] \end{cases} \tag{5.21}$$

which proves the proposition. □

To prove the main result of this section (Theorem 5.2.1), we need to pass to the limit as $T \to +\infty$, and for this, we need the following lemmas.

Lemma 5.2.7 *Let x_L be the solution to the following equation on $[0,L]$.*

$$\begin{cases} p(t)x''_L(t) + q(t)x'_L(t) - \beta x_L(t) = 0, \quad on\ [0,L] \\ x'_L(0) = 0, \qquad x_L(L) = 2\|u_0\| \end{cases} \tag{5.22}$$

Then $x_L \in H^3(0,L)$ and $\|x_L\|_{H^3(0,L)} \leq M$ where M is independent of L. x_L is a positive nondecreasing function, and for L_0 fixed, we have:

$$\lim_{L \geq L_0, L \to +\infty} \|x_L\|_{H^2(0,L_0)} = 0.$$

Proof. Let $y = x_L(L-t)$, $\tilde{p}(t) = p(L-t)$, and $\tilde{q}(t) = -q(L-t)$. Then we get:

$$\begin{cases} \tilde{p}(t)y''(t) + \tilde{q}y'(t) - \beta y(t) = 0 \text{ on } [0,L] \\ y(0) = 2\|u_0\|, \;\; y'(L) = 0 \end{cases} \tag{5.23}$$

By Lemmas 5.2.3 and 5.2.4 (with $H = \mathbb{R}$ and $\bar{A} = 0$), the first three claims in the lemma are proved. Since the embedding of $H^3(0,L)$ into $H^2(0,L)$ is compact, there exist a function $x \in H^3(0,+\infty)$ and a subnet of x_L such that for each L_0, we have $\lim \|x_L - x\|_{H^2(0,L_0)} = 0$. The function $t \mapsto x(t)$ is positive and nondecreasing as the limit of such functions, and since it belongs to $L^2(0,+\infty)$, it must be the constant zero. Now a classical unicity argument finishes the proof of the lemma. □

Lemma 5.2.8 *Suppose that ϕ is a function in $H^2(0,L;H)$ that satisfies the following relations:*

$$\begin{cases} \phi(0)=0, \ \|\phi(L)\| \leq 2\|u_0\| \\ p(t)(\phi'',\phi)(t)+q(t)(\phi',\phi)(t)-\beta\|\phi(t)\|^2 \geq 0 \ \text{a.e. on } [0,L]. \end{cases} \tag{5.24}$$

Then $\|\phi\|$ is a nondecreasing function that is bounded above by the solution x_L of the Equation (5.22).

Proof. If $\phi(L)=0$, then after integrating by parts the relation (5.24) on $[0,L]$, it follows immediately that ϕ is the constant zero. Otherwise, there exists $t_1 \in (0,L)$ such that $\|\phi(t_1)\| \geq \|\phi(L)\|$. Without loss of generality, we may assume that $\|\phi(t_1)\|$ is maximal and therefore $(\phi,\phi')(t_1)=0$. Integrating by parts on $[t_1,t]$ with $t>t_1$, we get:

$$\begin{aligned} \frac{1}{2}\rho(t)p(t)\frac{d}{dt}\|\varphi(t)\|^2 &= \rho(t)p(t)(\phi,\phi')(t) \\ &\geq \beta\int_{t_1}^{t}\rho(s)\|\phi(s)\|^2 ds + \int_{t_1}^{t}\rho(s)p(s)\|\phi'(s)\|^2 ds \geq 0 \end{aligned}$$

which is a contradiction since $\frac{d}{dt}\|\varphi(t)\|^2<0$ for t near t_1. By replacing L with an arbitrary t, it follows that $\|\varphi\|$ is nondecreasing, and the first claim is proved. Suppose that (t_0,L) is the interval where $\|\phi(t)\|>0$. The function $\|\phi\|$ belongs to $W^{2,1}_{\text{loc}}(t_0,L)$ and we have:

$$(\|\phi\|^2)' = 2(\phi,\phi') = 2\|\phi\|(\|\phi\|)' \Rightarrow \|\phi'\| \geq |(\|\phi\|)'|$$

$$(\|\phi\|^2)'' = 2\|\phi'\|^2 + 2(\phi,\phi'') = 2(\|\phi\|')^2 + 2\|\phi\|(\|\phi\|)''$$

and therefore $(\phi,\phi'') \leq \|\phi\|(\|\phi\|)''$ a.e. on (t_0,L). This implies that on (t_0,L) we have:

$$p(t)\|\phi\|''(t)+q(t)\|\phi\|'(t)-\beta\|\phi(t)\| \geq 0.$$

Let $w=\|\phi\|-x_L$. Then w satisfies the following relations:

$$\begin{cases} w \in W^{2,1}_{\text{loc}}(t_0,L), \ w(t_0) \leq 0, \ \ w(L) \leq 0 \\ p(t)w''(t)+q(t)w'(t)-\beta w(t) \geq 0, \ \text{a.e. on } (t_0,L). \end{cases} \tag{5.25}$$

Now by contradiction, assume that $w(t_1)>0$ for some $t_0<t_1<L$, where without loss of generality, we may assume that $w(t_1)$ is maximal. Then multiplying the relation (5.25) by $w(t)$ for $t>t_1$, and integrating by parts on $[t_1,t]$, with a similar reasoning as above, we get a contradiction. Therefore $w(t) \leq 0$, and this proves the theorem. □

Proposition 5.2.9 *For each fixed L, the net $(u_T)_{T\geq L}$ defined in Proposition 5.2.6 is Cauchy in $C^1([0,L];H)$.*

Proof. Suppose that u_T and $u_{T'}$ are solutions on $[0,T]$ and $[0,T']$ to the Equation (5.17) with the same initial value u_0. Suppose that $T' > T$ and let L and L' be such that $0 < L < L' < T < T'$. The function $\phi = u_T - u_{T'}$ satisfies the hypothesis of Lemma 5.2.8 on $[0,L']$, and therefore $\|\phi\|$ is bounded above by the solution $x_{L'}$ to (5.22) on $[0,L']$. By Lemma 5.2.7, for L fixed, we have $\lim_{L'\to+\infty}\|\phi(t)\| = 0$ uniformly on $[0,L]$. Moreover

$$\int_0^L \rho(t)p(t)\|w'(t)\|^2 dt + \beta\int_0^L \rho(t)\|\phi(t)\|^2 dt \le \rho(L)p(L)(\phi,\phi')(L)$$

and since by Lemma 5.2.4, $\phi'(L)$ is bounded, we deduce that the net $\{u_T\}_{T>L}$ is Cauchy in $H^1(0,L;H)$. Since u_T'' remains bounded in $L^2(0,T;H)$, by a similar proof as in Proposition 5.2.6, we conclude that this net is Cauchy in $C^1([0,L];H)$, for each fixed L. □

Now we complete the proof of Theorem 5.2.1:

Proof. We have already proved the uniqueness. To prove the existence, by Proposition 5.2.9 and Lemma 5.2.4, there exists a function u belonging to $H^2(0,+\infty;H)$ and $H_\rho^2(0,+\infty;H)$, which is the limit of the net (u_T) in $C^1([0,L];H)$, and moreover $w-\lim u_T'' = u''$ in $L^2(0,L;H)$ for each $L > 0$. Similar to the proof of Proposition 5.2.6, it follows that $u(t) \in D(A)$, a.e. on $(0,+\infty)$, and that u is the solution to (5.6). This completes the proof of the theorem. □

5.2.2 THE NON STRONGLY MONOTONE CASE

In this subsection, we assume that the assumptions (5.3), (5.4) and (5.5) are satisfied, and the operator A is maximal monotone and $0 \in A0$, but A is not strongly monotone, and prove the existence of the solution to (5.1).

Proposition 5.2.10 *For every $u_0 \in D(A)$, there is a unique continuously differentiable function u_T such that $u_T'' \in L^2(0,T;H)$ and satisfies the Equation* (5.17). *The function $\|u_T\|$ is nonincreasing and there exists M independent of T such that*

$$\begin{cases} \|u'\|_{H^1(0,T;H)} \le M \\ \|u'\|_{H_\rho^1(0,T;H)} \le M \end{cases} \tag{5.26}$$

Proof. The proof of the uniqueness is similar to the proof of Proposition 5.2.6. We prove the existence. For $\delta > 0$, let u_δ be the unique solution in $H^2(0,T;H)$ of the equation

$$\begin{cases} u_\delta(0) = u_0,\ u_\delta'(T) = 0,\ u_\delta(t) \in D(A), \text{ a.e. on}[0,T] \\ p(t)u_\delta''(t) + q(t)u_\delta'(t) - \delta u_\delta(t) \in Au_\delta(t), \text{ a.e. on } [0,T] \end{cases} \tag{5.27}$$

(the existence of u_δ follows from Proposition 5.2.6). By Remarks 2.6 and 2.7 from [VER2], there exist M independent of T and $\delta \le \delta_0$, such that

$$\begin{cases} \|u'_\delta\|_{H^1(0,T;H)} \le M \\ \|u'_\delta\|_{H^1_\rho(0,T;H)} \le M \end{cases} \tag{5.28}$$

and moreover the function $\|u_\delta\|$ is nonincreasing. For each γ and $\delta \le \delta_0$, consider the solutions u_γ and u_δ of (5.27), and set $w = u_\gamma - u_\delta$. By the monotonicity of A, we have:

$$-\int_0^T \rho(t)p(t)\|w'(t)\|^2 dt + \int_0^T \rho(t)(\delta u_\delta - \gamma u_\gamma, w)(t)dt \ge 0.$$

Therefore by using (5.28), we get $\|w'\|^2_{L^2_\rho(0,T;H)} \le K(\gamma+\delta)$. Now by letting $\gamma, \delta \to 0$, the proof is completed in a similar way as in Proposition 5.2.6. □

Theorem 5.2.11 *For each $u_0 \in D(A)$, there is a unique continuously differentiable function u, bounded on $\mathbb{R}^+$, such that $u' \in H^1(0,+\infty;H)$ and u satisfies* (5.6) *and* (5.7). *Moreover $\|u\|$ is nonincreasing and $u' \in H^1_\rho(0,+\infty;H)$.*

Proof. We first prove the uniqueness. Suppose that u_1 and u_2 are two solutions to (5.6) with the same initial value u_0, and let $w = u_1 - u_2$. By the monotonicity of A, we have:

$$\rho(t)p(t)(w',w)(t) \ge \int_0^t \rho(s)p(s)\|w'(s)\|^2 ds \ge 0.$$

Therefore the function $\|w(t)\|$ is nondecreasing. By contradiction, assume that w is not identically zero and let $(T,+\infty)$ be the interval where it is different from zero. Then after some computations (see [VER2]), we get the following inequality:

$$\|w(t)\| \le \|u_1 - u_2\|_{L^\infty} \frac{\int_0^t \exp(-\int_0^s \frac{q(\tau)}{p(\tau)} d\tau)ds}{\int_0^n \exp(-\int_0^s \frac{q(\tau)}{p(\tau)} d\tau)ds}, \quad \forall t \in [T,n].$$

For fixed t, letting $n \to +\infty$ in the above inequality, it follows from (5.5) that $w(t) = 0$, $\forall t \ge T$. Therefore $T = +\infty$, and the unicity follows.

For the existence, suppose that $L < T < T'$ and $w = u_T - u_{T'}$, where u_T and $u_{T'}$ are solutions to (5.17)(see Proposition 5.2.6) on $[0,T]$ and $[0,T']$ respectively. We have:

$$\begin{cases} p(t)(w'',w)(t) + q(t)(w',w)(t) \ge 0, \ \text{a.e. on } [0,T] \\ w(0) = 0, \ \|w(T)\| \le 2\|u_0\|. \end{cases} \tag{5.29}$$

Then with a similar argument as in the proof of the unicity, we get:

$$\|w(t)\| \le 2\|u_0\| \frac{\int_0^t \exp\left(-\int_0^s \frac{q(\tau)}{p(\tau)} d\tau\right)ds}{\int_0^T \exp\left(-\int_0^s \frac{q(\tau)}{p(\tau)} d\tau\right)ds}$$

which converges uniformly to zero on every compact interval, as T tends to infinity. It is easily shown that w' is bounded in $H^1(0,T;H)$; then a similar argument as in Proposition 5.2.6 shows that the net $(u_T)_{T>L}$ is Cauchy in $H^1(0,L;H)$ for every L, and hence also in $C^1([0,L];H)$. The proof is completed in a similar way as in Theorem 5.2.1. □

5.3 TWO POINT BOUNDARY VALUE PROBLEMS

In this section we consider the boundary value problem of the form:

$$\begin{cases} u''(t) \in Au(t) + f(t), & \text{a.e. on } (0,T) \\ u(0) = a, \quad u(T) = b \end{cases} \tag{5.30}$$

Definition 5.3.1 *The function $u : [0,T] \to H$ is called a solution of the two point boundary value problem* (5.30) *if $u \in W^{2,2}(0,T;H)$, and $u(t) \in D(A)$ for almost every $t \in (0,T)$, and u satifies the boundary conditions $u(0) = a$ and $u(T) = b$, and for almost every $t \in (0,T)$,*

$$u''(t) \in Au(t) + f(t).$$

We first study the existence problem for the case $a, b \in D(A)$.

Theorem 5.3.2 *If $A : D(A) \subset H \to H$ is a maximal monotone operator in a Hilbert space H, $a, b \in D(A)$ and $f \in L^2(0,T;H)$, then the problem* (5.30) *has a unique solution $u \in W^{2,2}(0,T;H)$. If u_1 and u_2 are solutions to* (5.30)*, then $t \mapsto \|u_1(t) - u_2(t)\|^2$ is a convex function and*

$$\|u_1(t) - u_2(t)\| \leq \max\{\|u_1(0) - u_2(0)\|, \|u_1(T) - u_2(T)\|\} \tag{5.31}$$

$$\int_0^T \gamma(t)\|u_1'(t) - u_2'(t)\|dt \leq \frac{1}{2}(\|u_1(0) - u_2(0)\|^2 + \|u_1(T) - u_2(T)\|^2) \tag{5.32}$$

where $t \in [0,T]$, and $\gamma(t) = \min\{t, T-t\}$

Proof. We start with the proof of uniqueness. Suppose that u_1 and u_2 are two solutions to (5.30). Then by the monotonicity of A, we get

$$(\|u_1 - u_2\|^2)'' \geq 2\|u_1' - u_2'\|^2 \quad \text{a.e. on } (0,T). \tag{5.33}$$

Therefore the function $t \mapsto \|u_1(t) - u_2(t)\|^2$ is convex on $[0,T]$, and (5.31) is satisfied. If u_1 and u_2 satisfy the boundary conditions, then the uniqueness follows. Multiplying (5.33) by $\gamma_\varepsilon(t) = \min\{t-\varepsilon, T-t-\varepsilon\}$ for $t \in (0,T)$, and $\varepsilon \in (0, \frac{T}{2})$ constant, and integrating from $t = \varepsilon$ to $t = T - \varepsilon$, we get

$$\begin{aligned} 2\int_\varepsilon^{T-\varepsilon} \gamma_\varepsilon\|u_1'(t) - u_2'(t)\|^2 dt \leq \gamma_\varepsilon(\|u_1 - u_2\|^2)'\Big|_\varepsilon^{T-\varepsilon} - \int_\varepsilon^{\frac{T}{2}} (\|u_1(t) - u_2(t)\|^2)' dt \\ + \int_{\frac{T}{2}}^{T-\varepsilon} (\|u_1(t) - u_2(t)\|^2)' dt \end{aligned}$$

or

$$2\int_{\varepsilon}^{T-\varepsilon}\gamma_\varepsilon\|u_1'(t)-u_2'(t)\|^2dt \le \gamma_\varepsilon(T-\varepsilon)(\|u_1-u_2\|^2)'(T-\varepsilon)$$
$$-\gamma_\varepsilon(\varepsilon)(\|u_1-u_2\|^2)'(\varepsilon)$$
$$+\|u_1-u_2\|^2(\varepsilon)+\|u_1-u_2\|^2(T-\varepsilon)$$

Letting $\varepsilon \to 0^+$, and using Fatou's lemma, we get (5.32).
Now we prove the existence result. Consider the following problem

$$\begin{cases} u_\lambda'' = A_\lambda u_\lambda + f, \ \text{a.e. on } (0,T) \\ u_\lambda(0) = a, \ \ u_\lambda(T) = b \end{cases} \tag{5.34}$$

We show that this problem has a unique solution $u_\lambda \in W^{2,2}(0,T;H)$ that converges uniformly on $[0,T]$ to a solution u of (5.30). We replace (5.34) with a boundary value problem with zero boundary conditions. To this end, we replace the function u_λ with $v_\lambda = u_\lambda - h$ where $h(t) = \frac{t}{T}b + \frac{T-t}{T}a, \ t \in [0,T]$. Then the above problem becomes

$$\begin{cases} v_\lambda'' - \frac{1}{\lambda}v_\lambda = \frac{1}{\lambda}h - \frac{1}{\lambda}J_\lambda(v_\lambda+h)+f, \ \text{a.e. on } (0,T) \\ v_\lambda(0) = v_\lambda(T) = 0. \end{cases} \tag{5.35}$$

For all $\alpha \in L^2(0,T;H)$, the problem

$$\begin{cases} v_\lambda'' - \frac{1}{\lambda}v_\lambda = \frac{1}{\lambda}h - \frac{1}{\lambda}J_\lambda(\alpha+h)+f, \ \text{a.e. on } (0,T) \\ v_\lambda(0) = v_\lambda(T) = 0, \end{cases} \tag{5.36}$$

has a unique solution $v_\lambda^\alpha \in W^{2,2}(0,T;H)$. Now consider the operator C : $L^2(0,T;H) \to L^2(0,T;H)$ defined by $C\alpha = v_\lambda^\alpha$. We show that C is a contraction. For $\alpha,\beta \in L^2(0,T;H)$, let $v_\lambda^\alpha, v_\lambda^\beta \in W^{2,2}(0,T;H)$ be the corresponding solutions to (5.36). We rewrite (5.36) in the form:

$$v_\lambda^\alpha = \frac{1}{\lambda}(\frac{1}{\lambda}I+B)^{-1}\mathscr{J}_\lambda(\alpha+h) - (I+\lambda B)^{-1}h - (\frac{1}{\lambda}I+B)^{-1}f \tag{5.37}$$

where $\mathscr{J}_\lambda$ is defined by $(\mathscr{J}_\lambda v)(t) = J_\lambda v(t)$ for all $v \in L^2(0,T;H)$, and the operator B is defined by

$$Bu = -u'', \ D(B) = \{u \in W^{2,2}(0,T;H), \ \ u(0) = u(T) = 0\}$$

We observe that B is linear and maximal monotone on $L^2(0,T;H)$. Therefore $-B$ generates a C_0 semigroup of linear nonexpansive mappings on $L^2(0,T;H)$. Moreover for each $\mu > 0$ and each $u \in L^2(0,T;H)$

$$((\mu I+B)^{-1}u)(t) = \frac{T}{\alpha\sinh\alpha}\sinh(\alpha-\frac{\alpha t}{T})\int_0^t \sinh\frac{\alpha s}{T}u(s)ds$$

$$+\frac{T}{\alpha \sinh \alpha} \sinh \frac{\alpha t}{T} \int_t^T \sinh(\alpha - \frac{\alpha s}{T}) u(s) ds,$$

where $\alpha = T\sqrt{\mu}$. Since

$$(Bu, u) \geq c\|u\|^2, \quad \forall u \in D(B)$$

we deduce that

$$\|(\mu I + B)^{-1} u - (\mu I + B)^{-1}(v)\| \leq \frac{1}{\mu + c}\|u - v\|, \quad \mu > 0.$$

By the nonexpansiveness of $\mathscr{J}_\lambda$ and (5.37) one obtains

$$\begin{aligned}\|C\alpha - C\beta\| &= \frac{1}{\lambda}\|(\frac{1}{\lambda}I + B)^{-1}\mathscr{J}_\lambda(\alpha + h) - (\frac{1}{\lambda}I + B)^{-1}\mathscr{J}_\lambda(\beta + h)\| \\ &\leq \frac{1}{1 + \lambda c}\|\alpha - \beta\|\end{aligned}$$

Therefore C is a contraction on $L^2(0, T; H)$ and hence has a unique fixed point $\alpha \in L^2(0, T; H)$ that clearly is v_λ^α. Hence the problem (5.35) and consequently (5.34) has a unique solution $u_\lambda \in W^{2,2}(0, T; H)$.

Now we prove that $u_\lambda \to u$ in $C([0, T]; H)$ where u satisfies (5.30). To this end, we multiply (5.34) by u_λ'' and integrate from 0 to T, then we get:

$$\int_0^T \|u_\lambda''\|^2 dt = (A_\lambda u_\lambda, u_\lambda')|_0^T - \int_0^T ((A_\lambda u_\lambda)', u_\lambda') dt + \int_0^T (f, u_\lambda'') dt. \tag{5.38}$$

We note that $(A_\lambda u_\lambda)'$ exists for almost every $t \in (0, T)$, because the function $t \mapsto A_\lambda u_\lambda(t)$ is Lipschitz continuous. By the monotonicity of A_λ, we have:

$$\left((A_\lambda u_\lambda)', u_\lambda'\right) \geq 0, \text{ a.e. on } (0, T).$$

Since

$$(A_\lambda b, u_\lambda'(T)) - (A_\lambda a, u_\lambda'(0)) = \int_0^T (\frac{t}{T}A_\lambda b + \frac{T-t}{T}A_\lambda a, u_\lambda'') dt + \frac{1}{T}(A_\lambda b - A_\lambda a, b - a)$$

Equality (5.38) implies that

$$\begin{aligned}\int_0^T \|u_\lambda''\|^2 dt &\leq \left[T^{\frac{1}{2}}\|A^0 a\| + T^{\frac{1}{2}}\|A^0 b\| + (\int_0^T \|f\|^2 dt)^{\frac{1}{2}}\right](\int_0^T \|u_\lambda''\|^2 dt)^{\frac{1}{2}} \\ &+ \frac{1}{T}(\|A^0 a\| + \|A^0 b\|)\|b - a\|\end{aligned} \tag{5.39}$$

This together with (5.34) implies the boundedness of $\{u_\lambda''\}$ and $\{A_\lambda u_\lambda\}$ in $L^2(0, T; H)$. Now we prove that u_λ is a Cauchy sequence in $C([0, T])$. Taking into account

$$(u_\lambda' - u_\mu', u_\lambda - u_\mu)' = (u_\lambda'' - u_\mu'', u_\lambda - u_\mu) + \|u_\lambda' - u_\mu'\|^2$$

almost everywhere on $(0,T)$, and using (5.34) and $x = \lambda A_\lambda x + J_\lambda x$, we get:

$$\int_0^T \|u'_\lambda - u'_\mu\|^2 dt = -\int_0^T (A_\lambda u_\lambda - A_\mu u_\mu, \lambda A_\lambda u_\lambda - \mu A_\mu u_\mu)dt$$
$$-\int_0^T (A_\lambda u_\lambda - A_\mu u_\mu, J_\lambda u_\lambda - J_\mu u_\mu)dt$$

Now the monotonicity of A_λ and the boundedness of $A_\lambda u_\lambda$ in $L^2(0,T;H)$ imply that:

$$\int_0^T \|u'_\lambda - u'_\mu\|^2 dt \leq K(\lambda + \mu). \tag{5.40}$$

Therefore

$$\|u_\lambda(t) - u_\mu(t)\|^2 \leq K'(\lambda + \mu) \tag{5.41}$$

where K, K' are positive constants. Then u_λ is uniformly convergent on $[0,T]$ to a function u and $u'_\lambda \to u'$ in $L^2(0,T;H)$.

Since u''_λ is bounded in $L^2(0,T;H)$, and $u''_\lambda \to u''$ in $\mathscr{D}'(0,T;H)$, we conclude that $u''_\lambda \rightharpoonup u''$ in $L^2(0,T;H)$. Moreover, we have the strong convergence of $u'_\lambda \to u'$ in $C(0,T;H)$ and $J_\lambda u_\lambda = \lambda A_\lambda u_\lambda + u_\lambda \to u$ in $C([0,T];H)$. Since $\bar{A}$ (the canonical extension of A) is demiclosed (by Proposition 3.3.4), and $\mathscr{J}_\lambda u_\lambda \to u$ and $u''_\lambda \rightharpoonup u''$ in $L^2(0,T;H)$, letting $\lambda \to 0^+$ in (5.34) rewritten in the form $u''_\lambda - f \in \bar{A}(\mathscr{J}_\lambda u_\lambda)$, we conclude that $u \in D(A)$ and u is a solution to (5.30). □

Remark 5.3.3 *The solution u to problem* (5.30) *can be approximated by the unique solution u_λ of the problem* (5.34), *because $u_\lambda \to u$ in $C([0,T];H)$. Moreover $u'_\lambda \to u'$ in $C([0,T];H)$ and $u''_\lambda \rightharpoonup u''$ in $L^2(0,T;H)$.*

In the sequel, we find an estimation for u'' in a weighted function space. Suppose that $\gamma : [0,T] \to \mathbb{R}$ is the function defined by $\gamma(t) = \min\{t, T-t\}$ for $t \in [0,T]$. Suppose that $L^2_\gamma(0,T;H)$ is the function space $L^2(0,T;H)$ with the weighted function γ. This means that the inner product of $L^2_\gamma(0,T;H)$ is defined by

$$(u,v)_\gamma = \int_0^T \gamma(t)(u(t),v(t))dt$$

and the norm $|\cdot|_\gamma$ is defined by

$$|u|_\gamma^2 = \int_0^T \gamma(t)\|u(t)\|^2 dt.$$

Similarly we define the weighted space $L^2_{\gamma^3}(0,T;H)$. Let $(\cdot,\cdot)_{\gamma^3}$ and $|\cdot|_{\gamma^3}$ be respectively the inner product and the norm for this space. We observe that

$$L^2(0,T;H) \subset L^2_\gamma(0,T;H) \subset L^2_{\gamma^3}(0,T;H)$$

and

$$u \in L^2_\gamma(0,T;H) \Leftrightarrow t^{\frac{1}{2}}(T-t)^{\frac{1}{2}}u(t) \in L^2(0,T;H)$$
$$u \in L^2_{\gamma^3}(0,T;H) \Leftrightarrow t^{\frac{3}{2}}(T-t)^{\frac{3}{2}}u(t) \in L^2(0,T;H)$$

Proposition 5.3.4 *With the assumptions of Theorem 5.3.2, if $u \in W^{2,2}(0,T;H)$ is a solution to* (5.30)*, then*

$$|u''|_{\gamma^3} \leq |f|_{\gamma^3} + 3|u'|_\gamma \tag{5.42}$$

Proof. Again we consider the approximate problem (5.34), and multiply the equation by $A_\lambda u_\lambda$, and use the estimate $(u'_\lambda, A_\lambda u_\lambda)' \geq (u''_\lambda, A_\lambda u_\lambda)$ to get

$$(u'_\lambda, A_\lambda u_\lambda)' \geq \|A_\lambda u_\lambda\|^2 + (f, A_\lambda u_\lambda) \tag{5.43}$$

Multiplying (5.43) by γ^3 and integrating on $[0,T]$ we get:

$$\int_0^T \gamma^3 \|A_\lambda u_\lambda + \frac{f}{2}\|^2 dt - \frac{1}{4}\int_0^T \gamma^3 \|f\|^2 dt \leq \int_0^T \gamma^3 (u'_\lambda, A_\lambda u_\lambda)' dt \tag{5.44}$$

Integrating by parts, and using the equality $\gamma(0) = \gamma(T) = 0$, we find:

$$\begin{aligned} &\int_0^T \gamma^3 \|A_\lambda u_\lambda + \frac{f}{2}\|^2 dt - \frac{1}{4}\int_0^T \gamma^3 \|f\|^2 dt \\ &\qquad \leq 3\Big(\int_0^T \gamma \|u'_\lambda\|^2 dt\Big)^{\frac{1}{2}} \Big(\int_0^T \gamma^3 \|A_\lambda u_\lambda + \frac{f}{2}\|^2 dt\Big)^{\frac{1}{2}} \\ &\qquad + \frac{3}{2}\Big(\int_0^T \gamma \|u'_\lambda\|^2 dt\Big)^{\frac{1}{2}} \Big(\int_0^T \gamma^3 \|f\|^2 dt\Big)^{\frac{1}{2}} \end{aligned}$$

Therefore

$$\Big(\int_0^T \gamma^3 \|A_\lambda u_\lambda + \frac{f}{2}\|^2 dt\Big)^{\frac{1}{2}} \leq \frac{1}{2}\Big(\int_0^T \gamma^3 \|f\|^2 dt\Big)^{\frac{1}{2}} + 3\Big(\int_0^T \gamma \|u'_\lambda\|^2 dt\Big)^{\frac{1}{2}}$$

Since $A_\lambda u_\lambda = u''_\lambda - f$ we get:

$$\Big(\int_0^T \gamma^3 \|u''_\lambda\|^2 dt\Big)^{\frac{1}{2}} \leq \Big(\int_0^T \gamma^3 \|f\|^2 dt\Big)^{\frac{1}{2}} + 3\Big(\int_0^T \gamma \|u'_\lambda\|^2 dt\Big)^{\frac{1}{2}} \tag{5.45}$$

Letting $\lambda \to 0^+$, and using the convergence $u''_\lambda \rightharpoonup u''$ in $L^2_{\gamma^3}(0,T;H)$ and $u'_\lambda \to u'$ in $L^2_\gamma(0,T;H)$ which follows from Remark 5.3.3, we get (5.42). □

Theorem 5.3.5 *If $A : D(A) \subset H \to H$ is maximal monotone on H and $a, b \in \overline{D(A)}$ and $f \in L^2(0,T;H)$, then the problem* (5.30) *has a unique solution $u \in C([0,T];H) \cap W^{2,2}_{\text{loc}}(0,T;H)$. Moreover, $u'' \in L^2_{\gamma^3}(0,T;H)$ and $u' \in L^2_\gamma(0,T;H)$.*

Proof. The proof of uniqueness is similar to the proof of Theorem 5.3.2. We prove the existence. Since $a, b \in \overline{D(A)}$, there are sequences a_n and b_n in $D(A)$ such that $a_n \to a$ and $b_n \to b$ in H. By Theorem 5.3.2 the problem

$$\begin{cases} u''_n \in Au_n + f \\ u_n(0) = a_n, \ \ u_n(T) = b_n \end{cases} \tag{5.46}$$

has a unique solution $u_n \in W^{2,2}(0,T;H)$. Inequality (5.31) for u_n and u_m shows that u_n is a Cauchy sequence in $C([0,T];H)$, therefore a function $u \in C([0,T];H)$ exists such that $u_n \to u$ in $C([0,T];H)$ and $u(0) = a$ and $u(T) = b$. By (5.32), $u_n' \to u'$ in $L^2_\gamma(0,T;H)$. Moreover since $f \in L^2(0,T;H) \subset L^2_{\gamma^3}(0,T;H)$, Proposition 5.3.4 implies the boundedness of u_n'' in $L^2_{\gamma^3}(0,T;H)$. Therefore $u \in W^{2,2}_{\text{loc}}(0,T;H)$ and $u' \in L^2_\gamma(0,T;H)$ and $u'' \in L^2_{\gamma^3}(0,T;H)$ and $u_n'' \rightharpoonup u''$ in $L^2_{\gamma^3}(0,T;H)$. Suppose that $\varepsilon \in (0,\frac{T}{2})$ and $\mathscr{A}$ is the operator defined by

$$\mathscr{A}u = \{v \in L^2(\varepsilon, T-\varepsilon; H),\ v(t) \in Au(t) \text{ a.e. on } (0,T)\}.$$

$\mathscr{A}$ is maximal monotone and therefore demiclosed. By rewriting the Equation (5.46) in the form $u_n'' - f \in \mathscr{A}u_n$ and noting that $u_n \to u$ and $u_n'' \rightharpoonup u''$ in $L^2(\varepsilon, T-\varepsilon; H)$, by taking the limit as $n \to +\infty$, we observe that u satisfies the equation $u'' \in Au + f$ almost everywhere on $(\varepsilon, T-\varepsilon)$. Since this holds for each $\varepsilon \in (0,\frac{T}{2})$, u is a solution to (5.30). We have already shown that u satisfies the boundary conditions and $u \in C([0,T];H) \cap W^{2,2}_{\text{loc}}(0,T;H)$ and $u' \in L^2_\gamma(0,T;H)$ and $u'' \in L^2_{\gamma^3}(0,T;H)$. □

5.4 EXISTENCE OF SOLUTIONS FOR THE NONHOMOGENEOUS CASE

Theorem 5.4.1 *Suppose that A is a maximal monotone operator in H and $f \in L^2_{\text{loc}}(0,+\infty;H)$ and $b \in \overline{D(A)}$ is an initial value such that Problem (5.2) for b has at least a solution $v \in C([0,\infty);H) \cap W^{2,2}_{\text{loc}}(0,\infty;H)$. Then Problem (5.2) has a solution $u \in C([0,\infty);H) \cap W^{2,2}_{\text{loc}}(0,\infty;H)$ for each initial data $a \in \overline{D(A)}$. Moreover $t^{\frac{1}{2}}u' \in L^2(0,\delta;H)$ and $t^{\frac{3}{2}}u'' \in L^2(0,\delta;H)$ for each $\delta > 0$. If $u,v \in C([0,\infty);H) \cap W^{2,2}_{\text{loc}}(0,\infty;H)$ are two bounded solutions of $u'' \in Au + f$, then the function $t \mapsto \|u(t) - v(t)\|^2$ is convex and nonincreasing and*

$$\int_0^\infty t\|u'(t) - v'(t)\|^2 dt \leq \frac{1}{2}\|u(0) - v(0)\|^2 \tag{5.47}$$

Proof. Suppose that v is a solution of (5.2) corresponding to the initial data $v(0) = b$. If $u \in C([0,\infty);H) \cap W^{2,2}_{\text{loc}}(0,\infty;H)$ is a solution of $u'' \in Au + f$, since $t \mapsto \|u(t) - v(t)\|^2$ is convex, we infer that

$$\sup_{t\in[0,T]} \|u(t)\| \leq \max\{\|u(0)\|, \|u(T)\|\} + M \tag{5.48}$$

where $M = 2\sup_{t\geq 0}\|v(t)\|$.

Step 1. Consider the case $a \in D(A)$. By Theorem 5.3.2, for every positive integer n, there is a unique solution $u_n \in W^{2,2}(0,n;H)$ to the problem:

$$\begin{cases} u_n''(t) \in Au_n(t) + f(t) \text{ a.e. on } (0,n) \\ u_n(0) = u_n(n) = a \end{cases}$$

Suppose that $0 < T < m \leq n$. Similar to the above, we can easily see that

$$2\|u_n'(t)-u_m'(t)\|^2 \leq (\|u_n(t)-u_m(t)\|^2)'' \quad \text{a.e. on } (0,n)$$

If u and v are two solutions to the Equation (5.2), then since $t \mapsto \|u(t)-v(t)\|^2$ is convex and bounded, we conclude that $\|u(t)-v(t)\|$ is nonincreasing. Therefore $\frac{d}{dt}\|u(t)-v(t)\|^2 \leq 0$. Now multiplying both sides of the above inequality by t, and integrating by parts on the interval $[0,T]$, we get:

$$\begin{aligned}
2\int_0^T t\|u'(t)-v'(t)\|^2 dt &\leq \int_0^T t(\|u(t)-v(t)\|^2)'' dt \\
&= t\frac{d}{dt}\|u(t)-v(t)\|^2|_0^T - \int_0^T \frac{d}{dt}\|u(t)-v(t)\|^2 dt \\
&\leq -\int_0^T \frac{d}{dt}\|u(t)-v(t)\|^2 dt \\
&= |u(0)-v(0)\|^2 - \|u(T)-v(T)\|^2 \\
&\leq \|u(0)-v(0)\|^2
\end{aligned}$$

which implies (5.47). By Theorem 5.3.2, one can show easily that $u \in W^{2,2}_{\text{loc}}(0,\infty;H)$ and verifies the equation $u'' \in Au+f$ on $(0,T)$ for each $T>0$. Then u is a solution of $u'' \in Au+f$ and since $\|u_n(t)\| \leq \|a\|+M$ for each $t \in [0,n]$, it also satisfies the boundary conditions in (5.2).

Step 2. We prove the theorem for $a \in \overline{D(A)}$. In this case $a_n \in D(A)$ exists such that $a_n \to a$. Let u_n be the solution of (5.2) with $u_n(0)=a_n$. Since $t \mapsto \|u_n(t)-v_n(t)\|^2$ is convex and bounded, it is nonincreasing. Therefore:

$$\|u_n(t)-u_m(t)\| \leq \|a_n-a_m\|$$

which shows that u_n converges uniformly on $[0,+\infty)$ to a function $u \in C([0,+\infty);H)$ with $u(0)=a$. By Theorem 5.3.2, $u \in W^{2,2}_{\text{loc}}(0,+\infty;\ H)$, and it is a solution to $u'' \in Au+f$. □

5.5 PERIODIC FORCING

Theorem 5.5.1 *Suppose that $A: D(A) \subset H \to H$ is maximal monotone and $f \in L^2_{\text{loc}}(0,+\infty;H)$ is periodic with period $T>0$. Let $u \in C(0,+\infty;H) \cap W^{2,2}_{\text{loc}}(0,+\infty;H)$ be a solution to (5.2) and $u_n(t)=u(t+nT)$, $t \in [0,+\infty)$, $n \in \mathbb{N}$. Then there exists a T-periodic solution $w \in C(0,+\infty;H) \cap W^{2,2}_{\text{loc}}(0,+\infty;H)$ to the Equation (5.2) with the following properties:*
(i) $u(t)-w(t) \rightharpoonup 0$ in H as $t \to +\infty$.
(ii) $t^{\frac{1}{4}}(u'(t)-w'(t)) \to 0$ in H as $t \to +\infty$.
(iii) $u_n'' \rightharpoonup w''$ in $L^2(0,T;H)$ as $n \to +\infty$.

Proof. First we prove the weak convergence of $(u_n(t))_n$ in H for each $t \in [0,T]$. The existence of a solution u which follows from Theorem 5.4.1 implies the existence of a unique solution $v \in C([s,\infty);H) \cap W^{2,2}_{\mathrm{loc}}(s,\infty;H)$ to the problem:

$$\begin{cases} v''(t) \in Av(t) + f(t), & \text{a.e. } t \in (s,\infty) \\ v(s) = a \end{cases} \tag{5.49}$$

for every $a \in \overline{D(A)}$. Then we define an evolution system $U(t,s) : \overline{D(A)} \to \overline{D(A)}$ by $U(t,s)a = v(t),\ 0 \le s \le t < +\infty$. By the uniqueness of the solution, we have

$$U(t,s)U(s,r) = U(t,r), \quad 0 \le r \le s \le t \tag{5.50}$$

Theorem 5.4.1 implies that

$$\|U(t,s)a - U(t,s)b\| \le \|a-b\|, \quad a,b \in \overline{D(A)}.$$

This means that $U(t,s)$ is nonexpansive. The T-periodicity of f implies that

$$U(t+T,s+T) = U(t,s).$$

This together with (5.50) implies that

$$U(t+nT,t) = U(t+T,t)^n, \quad n \in \mathbb{N}.$$

Then we have

$$u_n(t) = u(t+nT) = U(t+nT,t)u(t) = U(t+T,t)^n u(t), \tag{5.51}$$

Therefore $(u_n(t))_n$ is an orbit of the nonexpansive mapping $U(t+T,t)$. The boundedness of this sequence implies that this mapping has a fixed point (see [DJ1,DJ2,DJ3]). In particular, for $t=0$ we obtain the existence of a T-periodic solution w_1 for problem (5.2). By (5.47) we have

$$\int_0^\infty t\|u'(t) - w_1'(t)\|^2 dt \le \frac{1}{2}\,|u(0) - w_1(0)\|^2$$

Therefore

$$\|u_n' - w_1'\|_{L^2(0,T;H)} \le (2nT)^{\frac{-1}{2}} \|u(0) - w_1(0)\| \tag{5.52}$$

Then $u_n' \to w_1'$ in $L^2(0,T;H)$. Since

$$u_n(t) - u_{n+1}(t) = [u_n(t) - w_1(t)] - [u_n(t+T) - w_1(t+T)] = \int_t^{t+T} [w_1'(s) - u_n'(s)]ds$$

we obtain $u_{n+1}(t) - u_n(t) \to 0$ as $n \to \infty$. By the nonlinear ergodic theorem (see [DJ1,DJ2,DJ3]), this together with (5.51) implies that $(u_n(t))_n$ is weakly convergent to a fixed point of $U(t+T,t)$. Suppose that $w(t) = w - \lim_{n \to +\infty} u_n(t),\ t \ge 0$. Then clearly w is periodic with period T. As we can see, the inequality (5.52) implies the convergence $u_n' \to w_1'$ in $L^2(0,T;H)$. Since by Theorem 2.4 of [BRU2], $(u_n'')_n$ is bounded in $L^2(0,T;H)$, we conclude that $w \in W^{2,2}(0,T;H)$, w is a solution to (5.2), $u_n' \to w'$ and $u_n'' \rightharpoonup w''$ in $L^2(0,T;H)$. Therefore (iii) is proved. Since

$$u(t) - w(t) = u_n(0) - w(0) + \int_{nT}^{t} [u'(s) - w'(s)]ds$$

for $nT \leq t < nT + T$, and $u_n(0) \rightharpoonup w(0)$ and

$$\| \int_{nT}^{t} [u'(s) - w'(s)]ds \| \leq T^{\frac{1}{2}} \left(\int_0^T \|u_n' - w'\|^2 dt \right)^{\frac{1}{2}} \to 0$$

as $n \to +\infty$, this concludes (i). To prove (ii), (5.47) implies that

$$\alpha(t) = \int_t^{t+T} s\|u'(s) - w'(s)\|^2 ds \to 0, \tag{5.53}$$

as $t \to +\infty$. Moreover since $u_n'' \rightharpoonup w''$ in $L^2(0,T;H)$, a constant $c > 0$ exists such that

$$\int_t^{t+T} \|u''(s) - w''(s)\|^2 ds \leq c^2, \ \ t \geq T \tag{5.54}$$

Let $P_t = \{s \in [t, t+T]; \|u'(s) - w'(s)\| > \alpha(t)^{\frac{1}{4}} t^{\frac{-1}{4}}\}$. From (5.53)

$$|P_t| < \alpha(t)^{\frac{1}{2}} t^{\frac{-1}{2}} \tag{5.55}$$

where $|A|$ denotes the Lebesgue measure of A. By (5.54)

$$\|u'(t) - w'(t)\| \leq \|u'(s) - w'(s)\| + c(s-t)^{\frac{1}{2}}, \ \ s \in [t, t+T].$$

As a consequence of (5.55) we can choose $s \in [t, t + \alpha(t)^{\frac{1}{2}} t^{\frac{-1}{2}}] \setminus P_t$. Therefore

$$\|u'(t) - w'(t)\| \leq \alpha(t)^{\frac{1}{4}} t^{\frac{-1}{4}} + c\alpha(t)^{\frac{1}{4}} t^{\frac{-1}{4}}$$

which completes the proof of Part (ii). □

5.6 SQUARE ROOT OF A MAXIMAL MONOTONE OPERATOR

We consider problem (5.2) with $f(t) \equiv 0$ and $x \in \overline{D(A)}$. By Theorem 5.4.1, Equation (5.1) with $p(t) \equiv 1$ and $q(t) \equiv 0$ has a unique solution that we denote by $u(t)$. The nonexpansiveness of the trajectory of $u(t)$ implies that the mapping defining $u(t)$, which we denote by $S_{\frac{1}{2}}(t) : \mathbb{R} \times \overline{D(A)} \to \overline{D(A)}$ defined by $S_{\frac{1}{2}}(t)x := u(t)$ is a nonexpansive semigroup, which is called the square root semigroup. Assume that $A_{\frac{1}{2}}^0$ is the infinitesimal generator of $S_{\frac{1}{2}}(t)$ and $A_{\frac{1}{2}}$ is the unique maximal monotone extension of $A_{\frac{1}{2}}^0$. In the linear case $(A_{\frac{1}{2}})^2 = A$. $A_{\frac{1}{2}}$ is also called the square root of A.

Theorem 5.6.1 *[APR1] The nonlinear nonexpansive semigroup* $\{S_{\frac{1}{2}}(t);\, t \geq 0\}$ *has the following properties:*
(i) $S_{\frac{1}{2}}(t)$ *maps* $\overline{D(A)}$ *to* $D(A)$ *for all* $t > 0$.
(ii) For each $x \in \overline{D(A)}$, *the function* $t \mapsto S_{\frac{1}{2}}(t)x$ *is continuously differentiable on* $(0,+\infty)$ *and*

$$\frac{d}{dt}S_{\frac{1}{2}}(t)x + A^0_{\frac{1}{2}}S_{\frac{1}{2}}(t)x = 0, \quad \forall t > 0$$

(iii) If $x \in \overline{D(A)}$ *and* $y \in A^{-1}(0)$, *then*

$$\int_0^\infty t\|\frac{d}{dt}S_{\frac{1}{2}}(t)x\|^2 dt \leq \frac{1}{2}\|x-y\|^2$$

and

$$\|\frac{d^+}{dt}S_{\frac{1}{2}}(t)x\| \leq \frac{1}{t}\|x-y\|, \quad \forall t > 0$$

where $\frac{d^+}{dt}$ *is the right derivative.*
(iv) The second derivative $\frac{d^2}{dt^2}S_{\frac{1}{2}}(t)x$ *exists for almost every* $t \in (0,+\infty)$ *and*

$$\int_0^\infty t^3\|\frac{d^2}{dt^2}S_{\frac{1}{2}}(t)x\|^2 dt \leq \frac{3}{2}\|x-y\|^2$$

(v) If $x \in D(A^o_{\frac{1}{2}})$, *the function* $t \mapsto S_{\frac{1}{2}}(t)x$ *is continuously differentiable on* $[0,+\infty)$ *and* $t^{\frac{1}{2}}\frac{d^2}{dt^2}S_{\frac{1}{2}}(t)x \in L^2(0,+\infty;H)$.

5.7 ASYMPTOTIC BEHAVIOR

In this section, we study the asymptotic behavior of solutions to the following evolution equation:

$$\begin{cases} p(t)u''(t) + r(t)u'(t) \in Au(t) + f(t), \\ u(0) = x, \quad \sup_{t\geq 0}\|u(t)\| < +\infty \end{cases} \tag{5.56}$$

where A is a maximal monotone operator. We consider appropriate conditions on the real-valued functions p and r, and the function $f : (0,+\infty) \to H$, as well as the maximal monotone operator A. This section contains four subsetions. In the first one, we study the ergodic convergence of solutions, which is the convergence of the net of means of the solutions on the interval $[0,T]$, as $T \to +\infty$. The second subsection is devoted to the weak convergence of solutions to a zero of the maximal monotone operator A. In the third subsection, with suitable assumptions on p, r and f, or additional conditions on the monotone operator A, we prove the strong convergence of solutions to (5.56). The last subsection is devoted to an important special case of maximal monotone operators which are subdifferentials of proper, convex and lower semi-continuous functions. In this subsection, we investigate conditions by which ergodic, weak or strong convergence of solutions to a minimum point of the convex function is obtained. First we recall Véron's theorem on the existence of solutions to (5.56) for the homogeneous case $f \equiv 0$.

Theorem 5.7.1 *Let A be a maximal monotone operator in H. Assume that $f \equiv 0$, and suppose that p and r are two real valued functions satisfying*

$$p \in W^{2,\infty}(0,+\infty), \quad r \in W^{1,\infty}(0,+\infty) \tag{5.57}$$

$$\exists \alpha > 0, \quad \textit{such that} \quad \forall t \geq 0, \ \ p(t) \geq \alpha. \tag{5.58}$$

Then (5.56) *has a solution u such that u' and u'' belong to $L^2\big((0,+\infty);H\big)$. The solution is unique if moreover*

$$\int_0^{+\infty} e^{-\int_0^t \frac{r(s)}{p(s)}ds} dt = +\infty \tag{5.59}$$

Throughout this section, we always assume that the conditions (5.57) and (5.58) on p and r, which by Véron's theorem guarantee the existence of solutions to (5.56) are satisfied. We also concentrate on the homogeneous case of (5.56) (i.e. we assume that $f \equiv 0$). Throughout this section, we denote $M := \mathrm{Max}\{\|p\|_{W^{2,\infty}}, \ \|r\|_{W^{1,\infty}}, \ \sup_{t\geq 0}\|u(t)\|\}$, $u_h(t) := u(t+h) - u(t)$, and $\sigma_T(u) := \frac{1}{T}\int_0^T u(t)dt$.

5.7.1 ERGODIC CONVERGENCE

In this subsection, with the following assumptions on p and r, and without assuming $A^{-1}(0) \neq \varnothing$, we prove the almost weak convergence of solutions to (5.56) to a zero of A. With the above assumptions on the functions p and r, we also show that the existence of solutions to (5.56) implies that $A^{-1}(0) \neq \varnothing$, providing a converse to the existence theorem for the solutions to (5.56).

Definition 5.7.2 *We say that a sequence $\{x_n\}$ in a Banach space X is almost weakly convergent to $z \in X$ if $\frac{1}{n}\sum_{i=0}^{n-1} x_{k+i} \rightharpoonup z$ uniformly in k, as $n \to +\infty$.*

Lemma 5.7.3 *Suppose that $u(t)$ is a solution to* (5.56) *and $f \equiv 0$. If $q \in H$ satisfies the following inequality:*

$$\big(p(t)u''(t) + r(t)u'(t), u(t) - q\big) \geq 0,$$

then $\|u(t) - q\|$ is nonincreasing or eventually increasing (i.e. there exists $t_0 > 0$ such that $\|u(t) - q\|$ is increasing for $t \geq t_0$); therefore there exists $\lim_{t\to+\infty}\|u(t) - q\|$.

Proof. It follows from the assumption that

$$p(t)\frac{d^2}{dt^2}\|u(t) - q\|^2 + r(t)\frac{d}{dt}\|u(t) - q\|^2$$

$$
\begin{aligned}
&= 2p(t)\frac{d}{dt}\big(u'(t),u(t)-q\big)+2r(t)\big(u'(t),u(t)-q\big)\\
&= 2p(t)\big(u''(t),u(t)-q\big)+2p(t)\|u'(t)\|^2+2r(t)\big(u'(t),u(t)-q\big)\\
&= 2\big(p(t)u''(t)+r(t)u'(t),u(t)-q\big)+2p(t)\|u'(t)\|^2\geq 0
\end{aligned}
$$

Dividing both sides of the above inequality by $p(t)$ and multiplying by $e^{\int_0^t \frac{r(s)}{p(s)}ds}$, we get:

$$\frac{d}{dt}\big(e^{\int_0^t \frac{r(s)}{p(s)}ds}\frac{d}{dt}\|u(t)-q\|^2\big)\geq 0. \tag{5.60}$$

We consider two cases.

If $\frac{d}{dt}\|u(t)-q\|^2\leq 0$ for each $t>0$, then $\|u(t)-q\|$ is nonincreasing. Otherwise, there exists $t_0>0$ such that $\frac{d}{dt}\|u(t)-q\|^2_{|t=t_0}>0$. Integrating (5.60) from t_0 to t, we get for each $t\geq t_0$,

$$e^{\int_0^t \frac{r(s)}{p(s)}ds}\frac{d}{dt}\|u(t)-q\|^2\geq 2e^{\int_0^{t_0} \frac{r(s)}{p(s)}ds}\big(u'(t_0),u(t_0)-q\big)>0.$$

Then

$$\frac{d}{dt}\|u(t)-q\|^2>0,\quad \forall t\geq t_0,$$

which shows that $\|u(t)-q\|$ is eventually increasing. □

Corollary 5.7.4 *If $A^{-1}(0)\neq\emptyset$, $q\in A^{-1}(0)$ and $f\equiv 0$, then the conclusions of Lemma 5.7.3 hold.*

Proof. If $q\in A^{-1}(0)$, then by the monotonicity of A, q satisfies the inequality in Lemma 5.7.3. □

Theorem 5.7.5 *Let u be a solution to (5.56). Assume that p,r satisfy the assumptions (5.57) and (5.58). If either one of the following conditions hold:*
i) r and p' are monotone,
ii) $r'(t)\geq p''(t)$,
iii) $p''(t)\geq r'(t)$,
then $\sigma_T(u_h):=\frac{1}{T}\int_0^T u(t+h)dt\rightharpoonup c\in A^{-1}(0)$ as $T\to+\infty$, uniformly for $h\geq 0$. Moreover, c is the asymptotic center of the curve $(u(t))_{t\geq 0}$.

Proof. By the monotonicity of A, we have

$$
\begin{aligned}
&\big(p(t)u''(t)+r(t)u'(t),u(t)-u(s+h)\big)\geq\\
&\big(p(s+h)u''(s+h)+r(s+h)u'(s+h),u(t)-u(s+h)\big).
\end{aligned}
$$

Integrating from $s=0$ to $s=T$ and dividing by T, we get:

$$
\begin{aligned}
&\big(p(t)u''(t)+r(t)u'(t),u(t)-\sigma_T(u_h)\big)\\
&\geq\frac{1}{T}\int_0^T p(s+h)\frac{d}{ds}\big(u'(s+h),u(t)-u(s+h)\big)ds
\end{aligned}
$$

$$
\begin{aligned}
&-\frac{1}{T}\int_0^T \frac{r(s+h)}{2}\frac{d}{ds}\|u(t)-u(s+h)\|^2 ds \\
&= \frac{1}{T}p(T+h)\big(u'(T+h),u(t)-u(T+h)\big) \\
&\quad -\frac{1}{T}p(h)\big(u'(h),u(t)-u(h)\big)+\frac{p'(T+h)}{2T}\|u(t)-u(T+h)\|^2 \\
&\quad -\frac{1}{2T}p'(h)\|u(t)-u(h)\|^2-\frac{1}{2T}\int_0^T p''(s+h)\|u(t)-u(s+h)\|^2 ds \\
&\quad -\frac{r(T+h)}{2T}\|u(t)-u(T+h)\|^2+\frac{r(h)}{2T}\|u(t)-u(h)\|^2 \\
&\quad +\frac{1}{2T}\int_0^T r'(s+h)\|u(t)-u(s+h)\|^2 ds. \qquad (5.61)
\end{aligned}
$$

By (5.57) and (5.58), the third, fourth, sixth, and seventh terms on the right hand side of (5.61) converge to zero as $T\to+\infty$. Since we have:

$$
\begin{aligned}
\|u'(t+h)-u'(t)\| &= \|\int_t^{t+h} u''(s)ds\| \\
&\le \int_t^{t+h} \|u''(s)\|ds \\
&\le \sqrt{h}\|u''\|_{L^2((0,+\infty);H)}
\end{aligned}
$$

it follows that u', and hence also $\|u'\|$, is uniformly continuous on $(0,+\infty)$. Then since $u'\in L^2\big((0,+\infty);H\big)$, it follows that $\lim_{t\to+\infty}\|u'(t)\|=0$. This implies that the first and second terms on the right hand side of (5.61) converge to zero, uniformly for $h\ge 0$. Now assume that (i) holds. We will show that the fifth and the last terms on the right hand side of (5.61) are bigger than expressions tending to zero as $T\to+\infty$. If $p''(t)\le 0$, then:

$$
-\frac{1}{2T}\int_0^T p''(s+h)\|u(t)-u(s+h)\|^2 ds \ge 0. \qquad (5.62)
$$

Otherwise, $p''(t)\ge 0$ and by the boundedness of u, we have:

$$
-\frac{1}{2T}\int_0^T p''(s+h)\|u(t)-u(s+h)\|^2 ds \ge 4M^2\frac{1}{2T}(p'(h)-p'(T+h))\to 0 \quad (5.63)
$$

as $T\to+\infty$, uniformly for $h\ge 0$. If $r'(t)\ge 0$, then

$$
\frac{1}{2T}\int_0^T r'(s+h)\|u(t)-u(s+h)\|^2 ds \ge 0. \qquad (5.64)
$$

Otherwise, $r'(t)\le 0$ and we get:

$$
\frac{1}{2T}\int_0^T r'(s+h)\|u(t)-u(s+h)\|^2 ds \ge 4M^2\frac{1}{2T}(r(T+h)-r(h))\to 0 \qquad (5.65)
$$

as $T \to +\infty$, uniformly for $h \geq 0$. Suppose q is a weak cluster point of $\sigma_T(u_h)$. Then for any two sequences T_n and h_n of positive real numbers such that $T_n \to +\infty$ and $\sigma_{T_n}(u_{h_n}) \rightharpoonup q$, by replacing T by T_n and h by h_n in (5.61)–(5.65), and letting $n \to +\infty$, we get:

$$\big(p(t)u''(t) + r(t)u'(t), u(t) - q\big) \geq 0 \tag{5.66}$$

If $p(t)$ and $r(t)$ satisfy condition (ii), then we get again (5.66) from (5.61)–(5.65). If $p(t)$ and $r(t)$ satisfy condition (iii), then we have

$$\begin{aligned}
&-\frac{1}{2T}\int_0^T p''(s+h)\|u(t)-u(s+h)\|^2 ds \\
&+\frac{1}{2T}\int_0^T r'(s+h)\|u(t)-u(s+h)\|^2 ds \\
&=\frac{1}{2T}\int_0^T (r'(s+h)-p''(s+h))\|u(t)-u(s+h)\|^2 ds \\
&\geq 4M^2\frac{1}{2T}\int_0^T (r'(s+h)-p''(s+h))ds \\
&=\frac{4M^2}{2T}(r(T+h)-r(h)-p'(T+h)+p'(h)),
\end{aligned}$$

which converges to zero as $T \to +\infty$. Then we get again (5.66) from (5.61) to (5.65). Now by Lemma 5.7.3, there exists $\lim_{t\to+\infty}\|u(t)-q\|$. If c is another weak cluster point of $\sigma_T(u_h)$, then there exists $\lim_{t\to+\infty}(\|u(t)-q\|^2 - \|u(t)-c\|^2)$. This implies that there exists $\lim_{t\to+\infty}(u(t), c-q)$, then there exists $\lim_{T\to+\infty}(\sigma_T(u_h), c-q)$ uniformly for $h \geq 0$. It follows that $(c, c-q) = (q, c-q)$. Therefore $c = q$, and hence $\sigma_T(u_h) \rightharpoonup c$ as $T \to +\infty$, uniformly for $h \geq 0$, which shows the almost weak convergence of $u(t)$ to c as $t \to +\infty$. Now we prove that $c \in A^{-1}(0)$. Let $y \in Ax$. By the monotonicity of A, we have:

$$\begin{aligned}
\big(x-u(t), y\big) &= \big(x-u(t), y-Au(t)\big) + \big(x-u(t), Au(t)\big) \\
&\geq \big(x-u(t), p(t)u''(t) + r(t)u'(t)\big)
\end{aligned}$$

Integrating from $t = 0$ to T, dividing by T, and letting $T \to +\infty$, by a similar proof as above we get: $(x-c, y) \geq 0$. Now the maximality of A implies that $c \in A^{-1}(0)$. Finally, we show that c is the asymptotic center of the curve $u(t)$. Let $x \in H$, with $x \neq c$. Then:

$$\|u(t)-x\|^2 = \|u(t)-c\|^2 + \|x-c\|^2 + 2\big(u(t)-c, c-x\big).$$

Integrating from 0 to T, and dividing by T, then taking limsup when $T \to +\infty$, since $\sigma_T \rightharpoonup c$ and $\lim_{n\to+\infty}\|u(t)-c\|$ exists, we get:

$$\begin{aligned}
\limsup_{t\to+\infty}\|u(t)-x\|^2 &\geq \limsup_{t\to+\infty}\|u(t)-c\|^2 + \|x-c\|^2 \\
&> \limsup_{t\to+\infty}\|u(t)-c\|^2.
\end{aligned}$$

Hence c is the asymptotic center of $u(t)$ as desired. The proof is now complete. □

5.7.2 WEAK CONVERGENCE

In this subsection, we prove the weak convergence of solutions to (5.56) to a zero of A. The proof is based on Lorentz' Tauberian theorem which states that an almost weakly convergent sequence $\{x_n\}$ is weakly convergent if and only if $x_{n+1} - x_n \rightharpoonup 0$ as $n \to +\infty$.

Theorem 5.7.6 *Let u be a solution to* (5.56) *with $f \equiv 0$. If* (5.57), (5.58) *and the assumptions of Theorem 5.7.5 are satisfied, then $u(t) \rightharpoonup c \in A^{-1}(0)$ as $t \to +\infty$, where c is the asymptotic center of the curve $(u(t))_{t\geq 0}$.*

Proof. Since $u' \in L^2\big((0,+\infty);H\big)$, then u is asymptotically regular (i.e. $u(t+h) - u(t) \to 0$, as $t \to +\infty$, $\forall h \geq 0$). Now the result follows from Theorem 5.7.5 and G. G. Lorentz' Tauberian condition for almost convergence (see [LOR]). □

Remark 5.7.7 *The conclusions of Theorems 5.7.5 and 5.7.6 still hold if the assumptions (i), (ii) or (iii) in Theorem 5.7.5 are satisfied only for large enough t(i.e. for $t \geq t_0$).*

Remark 5.7.8 *Since u' and u'' belong to $L^2(0,+\infty;H)$, then as stated in the proof of Theorem 5.7.5, we have that $\lim_{t\to+\infty}\|u'(t)\| = 0$. Since $p, r \in L^\infty\big((0,+\infty);H\big)$, then $pu'' + ru' \in L^2\big((0,+\infty);H\big)$. Therefore there is a sequence $t_n \to +\infty$ such that $p(t_n)u''(t_n) + r(t_n)u'(t_n) \to 0$ as $n \to +\infty$. Now the demiclosedness of A shows that $A^{-1}(0) \neq \emptyset$. However, the weak convergence of $u(t)$ cannot be directly derived from this fact, and we need the arguments in Theorems 5.7.5 and 5.7.6.*

Example 5.7.9 *Consider the following second order evolution equation:*

$$\begin{cases} \big(q(t)u'(t)\big)' \in Au(t) \text{ a.e. on } \mathbb{R}^+ \\ u(0) = u_0, \quad \sup_{t\geq 0}\|u(t)\| < +\infty \end{cases} \tag{5.67}$$

where A is a maximal monotone operator and $q(t) \in W^{2,\infty}\big((0,+\infty);H\big)$. Then (5.57) *and condition (ii) of Theorem 5.7.5 are satisfied.* (5.58) *is satisfied if $q(t) \geq \alpha > 0, \forall t \geq 0$. Then $u(t)$ converges weakly to a zero of A as $t \to +\infty$. In this case, condition* (5.59) *is equivalent to the following condition:*

$$\int_0^\infty \frac{1}{q(t)} = +\infty. \tag{5.68}$$

For example, $q(t) = \sin t + 2$ satisfies all of the above conditions.

Example 5.7.10 *Consider bounded solutions to the following ordinary differential equation:*

$$\frac{3}{2}u'' + \frac{2}{t+1}u' = u^3, \quad u(0) = 1.$$

One can easily verify that the solution is given by $u(t) = \frac{1}{t+1}$, which converges to zero as $t \to +\infty$, as predicted by Theorem 5.7.6.

5.7.3 STRONG CONVERGENCE

In this subsection, we prove the strong convergence of solutions to (5.56) with $f \equiv 0$, to a zero of the maximal monotone operator A. In particular, as in the previous subsection, we show that the existence of solutions implies that $A^{-1}(0) \neq \emptyset$. We assume that p and r satisfy (5.57) and (5.58), and then by Theorem 5.7.1 we have $u', u'' \in L^2\big((0,+\infty);H\big)$.

Lemma 5.7.11 *Let u be a solution to* (5.56). *Then* $\liminf_{t\to+\infty} p(t)\frac{d}{dt}\|v(t,h)\|^2 \leq 0$ *and* $\limsup_{t\to+\infty} r(t)\|v(t,h)\|^2 \leq 0$, *where* $v(t,h) = u(t+h) - u(t)$.

Proof. Assume by contradiction that $\liminf_{t\to+\infty} p(t)\frac{d}{dt}\|v(t,h)\|^2 \geq c > 0$.
Since $0 < \alpha \leq p(t) \leq M$, integrating from $t = s$ to T, we get:

$$\|v(T,h)\|^2 - \|v(s,h)\|^2 > \frac{c}{M}(T-s).$$

Letting $T \to +\infty$, we get a contradiction since u is bounded.
On the other hand,

$$\begin{aligned} |r(t)|\|v(t,h)\|^2 &= |r(t)|\|\int_t^{t+h} u'(s)ds\|^2 \\ &\leq Mh(\int_t^{t+h} \|u'(s)\|^2 ds) \to 0 \end{aligned}$$

as $t \to +\infty$, since $u' \in H^1\big((0,+\infty);H\big)$. □

Theorem 5.7.12 *Let $u(t)$ be a solution to* (5.56) *with* $f \equiv 0$. *Assume that the following conditions (i) and (ii) are satisfied:*
$(i)\ \int_0^\infty e^{-\int_0^s \frac{r(\tau)}{2p(\tau)}d\tau} ds < +\infty$,
$(ii) M_1 := \int_0^\infty [\int_0^t e^{-\int_s^t \frac{r(\tau)}{p(\tau)}d\tau} R(s)ds]^{1/2} dt < +\infty$,
where $R(s) := \mathrm{Max}\{0, \sup_{t\geq s} r'(t)\}$. *Then* $u(t) \to p \in H$, *as* $t \to +\infty$.

Proof. Take $v(t) = u(t+h) - u(t)$. By the monotonicity of A and (5.56), we get:

$$\begin{aligned} & p(t)\big(v''(t),v(t)\big) + \big(p(t+h)-p(t)\big)\big(u''(t+h),v(t)\big) + \\ & r(t)\big(v'(t),v(t)\big) + \big(r(t+h)-r(t)\big)\big(u'(t+h),v(t)\big) \geq 0 \\ \Rightarrow\ & \frac{1}{2}p(t)\frac{d^2}{dt^2}\|v(t)\|^2 + \big(p(t+h)-p(t)\big)\big(u''(t+h),v(t)\big) \\ & + \frac{1}{2}r(t)\frac{d}{dt}\|v(t)\|^2 + \big(r(t+h)-r(t)\big)\big(u'(t+h),v(t)\big) \geq 0. \end{aligned}$$

Integrating by parts from s to t, we get:

$$\frac{1}{2}p(t)\frac{d}{dt}\|v(t)\|^2 - \frac{1}{2}p(s)\frac{d}{ds}\|v(s)\|^2 - \frac{1}{2}\int_s^t p'(\tau)\frac{d}{d\tau}\|v(\tau)\|^2 d\tau +$$

$$\int_s^t (p(\tau+h)-p(\tau))(u''(\tau+h),v(\tau))d\tau+\frac{1}{2}r(t)\|v(t)\|^2-\frac{1}{2}r(s)\|v(s)\|^2-$$
$$\frac{1}{2}\int_s^t r'(\tau)\|v(\tau)\|^2d\tau+\int_s^t (r(\tau+h)-r(\tau))(u'(\tau+h),v(\tau))d\tau\geq 0$$

By Lemma 5.7.11, there is a sequence $t_n\to+\infty$ such that

$$\lim_{n\to+\infty} p(t_n)(v(t_n),v'(t_n))\leq 0.$$

Now replacing t by t_n in the above inequality, and letting $n\to+\infty$, we get:

$$\frac{1}{2}p(s)\frac{d}{ds}\|v(s)\|^2+\frac{1}{2}r(s)\|v(s)\|^2\leq$$
$$-\frac{1}{2}\int_s^\infty p'(\tau)\frac{d}{d\tau}\|v(\tau)\|^2d\tau+\int_s^\infty (p(\tau+h)-p(\tau))(u''(\tau+h),v(\tau))d\tau$$
$$-\frac{1}{2}\int_s^\infty r'(\tau)\|v(\tau)\|^2d\tau+\int_s^\infty (r(\tau+h)-r(\tau))(u'(\tau+h),v(\tau))d\tau.$$

Dividing by h^2 and letting $h\to 0$, by an application of Fatou's Lemma, it follows from the assumptions that:

$$p(s)\frac{d}{ds}\|u'(s)\|^2+r(s)\|u'(s)\|^2\leq\int_s^\infty r'(\tau)\|u'(\tau)\|^2d\tau\leq R(s)\int_s^\infty \|u'(\tau)\|^2d\tau. \tag{5.69}$$

Then:

$$\frac{d}{ds}\|u'(s)\|^2e^{\int_{t_0}^s\frac{r(\tau)}{p(\tau)}d\tau}+\frac{r(s)}{p(s)}\|u'(s)\|^2e^{\int_{t_0}^s\frac{r(\tau)}{p(\tau)}d\tau}\leq\frac{e^{\int_{t_0}^s\frac{r(\tau)}{p(\tau)}d\tau}}{p(s)}R(s)\int_s^\infty \|u'(\tau)\|^2d\tau.$$

Therefore:

$$\frac{d}{ds}\left(\|u'(s)\|^2e^{\int_{t_0}^s\frac{r(\tau)}{p(\tau)}d\tau}\right)\leq\frac{e^{\int_{t_0}^s\frac{r(\tau)}{p(\tau)}d\tau}}{p(s)}R(s)\int_s^\infty \|u'(\tau)\|^2d\tau.$$

Integrating from t_0 to t with respect to s, we get:

$$\|u'(t)\|^2\leq\|u'(t_0)\|^2e^{-\int_{t_0}^t\frac{r(\tau)}{p(\tau)}d\tau}+\frac{e^{-\int_{t_0}^t\frac{r(\tau)}{p(\tau)}d\tau}}{\alpha}\int_{t_0}^t (e^{\int_{t_0}^s\frac{r(\tau)}{p(\tau)}d\tau}R(s)\int_s^\infty \|u'(\tau)\|^2d\tau)ds$$
$$=\|u'(t_0)\|^2e^{-\int_{t_0}^t\frac{r(\tau)}{p(\tau)}d\tau}+\frac{1}{\alpha}\int_{t_0}^t (e^{-\int_s^t\frac{r(\tau)}{p(\tau)}d\tau}R(s)\int_s^\infty \|u'(\tau)\|^2d\tau)ds.$$

Therefore:

$$\|u'(t)\|\leq\|u'(t_0)\|e^{-\int_{t_0}^t\frac{r(\tau)}{2p(\tau)}d\tau}+\frac{1}{\sqrt{\alpha}}\Big[\int_{t_0}^t (e^{-\int_s^t\frac{r(\tau)}{p(\tau)}d\tau}R(s)\int_s^\infty \|u'(\tau)\|^2d\tau)ds\Big]^{\frac{1}{2}}. \tag{5.70}$$

Hence:

$$\|u(T')-u(T)\| \leq \int_T^{T'} \|u'(t)\|dt \leq \|u'(t_0)\| \int_T^{T'} e^{-\int_{t_0}^t \frac{r(\tau)}{2p(\tau)}d\tau}dt$$

$$+\frac{1}{\sqrt{\alpha}}\int_T^{T'}\Big[\int_{t_0}^t (e^{-\int_s^t \frac{r(\tau)}{p(\tau)}d\tau}R(s)\int_s^\infty \|u'(\tau)\|^2 d\tau)ds\Big]^{\frac{1}{2}}dt.$$

Assume $u(T_n') \rightharpoonup p$. Then we get:

$$\|u(T)-p\| \leq \|u'(t_0)\| \int_T^{\infty} e^{-\int_{t_0}^t \frac{r(\tau)}{2p(\tau)}d\tau}dt$$

$$+\frac{1}{\sqrt{\alpha}}\int_T^{\infty}\Big[\int_{t_0}^t (e^{-\int_s^t \frac{r(\tau)}{p(\tau)}d\tau}R(s)\int_s^\infty \|u'(\tau)\|^2 d\tau)ds\Big]^{\frac{1}{2}}dt.$$

Given $\varepsilon > 0$, choose t_0 big enough such that: $\int_s^\infty \|u'(\tau)\|^2 d\tau \leq \varepsilon^2, \ \ \forall s \geq t_0$. Then we have:

$$\|u(t)-p\| \leq \|u'(t_0)\| \int_t^\infty e^{-\int_{t_0}^s \frac{r(\tau)}{2p(\tau)}d\tau}ds + \frac{M_1\varepsilon}{\sqrt{\alpha}}. \tag{5.71}$$

Therefore $\limsup_{t\to+\infty}\|u(t)-p\| \leq \frac{M_1\varepsilon}{\sqrt{\alpha}}$. Since $\varepsilon > 0$ is arbitrary, we conclude that $u(t) \to p$ as $t \to +\infty$. □

Corollary 5.7.13 *Let $u(t)$ be a solution to* (5.56) *with $f \equiv 0$. Assume that $r(t) \geq 0$. In addition, assume that (i) and the following stronger condition (ii') is satisfied:*
$(ii')M_1' := \int_0^{+\infty} e^{\int_0^s \frac{r(\tau)}{p(\tau)}d\tau}R(s)ds < +\infty$.
Then $u(t) \to p \in A^{-1}(0)$ as $t \to +\infty$.

Proof. First of all, since: $-\int_s^t \frac{r(\tau)}{p(\tau)}d\tau = -2\int_0^t \frac{r(\tau)}{2p(\tau)}d\tau + \int_0^s \frac{r(\tau)}{p(\tau)}d\tau$, it is clear that (i) and (ii') imply (ii). Therefore by Theorem 5.7.12, $u(t) \to p \in H$ as $t \to +\infty$. Now since $r(t) \geq 0$, it follows from $(i), (ii')$ and (5.70) that $\|u'(t)\| \to 0$ as $t \to +\infty$. Since $u'' \in L^2((0,+\infty);H)$, there is a sequence $t_n \to +\infty$ such that $u''(t_n) \to 0$ as $n \to +\infty$. Since $p(t)$ and $r(t)$ are bounded, the closedness of A implies that $p \in A^{-1}(0)$. □

Corollary 5.7.14 *Let $u(t)$ be a solution to* (5.56) *with $f \equiv 0$. Assume that $r'(t) \leq 0$. If (i) holds, then $u(t) \to p \in A^{-1}(0)$ as $t \to +\infty$ and $\|u(t)-p\| = O(\int_t^\infty e^{-\int_0^s \frac{r(\tau)}{2p(\tau)}d\tau}ds)$.*

Proof. Since $r'(t) \leq 0$, then (i) implies that $r(t) \geq 0$, and $R(s) = 0$ so that (ii') is clearly satisfied. The conclusion follows now from Corollary 5.7.13. The rate of convergence is concluded by (5.71), because $M_1 = 0$. □

Example 5.7.15 *Let $p(t) \equiv 1$, $r(t) = \frac{4}{t+1}$. Then the assumptions of Corollary 5.7.14 are satisfied; therefore $u(t) \to p \in A^{-1}(0)$ as $t \to +\infty$. Moreover, the proof of Theorem 5.7.12 shows that we have the following rate of convergence: $|u(t)-p| = O(\frac{1}{t+1})$.*

Example 5.7.16 *Let $p(t) \equiv 1$ and $r(t) = 1 - e^{-t} \geq 0$. Then $r'(t) = e^{-t} > 0$ and $r'(t) \to 0$ as $t \to +\infty$. In this case, (i) and (ii) are satisfied, but (ii′) does not hold. Therefore, it follows from Theorem 5.7.12 that $u(t) \to p \in H$, as $t \to +\infty$.*

Problem 5.7.17 *Say for $p(t) \equiv 1$, is it possible to get the strong convergence of $u(t)$ by assuming that $\limsup_{t\to+\infty} r'(t) \leq 0$, and is there any relationship between this condition and condition (ii) in Theorem 5.7.12?*

5.7.4 SUBDIFFERENTIAL CASE

In this subsection, we consider the evolution Equation (5.56) with $f(t) \equiv 0$ when the monotone operator A is the subdifferential $\partial\varphi$ of a proper, convex and lower semicontinuous function $\varphi : H \to (-\infty, +\infty]$. We prove a weak convergence theorem with suitable assumptions on $p(t)$ and $r(t)$, as well as a strong convergence theorem with additional assumptions on φ.

Lemma 5.7.18 *Suppose that $u(t)$ is a solution to (5.56) with $f \equiv 0$. Then for each $p \in A^{-1}(0)$, $\lim_{t\to+\infty} \|u(t) - p\|^2$ exists and $\liminf_{t\to+\infty} \frac{d}{dt}\|u(t) - p\|^2 \leq 0$. In addition, if either (5.59) is satisfied or A is strongly monotone, then $\|u(t) - p\|^2$ is nonincreasing.*

Proof. The existence of $\lim_{t\to+\infty} \|u(t) - p\|^2$ follows from Lemma 5.7.3.
By contradiction, assume that $\liminf_{t\to+\infty} \frac{d}{dt}\|u(t) - p\|^2 > 0$. Then there exists $t_0 > 0$ and $\lambda > 0$, such that for each $t \geq t_0$,

$$\frac{d}{dt}\|u(t) - p\|^2 \geq \lambda.$$

Integrating from $t = t_0$ to $t = T$, we get

$$\|u(T) - p\|^2 - \|u(t_0) - p\|^2 \geq \lambda T - \lambda t_0.$$

Letting $T \to +\infty$, we deduce that u is not bounded, a contradiction. If in addition (5.59) is satisfied, assume that $\|u(t) - p\|$ is eventually increasing. Then there exists $t_0 > 0$ such that $(u'(t_0), u(t_0) - p) > 0$. Integrating (5.60) from t_0 to t, then dividing both sides by $e^{\int_0^t \frac{r(s)}{p(s)}ds}$, and integrating from $t = t_0$ to $t = T$, we get:

$$\|u(T) - p\|^2 - \|u(t_0) - p\|^2 \geq 2e^{\int_0^{t_0} \frac{r(s)}{p(s)}ds}\left(u'(t_0), u(t_0) - p\right)\int_{t_0}^{T} e^{-\int_0^t \frac{r(s)}{p(s)}ds}dt.$$

Letting $T \to +\infty$, we obtain a contradiction to Assumption (5.59). This implies that $\|u(t) - p\|$ is nonincreasing.
Finally, assume that A is strongly monotone, and let $p \in A^{-1}(0)$. Then we have

$$\left(p(t)u''(t) + r(t)u'(t), u(t) - p\right) \geq \beta\|u(t) - p\|^2.$$

This implies that

$$p(t)\frac{d^2}{dt^2}\|u(t) - p\|^2 + r(t)\frac{d}{dt}\|u(t) - p\|^2 \geq 2\beta\|u(t) - p\|^2.$$

Suppose to the contrary that $\|u(t)-p\|$ is increasing for $t \geq T_0 > 0$. Let K (resp. M) be an upper bound for $p(t)$ (resp. $|r(t)|$). Integrating both sides of this inequality from $t = T_0$ to $t = T$, we get:

$$
\begin{aligned}
&2\beta \int_{T_0}^{T} \|u(t)-p\|^2 dt \\
&\leq K\left(\frac{d}{dT}\|u(T)-p\|^2 - 2\big(u'(T_0), u(T_0)-p\big) + \int_{T_0}^{T} \frac{r(t)}{p(t)} \frac{d}{dt}\|u(t)-p\|^2 dt\right) \\
&\leq K\left(\frac{d}{dT}\|u(T)-p\|^2 - 2\big(u'(T_0), u(T_0)-p\big) + \frac{M}{\alpha}\big(\|u(T)-p\|^2 - \|u(T_0)-p\|^2\big)\right).
\end{aligned}
$$

Since $\|u(t)-p\|$ is increasing for $t \geq T_0 > 0$, we have:

$$
\begin{aligned}
&2\beta \|u(T_0)-p\|^2 (T-T_0) \leq \\
&K\left(\frac{d}{dT}\|u(T)-p\|^2 - 2\big(u'(T_0), u(T_0)-p\big) + \frac{M}{\alpha}\big(\|u(T)-p\|^2 - \|u(T_0)-p\|^2\big)\right).
\end{aligned}
$$

Taking $\liminf$ as $T \to +\infty$ of both sides in the above inequality, by the first part of this Lemma we deduce that $u(t)$ is unbounded, a contradiction. □

In the following, we prove a mean ergodic theorem when A is the subdifferential of a proper, convex and lower semicontinuous function.

Theorem 5.7.19 *Suppose that $u(t)$ is a solution to* (5.56) *with $f \equiv 0$ and $A = \partial\varphi$, where $\varphi : H \to (-\infty, +\infty]$ is a proper, convex and lower semicontinuous function. If* (5.59) *is satisfied, then $\sigma_T := \frac{1}{T}\int_0^T u(t)dt \rightharpoonup p \in A^{-1}(0)$, as $T \to +\infty$.*

Proof. By the subdifferential inequality and Equation (5.56), we get for each $p \in A^{-1}(0)$,

$$
\begin{aligned}
\varphi(u(t)) - \varphi(p) &\leq \big(p(t)u''(t) + r(t)u'(t), u(t)-p\big) \\
&\leq \frac{p(t)}{2}\frac{d^2}{dt^2}\|u(t)-p\|^2 + \frac{r(t)}{2}\frac{d}{dt}\|u(t)-p\|^2 \\
&= \frac{p(t)}{2} e^{-\int_0^t \frac{r(s)}{p(s)}ds} \frac{d}{dt}\big(e^{\int_0^t \frac{r(s)}{p(s)}ds}\frac{d}{dt}\|u(t)-p\|^2\big).
\end{aligned}
$$

Let K be an upper bound for $\frac{p(t)}{2}$. Integrating the above inequality from $t = 0$ to $t = T$, and using integration by parts, we get:

$$
\begin{aligned}
&\int_0^T (\varphi(u(t)) - \varphi(p))dt \\
&\leq K\big(\frac{d}{dT}\|u(T)-p\|^2 - 2(u'(0), u(0)-p) + \int_0^T \frac{r(t)}{p(t)}\frac{d}{dt}\|u(t)-p\|^2 dt\big) \\
&\leq K\big(-2(u'(0), u(0)-p) + \int_0^T \frac{r(t)}{p(t)}\frac{d}{dt}\|u(t)-p\|^2 dt\big) \qquad (5.72)
\end{aligned}
$$

(the second inequality holds by Lemma 5.7.18). Let R be an upper bound for $|r(t)|$, which exists by (5.57). Since $\|u(t)-p\|$ is nonincreasing (by Lemma 5.7.18) we get from (5.72),

$$\begin{aligned}
&\limsup_{T\to+\infty}\frac{1}{T}\int_0^T\big(\varphi(u(t))-\varphi(p)\big)dt\\
&\leq \limsup_{T\to+\infty}\frac{K}{T}\int_0^T\frac{r(t)}{p(t)}\frac{d}{dt}\|u(t)-p\|^2dt\\
&\leq \frac{-KR}{\alpha}\limsup_{T\to+\infty}\frac{1}{T}\left[\|u(T)-p\|^2-\|u(0)-p\|^2\right]=0 \qquad (5.73)
\end{aligned}$$

Since $p\in A^{-1}(0)$ and $A=\partial\varphi$, p is a minimum point of φ. Convexity of φ implies,

$$0\leq\varphi(\sigma_T)-\varphi(p)\leq\frac{1}{T}\int_0^T\varphi(u(t))dt-\varphi(p).$$

Taking the lim sup as $T\to+\infty$ in the above inequality we get by (5.73)

$$\limsup_{T\to+\infty}\varphi(\sigma_T)\leq\varphi(p).$$

Assume that $\sigma_{T_n}\rightharpoonup q$ for some sequence $\{T_n\}$ converging to $+\infty$ as $n\to+\infty$. Since φ is lower semicontinuous, we have

$$\liminf_{n\to+\infty}\varphi(\sigma_{T_n})\geq\varphi(q).$$

Therefore

$$\varphi(p)\geq\limsup_{T\to+\infty}\varphi(\sigma_T)\geq\liminf_{n\to+\infty}\varphi(\sigma_{T_n})\geq\varphi(q).$$

Hence, $q\in A^{-1}(0)$ and by Lemma 5.7.18 $\lim_{t\to+\infty}\|u(t)-q\|^2$ exists. Now if p is another weak cluster point of σ_T, then $\lim_{t\to+\infty}(\|u(t)-p\|^2-\|u(t)-q\|^2)$ exists. It follows that $\lim_{t\to+\infty}(u(t),p-q)$ exists, hence $\lim_{T\to+\infty}(\sigma_T,p-q)$ exists. This implies that $p=q$ and therefore $\sigma_T\rightharpoonup p\in A^{-1}(0)$, as $T\to+\infty$. □

Proposition 5.7.20 *Let $u(t)$ be a solution to* (5.56) *with $f\equiv 0$. Assume that* (5.57) *and* (5.58) *hold. If $\int_0^\infty R(t)dt<+\infty$, then $\lim_{t\to+\infty}\varphi(u(t))$ exists.*

Proof. By Lemma 5.7.3, (5.56) and (5.69), we have:

$$\begin{aligned}
\frac{d}{dt}\varphi(u(t))&=\big(\partial\varphi(u(t)),u'(t)\big)\\
&=\big(p(t)u''(t)+r(t)u'(t),u'(t)\big)\\
&=\frac{1}{2}p(t)\frac{d}{dt}\|u'(t)\|^2+r(t)\|u'(t)\|^2\\
&\leq\frac{1}{2}\int_t^\infty r'(s)\|u'(s)\|^2ds+\frac{1}{2}r(t)\|u'(t)\|^2
\end{aligned}$$

$$\leq \frac{1}{2}R(t)\int_t^{\infty}\|u'(s)\|^2ds + \frac{1}{2}r(t)\|u'(t)\|^2.$$

Therefore:

$$\begin{aligned}\varphi\big(u(T')\big)-\varphi\big(u(T)\big) &\leq \frac{1}{2}\int_T^{T'} R(t)\int_t^{\infty}\|u'(s)\|^2dsdt + \frac{1}{2}\int_T^{T'} r(t)\|u'(t)\|^2dt \\ &\leq \frac{1}{2}\Big(\int_T^{\infty}\|u'(t)\|^2dt\Big)\Big[\int_T^{T'} R(t)dt + M\Big]\end{aligned}$$

This implies that: $\limsup_{t\to+\infty}\varphi\big(u(t)\big) \leq \varphi\big(u(T)\big) + C\int_T^{\infty}\|u'(s)\|^2ds$, for some constant C. Now since $u' \in L^2\big((0,+\infty);H\big)$, letting $T \to +\infty$, we get: $\limsup_{t\to+\infty}\varphi\big(u(t)\big) \leq \liminf_{t\to+\infty}\varphi\big(u(t)\big)$, which completes the proof of the proposition. □

Remark 5.7.21 *In Proposition 5.7.20, if $r(t) \leq 0$ and $r'(t) \leq 0$, then $\varphi(u(t))$ is nonincreasing.*

Lemma 5.7.22 *Suppose that $u(t)$ is a solution to* (5.56) *with $f \equiv 0$, and $q \in H$. Then $\liminf_{t\to+\infty} t\frac{d}{dt}\|u(t)-q\|^2 \leq 0$.*

Proof. Assume by contradiction that $\liminf_{t\to+\infty} t\frac{d}{dt}\|u(t)-q\|^2 > 0$. Then There exist $t_0 > 0$ and $c > 0$ such that for each $t > t_0$, we have $t\frac{d}{dt}\|u(t)-q\|^2 \geq c > 0$. Dividing both sides by t and then integrating from $t = t_0$ to T, we get

$$\|u(T)-q\|^2 - \|u(t_0)-q\|^2 \geq c(\ln T - \ln t_0).$$

Since $u(t)$ is bounded, we get a contradiction by letting $T \to +\infty$. □

Theorem 5.7.23 *Let $u(t)$ be a solution to* (5.56) *with $f \equiv 0$. Suppose that the assumptions of Proposition 5.7.20 are satisfied. Then $u(t)$ converges weakly to an element of $A^{-1}(0)$, as $t \to +\infty$.*

Proof. By Remark 5.7.8, we know that $A^{-1}(0) \neq \varnothing$. Let $q \in A^{-1}(0)$. By the subdifferential inequality and (5.56), we get

$$\begin{aligned}\varphi\big(u(t)\big)-\varphi(q) &\leq \big(p(t)u''(t)+r(t)u'(t),u(t)-q\big) \\ &\leq \frac{1}{2}p(t)\frac{d^2}{dt^2}\|u(t)-q\|^2 + \frac{1}{2}r(t)\frac{d}{dt}\|u(t)-q\|^2 \end{aligned} \tag{5.74}$$

Integrating from $t = 0$ to T, we get:

$$\int_0^T \big(\varphi(u(t))-\varphi(q)\big)dt \leq \frac{p(T)}{2}\frac{d}{dT}\|u(T)-q\|^2 - p(0)\big(u'(0),u(0)-q\big)$$
$$-\frac{1}{2}\int_0^T p'(t)\frac{d}{dt}\|u(t)-q\|^2 + \frac{1}{2}\int_0^T r(t)\frac{d}{dt}\|u(t)-q\|^2dt$$

By Lemma 5.7.3, $\|u(t)-q\|$ is nonincreasing or eventually increasing. If $\|u(t)-q\|$ is nonincreasing, then

$$\int_0^T (\varphi(u(t))-\varphi(q))dt \leq -p(0)\big(u'(0),u(0)-q\big)+M(\|u(0)-q\|^2-\|u(T)-q\|^2). \tag{5.75}$$

If $\|u(t)-q\|$ is eventually increasing, then there exists t_0 such that for each $t\geq t_0$, $\frac{d}{dt}\|u(t)-q\|>0$. Integrating (5.74) from $t=t_0$ to T, we get:

$$\int_{t_0}^T (\varphi(u(t))-\varphi(q))dt \leq \frac{p(T)}{2}\frac{d}{dT}\|u(T)-q\|^2-p(t_0)\big(u'(t_0),u(t_0)-q\big)$$
$$-\frac{1}{2}\int_{t_0}^T p'(t)\frac{d}{dt}\|u(t)-q\|^2+\frac{1}{2}\int_{t_0}^T r(t)\frac{d}{dt}\|u(t)-q\|^2dt$$

Therefore:

$$\begin{aligned}\int_{t_0}^T (\varphi(u(t))-\varphi(q))dt &\leq \frac{M}{2}\frac{d}{dT}\|u(T)-q\|^2-p(t_0)\big(u'(t_0),u(t_0)-q\big)\\ &\quad +M(\|u(T)-q\|^2-\|u(t_0)-q\|^2).\end{aligned} \tag{5.76}$$

Taking liminf of (10.14) and (8.85) as $T\to+\infty$, by Lemma 5.7.18, we get

$$\int_{t_0}^{\infty} (\varphi(u(t))-\varphi(q))dt<+\infty.$$

Therefore $\liminf_{t\to+\infty}\varphi\big(u(t)\big)\leq\varphi(q)$. By Proposition 5.7.20, $\lim_{t\to+\infty}\varphi\big(u(t)\big)=\varphi(q)=\text{Min}\{\varphi(z);z\in H\}$. If $u(t_n)\rightharpoonup s$ as $n\to+\infty$, then

$$\varphi(s)\leq\liminf_{n\to+\infty}\varphi(u(t_n))=\lim_{t\to+\infty}\varphi(u(t))=\varphi(q).$$

Hence $s\in A^{-1}(0)$, and therefore by Corollary 5.8.8, there exists $\lim_{t\to+\infty}\|u(t)-s\|$. Now the proof is completed by a similar argument as in Theorem 5.7.5. □

Theorem 5.7.24 *Let $u(t)$ be a solution to (5.56) with $f\equiv 0$. Suppose (5.57) and (5.58) hold and that $r(t)\leq 0$ and $r'(t)\leq 0$. Let $\varphi: H\to]-\infty,+\infty]$ be a proper, convex and lower semicontinuous function satisfying the following conditions: $D(\varphi)=-D(\varphi)$, and*
$\varphi(x)-\varphi(0)\geq a(\|x\|)\big(\varphi(-x)-\varphi(0)\big),\quad \forall x\in D(\varphi),$
where $a:\mathbb{R}^+\to(0,1)$ is a continuous function. Then $u(t)\to q\in A^{-1}(0)$ as $t\to+\infty$, which is a minimum point of φ.

Proof. By Remark 5.7.8, we know that $A^{-1}(0)\neq\emptyset$, and therefore φ has a minimum point. Without loss of generality, we may assume that $\varphi(0)=0$ and 0 is a minimum point of φ. For $t\leq s$, by the assumptions, Proposition 5.7.20 and Remark 5.7.21, we get:

$$\varphi\big(u(t)\big)\geq\varphi\big(u(s)\big)\geq a(\|u(s)\|)\varphi\big(-u(s)\big)+\big(1-a(\|u(s)\|)\big)\varphi(0)$$

$$\begin{aligned}
&\geq \varphi\big(-a(\|u(s)\|)u(s)\big)\\
&\geq \varphi\big(u(t)\big)+\big(\partial\varphi(u(t)),-a(\|u(s)\|)u(s)-u(t)\big)\\
&= \varphi\big(u(t)\big)-\big(p(t)u''(t)+r(t)u'(t),a(\|u(s)\|)u(s)+u(t)\big).
\end{aligned}$$

Therefore:

$$\big(p(t)u''(t)+r(t)u'(t),a(\|u(s)\|)u(s)+u(t)\big)\geq 0.$$

Let:

$$g(t)=\big(1+a(\|u(s)\|)\big)\big(\|u(t)\|^2-\|u(s)\|^2\big)-a(\|u(s)\|)\|u(t)-u(s)\|^2$$

then

$$\begin{aligned}
g'(t)&=2\big(u'(t),u(t)+a(\|u(s)\|)u(s)\big),\\
g''(t)&=2\big(u''(t),u(t)+a(\|u(s)\|)u(s)\big)+2\|u'(t)\|^2.
\end{aligned}$$

Hence $p(t)g''(t)+r(t)g'(t)\geq 0$. Now the same argument as in Lemma 5.7.3, with $\|u(t)-q\|^2$ replaced by $g(t)$, shows that $g(t)$ is either nonincreasing or eventually increasing. Since $r(t)\leq 0$, condition (5.59) is satisfied. Therefore by Lemma 5.7.18, we conclude that $g(t)$ is nonincreasing. Then $g(t)\geq g(s)=0$ for $t\leq s$. It follows that:

$$\begin{aligned}
\|u(t)-u(s)\|^2&\leq \frac{1+a(\|u(s)\|)}{a(\|u(s)\|)}(\|u(t)\|^2-\|u(s)\|^2)\\
&< \frac{2}{a(\|u(s)\|)}(\|u(t)\|^2-\|u(s)\|^2),\quad \forall s\geq t
\end{aligned}$$

By Corollary 5.8.8, there exists $\lim_{s\to+\infty}\|u(s)\|$. If $\|u(s)\|\to 0$ as $s\to+\infty$, then $u(s)\to 0$ and this yields the theorem. Otherwise, if $\|u(s)\|\to r>0$, from the continuity of a, we have $\lim_{s\to+\infty}a(\|u(s)\|)=a(\lim_{s\to+\infty}\|u(s)\|)=a(r)>0$. Therefore $\{u(t)\}$ is a cauchy sequence in H, hence $u(t)\to q$ as $t\to+\infty$, and $q\in A^{-1}(0)$ by Theorem 5.7.23. □

5.8 ASYMPTOTIC BEHAVIOR FOR SOME SPECIAL NONHOMOGENEOUS CASES

In this section, we study the weak and strong convergence of solutions to (5.56) when $p(t)\equiv 1$ and $r(t)\equiv c$, where c is a constant real number. Our investigations in this section include the nonhomogeneous case.

We consider the following second order nonhomogeneous evolution equation

$$\begin{cases} u''(t)+cu'(t)\in Au(t)+f(t) \ \text{ a.e. on } \mathbb{R}^+\\ u(0)=u_0,\quad \sup_{t\geq 0}\|u(t)\|<+\infty \end{cases}\tag{5.77}$$

where A is a maximal monotone operator, and $f:\mathbb{R}^+\to H$ satisfies the following condition:

$$\text{There exists } t_0>0 \text{ such that } \int_{t_0}^{\infty} t\|f(t)-f_\infty\|dt<+\infty,\tag{5.78}$$

for some $f_\infty \in H$. By replacing $f(t)$ by $f(t) - f_\infty$ and A by $A + f_\infty$, we may assume without loss of generality that $f_\infty = 0$.

We study the asymptotic behavior of (5.77) in two subsections: first when the constant c is non-positive, and then in the next subsection when c is positive.

5.8.1 CASE $C \leq 0$

In this case, (5.77) satisfies the assumptions in Véron's existence Theorem (Theorem 5.7.1) when $f \equiv 0$.

Lemma 5.8.1 *Suppose $f : [0, +\infty) \to \mathbb{R}$ is bounded from above and absolutely continuous on every compact subinterval, then* $\liminf_{t \to +\infty} f'(t) \leq 0$.

Proof. To get a contradiction, assume that $\liminf_{t \to +\infty} f'(t) > 0$. Then there exist $\lambda > 0$ and $t_0 > 0$ such that for each $t \geq t_0$, $f'(t) \geq \lambda$. Integrating this inequality from t_0 to T, we get: $f(T) - f(t_0) = \int_{t_0}^{T} f'(t)dt \geq \lambda(T - t_0)$. Letting $T \to +\infty$, we get a contradiction. □

Lemma 5.8.2 *Assume $f : [0, +\infty) \to \mathbb{R}$ is bounded from above and f and f' are absolutely continuous on every compact subinterval. If*

$$f''(t) + cf'(t) \geq -Mg(t) \tag{5.79}$$

with $g(t) \geq 0$, and g satisfying $\int_{t_0}^{+\infty} tg(t)dt < +\infty$, for some $t_0 > 0$, then $f(T_2) \leq f(T_1) + \eta(T_1)$, $\forall T_2 \geq T_1 \geq t_0$, where $\lim_{T_1 \to +\infty} \eta(T_1) = 0$, and $\lim_{t \to +\infty} f(t)$ exists.

Proof. Multiplying both sides of (5.79) by e^{ct}, we get

$$\frac{d}{dt}(e^{ct} f'(t)) \geq -Me^{ct} g(t)$$

Integrating from $t = S$ to $t = T$, we get

$$e^{cT} f'(T) - e^{cS} f'(S) \geq -M \int_S^T e^{ct} g(t)dt \geq -Me^{cS} \int_S^T g(t)dt$$

This implies that

$$e^{cS} f'(S) \leq e^{cT} f'(T) + Me^{cS} \int_S^T g(t)dt$$

Taking $\liminf$ as $T \to +\infty$, by Lemma 5.8.1, we get

$$f'(S) \leq M \int_S^\infty g(t)dt$$

Integrating the above inequality from $S = T_1$ to $S = T_2$, and applying Fubini's Theorem, we obtain

$$f(T_2) \leq f(T_1) + M \int_{T_1}^{T_2} dS \int_S^\infty g(t)dt$$

$$
\begin{aligned}
&\leq f(T_1)+M\int_{T_1}^{+\infty} dS \int_{S}^{\infty} g(t)dt \\
&= f(T_1)+M\int_{T_1}^{\infty} dt \int_{T_1}^{t} g(t)dS \\
&= f(T_1)+M\int_{T_1}^{\infty} (t-T_1)g(t)dt \\
&\leq f(T_1)+M\int_{T_1}^{+\infty} tg(t)dt \\
&= f(T_1)+\eta(T_1)
\end{aligned}
$$

where $\eta(T_1) = M\int_{T_1}^{+\infty} tg(t)dt \to 0$ as $T_1 \to +\infty$. This implies that there exists $\lim_{T\to+\infty} f(T)$. □

Theorem 5.8.3 *Assume that u is a solution to* (5.77) *and* $f(t)$ *satisfies the condition* (5.78), *then* $\sigma_T := \frac{1}{T}\int_0^T u(t)dt$ *converges weakly as* $T \to +\infty$ *to some* $p \in L := \{q \in H: \lim_{n\to+\infty} \|u(t)-q\|$ *exists*$\}$, *which is also the asymptotic center of the curve* $(u(t))_{t\geq 0}$.

Proof. We show that the curve u is almost nonexpansive. Let $h \geq 0$ be fixed. By the monotonicity of A, we get from (5.77) that

$$
\begin{aligned}
&\Big(\frac{d^2u}{dt^2}(t+h)-\frac{d^2u}{dt^2}(t),u(t+h)-u(t)\Big)+c\Big(\frac{du}{dt}(t+h)-\frac{du}{dt}(t),u(t+h)-u(t)\Big) \\
&\qquad \geq \big(f(t+h)-f(t),u(t+h)-u(t)\big) \\
&\qquad \geq -\|f(t+h)-f(t)\|\|u(t+h)-u(t)\| \\
&\qquad \geq -\frac{M}{2}\|f(t+h)-f(t)\| \\
&\qquad \geq -\frac{M}{2}\|f(t+h)\|-\frac{M}{2}\|f(t)\|
\end{aligned}
$$

where $M := \sup_{t\geq 0}\|u(t)\|$. This implies that

$$
\begin{aligned}
&\frac{1}{2}\frac{d^2}{dt^2}\|u(t+h)-u(t)\|^2+\frac{c}{2}\frac{d}{dt}\|u(t+h)-u(t)\|^2 \\
&\qquad \geq \|u'(t+h)-u'(t)\|^2 \\
&\qquad -\frac{M}{2}\|f(t+h)\|-\frac{M}{2}\|f(t)\|
\end{aligned}
$$

From Lemma 5.8.2 we conclude that there exists $\lim_{T\to+\infty}\|u(T+h)-u(T)\|$ and $u(t)$ is almost nonexpansive. By Theorem 4.5.23, we get: $\frac{1}{t}\int_0^t u(s)ds \rightharpoonup p$ as $t \to +\infty$, where p is the asymptotic center of u and $\lim_{t\to+\infty}\|u(t)-p\|^2$ exists. □

Lemma 5.8.4 *Assume that u is a solution to* (5.77), *and f satisfies* (5.78). *If p is the asymptotic center of u, then we have:*

$$
\Big(\frac{d^2u}{dt^2}(t)+c\frac{du}{dt}(t)-f(t)+f_\infty,u(t)-p\Big)\geq 0, \quad \text{for a.e. } t\in\mathbb{R}^+. \tag{5.80}
$$

Proof. Without loss of generality, we assume that $f_\infty = 0$. Let u be a solution to (5.77), and let $t \in \mathbb{R}^+$ be fixed so that $u(t)$ satisfies (5.77). Then by the monotonicity of A we have:

$$\begin{aligned}
&\big(u''(t)+cu'(t)-f(t),u(t)-\frac{1}{T}\int_{t_0}^{T}u(s)ds\big)\\
&=\frac{1}{T}\int_{t_0}^{T}\big[\big(u''(t)+cu'(t)-f(t)-u''(s)-cu'(s)+f(s),u(t)-u(s)\big)\\
&\quad+\big(u''(s)+cu'(s)-f(s),u(t)-u(s)\big)\big]ds\\
&\geq\frac{1}{T}\int_{t_0}^{T}\big(u''(s)+cu'(s)-f(s),u(t)-u(s)\big)ds\\
&\geq\frac{1}{T}\int_{t_0}^{T}\big[\frac{d}{ds}\big(u'(s),u(t)-u(s)\big)+\|u'(s)\|^2\\
&\quad-\frac{c}{2}\frac{d}{ds}\|u(t)-u(s)\|^2-\|f(s)\|\|u(t)-u(s)\|\big]ds\\
&\geq\frac{1}{T}\big[\big(u'(T),u(t)-u(T)\big)-\big(u'(t_0),u(t)-u(t_0)\big)\big]\\
&\quad-\frac{1}{T}\big[\frac{c}{2}\|u(t)-u(T)\|^2-\frac{c}{2}\|u(t)-u(t_0)\|^2\big]\\
&\quad-\frac{1}{T}\int_{t_0}^{T}\|f(s)\|\|u(t)-u(s)\|ds
\end{aligned}$$

Letting $T \to +\infty$, by Lemma 5.8.1, we get:

$$\big(u''(t)+cu'(t)-f(t),u(t)-p\big)\geq\limsup_{T\to+\infty}\frac{-1}{2T}\frac{d}{dT}\|u(t)-u(T)\|^2\geq 0$$

This proves the lemma. □

Theorem 5.8.5 *Let u be a solution to* (5.77) *and f satisfies* (5.78). *Then $u(t)\rightharpoonup p$ as $t\to+\infty$, where p is an element of $A^{-1}(-f_\infty)$, as well as the asymptotic center of u.*

Proof. First we show that u is asymptotically regular. For all $h \geq 0$ we have

$$\|u(t+h)-u(t)\|\leq\int_{t}^{t+h}\|u'(s)\|ds\leq\sqrt{h}\Big(\int_{t}^{t+h}\|u'(s)\|^2ds\Big)^{\frac{1}{2}}.$$

From (5.80), we deduce that

$$\|u'(s)\|^2\leq\frac{d}{ds}\big(u'(s),u(s)-p\big)+\frac{c}{2}\frac{d}{ds}\|u(s)-p\|^2+M\|f(s)-f_\infty\|$$

where $M := \sup_{t\geq 0}\|u(t)-p\|$. Integrating on $[t,t+h]$, we get

$$\int_{t}^{t+h}\|u'(s)\|^2ds\leq\big(u'(t+h),u(t+h)-p\big)-\big(u'(t),u(t)-p\big)$$

$$+\frac{c}{2}\|u(t+h)-p\|^2-\frac{c}{2}\|u(t)-p\|^2$$
$$+M\int_t^{t+h}\|f(s)-f_\infty\|ds$$

Since there exists

$$\lim_{t\to+\infty}\|u(t+h)-u(t)\|^2$$

it is sufficient to prove that

$$\liminf_{t\to+\infty}\|u(t+h)-u(t)\|^2=0$$

By Assumption (5.78) and Lemma 8.116, we get

$$\begin{aligned}\liminf_{t\to+\infty}\|u(t+h)-u(t)\|^2 &\leq h\liminf_{t\to+\infty}\int_t^{t+h}\|u'(s)\|^2ds\\ &\leq \liminf_{t\to+\infty}\left[\left(u'(t+h),u(t+h)-p\right)-\left(u'(t),u(t)-p\right)\right]\\ &\quad -\frac{c}{2}\lim_{t\to+\infty}\|u(t)-p\|^2+\frac{c}{2}\lim_{t\to+\infty}\|u(t+h)-p\|^2\\ &=\frac{1}{2}\liminf_{t\to+\infty}\frac{d}{dt}\left[\|u(t+h)-p\|^2-\|u(t)-p\|^2\right]\leq 0\end{aligned}$$

Therefore

$$\lim_{t\to+\infty}\|u(t+h)-u(t)\|=0$$

By Theorem 4.5.25, $u(t)\rightharpoonup p$ where p is the asymptotic center of u. Now we prove that $p\in A^{-1}(-f_\infty)$. Suppose that $[x,y]\in A$. By the monotonicity of A and Equation (5.77), we get:

$$\begin{aligned}\left(x-\frac{1}{T}\int_{t_0}^{T}u(t)dt,y+f_\infty\right) &=\frac{1}{T}\int_{t_0}^{T}\left(x-u(t),y+f_\infty\right)dt\\ &\geq\frac{1}{T}\int_{t_0}^{T}\left(x-u(t),u''(t)+cu'(t)-f(t)+f_\infty\right)dt\\ &\geq\frac{1}{T}\int_{t_0}^{T}\Big[\frac{d}{dt}\left(x-u(t),u'(t)\right)-\frac{c}{2}\frac{d}{dt}\|u(t)-x\|^2\\ &\quad-\|f(t)-f_\infty\|\|u(t)-x\|\Big]dt\\ &=\frac{1}{T}\Big[\left(x-u(T),u'(T)\right)-\left(x-u(t_0),u'(t_0)\right)\\ &\quad-\frac{c}{2}\|u(T)-x\|^2+\frac{c}{2}\|u(t_0)-x\|^2\Big]\\ &\quad-\frac{M}{T}\int_{t_0}^{T}\|f(t)-f_\infty\|dt\end{aligned}$$

where $M=\sup_{t\geq t_0}\|u(t)-x\|$. Taking limsup as $T\to+\infty$ and using Lemma 5.8.1, we get $(x-p,y+f_\infty)\geq 0$. Now the maximality of A implies that $p\in A^{-1}(-f_\infty)$. □

Theorem 5.8.6 *Assume that the operator A in (5.77) is strongly monotone, and f satisfies (5.78). Let u be a solution to (5.77). Then $u(t)$ converges strongly as $t \to +\infty$ to the asymptotic center p of u.*

Proof. Without loss of generality, we assume that $f_\infty = 0$. Assume $(y_2 - y_1, x_2 - x_1) \geq \alpha \|x_2 - x_1\|^2$, for all $[x_i, y_i] \in A, i = 1,2$ and for some $\alpha > 0$. Then by using the strong monotonicity of A, and a similar proof as in Lemma 5.8.4, we get:

$$\begin{aligned}
&\big(u''(t) + cu'(t) - f(t), u(t) - \frac{1}{T}\int_{t_0}^{T} u(s)ds\big) \\
&\geq \alpha \frac{1}{T}\int_{t_0}^{T} \|u(t) - u(s)\|^2 ds \\
&\quad - \frac{1}{2T}\frac{d}{dT}\|u(t) - u(T)\|^2 - \frac{1}{T}\big(u'(t_0), u(t) - u(t_0)\big) \\
&\quad - \frac{1}{T}\big[\frac{c}{2}\|u(t) - u(T)\|^2 - \frac{c}{2}\|u(t) - u(t_0)\|^2\big] \\
&\quad - \frac{1}{T}\int_{t_0}^{T} \|f(s)\|\|u(t) - u(s)\|ds
\end{aligned}$$

Taking limsup as $T \to +\infty$ in the above inequality and using Lemma 5.8.1, Theorem 5.8.3 and Theorem 5.8.5, we get:

$$\begin{aligned}
\big(u''(t) + cu'(t) - f(t), u(t) - p\big) &\geq \alpha \liminf_{T \to +\infty} \frac{1}{T}\int_{t_0}^{T} \|u(t) - u(s)\|^2 ds \\
&\geq \alpha \liminf_{s \to +\infty} \|u(t) - u(s)\|^2 \\
&\geq \alpha \|u(t) - p\|^2.
\end{aligned}$$

Now integrating both sides of the above inequality on $[\zeta, t]$ where $\zeta > t_0$, we get

$$\begin{aligned}
\alpha \int_{\zeta}^{t} \|u(s) - p\|^2 ds &\leq \int_{\zeta}^{t} \big[\frac{d}{ds}\big(u'(s), u(s) - p\big) - \|u'(s)\|^2 \\
&\quad + \frac{c}{2}\frac{d}{ds}\|u(s) - p\|^2 - \big(f(s), u(s) - p\big)\big]ds \\
&\leq \big(u'(t), u(t) - p\big) - \big(u'(\zeta), u(\zeta) - p\big) \\
&\quad + \frac{c}{2}\|u(t) - p\|^2 - \frac{c}{2}\|u(\zeta) - p\|^2 + M\int_{\zeta}^{t} \|f(s)\|ds
\end{aligned}$$

where $M := \sup_{t \geq 0} \|u(t) - p\|$. We define $F(t) := \alpha \int_{\zeta}^{t} \|u(s) - p\|^2 ds$. Integrating the above inequality on $[T, 2T]$ and dividing by T, since F is nondecreasing, we get:

$$\begin{aligned}
F(T) &\leq \frac{1}{T}\int_{T}^{2T} F(t)dt \\
&\leq M(\zeta) + \frac{1}{T}\int_{T}^{2T} \big(u'(t), u(t) - p\big)dt
\end{aligned}$$

$$= M(\zeta) + \frac{1}{2T}\left(\|u(2T)-p\|^2 - \|u(T)-p\|^2\right)$$

where $M(\zeta) := -\left(u'(\zeta), u(\zeta)-p\right) - \frac{c}{2}\|u(\zeta)-p\|^2 + M\int_{\zeta}^{+\infty}\|f(s)\|ds$. This implies that

$$\int_{\zeta}^{+\infty}\|u(s)-p\|^2 ds < +\infty$$

Hence $\liminf_{t\to+\infty}\|u(t)-p\|^2 = 0$. Since there exists $\lim_{t\to+\infty}\|u(t)-p\|^2$, the proof of the theorem is now complete. □

Now we prove the strong convergence of solutions to (5.77) without even the monotonicity assumption on A, but assuming A to satisfy some positivity condition and the following condition (b):

$$(x_2,y_1)+(x_1,y_2)+b(\|x_1\|,\|x_2\|)\{(x_1,y_1)+(x_2,y_2)\} \geq 0, \tag{5.81}$$

where $[x_1,y_1]\in A$, $[x_2,y_2]\in A$ and $b:\mathbb{R}^+\times\mathbb{R}^+\to\mathbb{R}^+$ is a continuous function. Condition (5.81) is a weaker version of the following condition (a):

$$|(x_1,y_2)+(x_2,y_1)| \leq a(\|x_1\|,\|x_2\|)\{(x_1,y_1)+(x_2,y_2)\}, \tag{5.82}$$

where $[[x_1,y_1]\in A$, $[x_2,y_2]\in A$ and $a:\mathbb{R}^+\times\mathbb{R}^+\to\mathbb{R}^+$ is a continuous function. The condition (5.82) has been applied by Mitidieri [MIT] to prove the strong convergence of solutions to (5.77) when $c=0$ in the homogeneous case ($f\equiv 0$). Since these conditions are no longer invariant by translation of A with f_∞, we will assume in our next theorem that f satisfies (5.78) with $f_\infty = 0$.

Theorem 5.8.7 *Assume that the (not necessarily monotone) operator A in* (5.77) *satisfies $(y,x)\geq 0$, $\forall[x,y]\in A$, as well as condition (b), and f satisfies* (5.78) *with $f_\infty=0$. Let u be a solution to* (5.77). *Then $u(t)$ converges strongly as $t\to+\infty$ to the asymptotic center p of u.*

Proof. For $h\geq 0$, let $M := \max\{1, \sup_{t\geq 0} b(\|u(t)\|,\|u(t+h)\|)\}$ and

$$H(t) := \frac{M}{2}\|u(t)\|^2 + \frac{M}{2}\|u(t+h)\|^2 + \left(u(t),u(t+h)\right)$$

By the assumption on A, we have

$$\begin{aligned}
H''(t) &= M\left(u''(t),u(t)\right) + M\|u'(t)\|^2 + M\left(u''(t+h),u(t+h)\right) + M\|u'(t+h)\|^2 \\
&\quad + \left(u''(t),u(t+h)\right) + \left(u''(t+h),u(t)\right) + 2(u'(t),u'(t+h)) \\
&= M\left(u''(t)+cu'(t)-f(t),u(t)\right) - Mc\left(u'(t),u(t)\right) + M\left(f(t),u(t)\right) \\
&\quad + M\|u'(t)\|^2 + M\left(u''(t+h)+cu'(t+h)-f(t+h),u(t+h)\right) \\
&\quad - Mc\left(u'(t+h),u(t+h)\right) + M\left(f(t+h),u(t+h)\right) + M\|u'(t+h)\|^2 \\
&\quad + \left(u''(t)+cu'(t)-f(t),u(t+h)\right) - c\left(u'(t),u(t+h)\right) + \left(f(t),u(t+h)\right)
\end{aligned}$$

$$
\begin{aligned}
&+\big(u''(t+h)+cu'(t+h)-f(t+h),u(t)\big)-c\big(u'(t+h),u(t)\big)\\
&+\big(f(t+h),u(t)\big)+2\big(u'(t),u'(t+h)\big)\\
\geq{}& -Mc\big(u'(t),u(t)\big)+M\big(f(t),u(t)\big)-Mc\big(u'(t+h),u(t+h)\big)\\
&+M\big(f(t+h),u(t+h)\big)-c\big(u'(t),u(t+h)\big)+\big(f(t),u(t+h)\big)\\
&-c\big(u'(t+h),u(t)\big)+\big(f(t+h),u(t)\big)\\
\geq{}& -\frac{Mc}{2}\frac{d}{dt}\|u(t)\|^2-\frac{Mc}{2}\frac{d}{dt}\|u(t+h)\|^2-c\frac{d}{dt}\big(u(t),u(t+h)\big)\\
&-M\|f(t)\|\|u(t)\|-M\|f(t+h)\|\|u(t+h)\|\\
&-\|f(t)\|\|u(t+h)\|-\|f(t+h)\|\|u(t)\|\\
={}& -c\frac{d}{dt}H(t)-\|f(t)\|\big(M\|u(t)\|+\|u(t+h)\|\big)\\
&-\|f(t+h)\|\big(M\|u(t+h)\|+\|u(t)\|\big)\\
\geq{}& -K\|f(t)\|-K\|f(t+h)\|-cH'(t)
\end{aligned}
$$

where $K := 2M\sup_{t\geq 0}\|u(t)\|$. This implies that

$$H''(t)+cH'(t)\geq -K\|f(t)\|-K\|f(t+h)\| \quad \text{a.e. on } \mathbb{R}^+$$

From Lemma 5.8.2, we deduce that

$$H(t)\leq H(s)+\eta(s) \quad \forall t\geq s\geq t_0$$

where $\lim_{s\to+\infty}\eta(s)=0$. This implies that

$$
\begin{aligned}
\big(u(t),u(t+h)\big) &\leq \big(u(s),u(s+h)\big)+\eta(s)\\
&+\frac{M}{2}\big(\|u(s)\|^2-\|u(t)\|^2+\|u(s+h)\|^2-\|u(t+h)\|^2\big)
\end{aligned} \tag{5.83}
$$

Multiplying (5.77) by $u(t)$, and using the positivity of A, we get

$$\big(u''(t)+cu'(t),u(t)\big)\geq \big(f(t),u(t)\big) \tag{5.84}$$

This implies that $\frac{d^2}{dt^2}\|u(t)\|^2+c\frac{d}{dt}\|u(t)\|^2\geq -\frac{K}{M}\|f(t)\|$. By Lemma 5.8.2, there exists

$$\lim_{t\to+\infty}\|u(t)\|^2$$

From (5.83), we obtain

$$
\begin{aligned}
\|u(s+h)-u(s)\|^2 &\leq \|u(t+h)-u(t)\|^2+2\eta(s)\\
&+(M+1)\big(\|u(s)\|^2-\|u(t)\|^2+\|u(s+h)\|^2-\|u(t+h)\|^2\big).
\end{aligned}
$$

This implies that for all $\delta>0$ there exists $t_1>0$ such that for all $t\geq s\geq t_1$ and $h\geq 0$, we have

$$\|u(s+h)-u(s)\|^2\leq \|u(t+h)-u(t)\|^2+\delta \tag{5.85}$$

From (5.84) we get:

$$\|u'(t)\|^2 \le \frac{d}{dt}(u'(t),u(t)) - \frac{c}{2}\frac{d}{dt}\|u(t)\|^2 + M\|f(t)\|$$

Now integrating from s to $s+h$ and with a similar proof to that in Theorem 5.8.5, we conclude that $u(t)$ is asymptotically regular. From (5.85), by letting $t \to +\infty$ and using the asymptotic regularity of u, we get:

$$\|u(s+h)-u(s)\| \le \delta \quad \forall s \ge t_0, \ \forall h \ge 0.$$

Therefore $(u(t))_{t\ge 0}$ is a Cauchy net in H, and hence $u(t) \to p$ as $t \to +\infty$. □

Now we state the following two Corollaries.

Corollary 5.8.8 *Assume that the operator A in* (5.77) *is monotone and satisfies the Condition (b), and f satisfies* (5.78) *with $f_\infty = 0$. If u is a solution to* (5.77)*, then $u(t)$ converges strongly as $t \to +\infty$ to the asymptotic center p of u.*

Proof. Let $[x_1,y_1],[x_2,y_2] \in A$. By Condition (5.81), we have:

$$-(x_1,y_2)-(x_2,y_1) \le b(\|x_1\|,\|x_2\|)\{(x_1,y_1)+(x_2,y_2)\}$$

On the other hand, by the monotonicity of A, we have:

$$(x_1,y_2)+(x_2,y_1) \le (x_1,y_1)+(x_2,y_2)$$

Adding the above two inequalities, we get:

$$(x_1,y_1)+(x_2,y_2) \ge 0.$$

This implies that A is positive. Then the result follows from Theorem 5.8.7. □

Corollary 5.8.9 *Let u be a solution to* (5.77) *with $A = \partial\varphi$, where φ is a proper, convex and lower semicontinuous function on H satisfying the following Condition (a_1): $D(\varphi) = -D(\varphi)$ and $\varphi(x)-\varphi(0) \ge a(\|x\|)\big(\varphi(-x)-\varphi(0)\big)$, $\forall x \in D(\varphi)$, where $a: \mathbb{R}^+ \to (0,+\infty)$ is a continuous function. Assume that f satisfies* (5.78) *with $f_\infty =$ 0. Then $u(t)$ converges strongly as $t \to +\infty$ to the asymptotic center p of u.*

Proof. We prove that the assumption on φ implies the condition (b), and then the result follows from Corollary 5.8.8. Without loss of generality we may assume that $\varphi(0) = 0$. From the convexity of φ and the assumption, we have:

$$\begin{aligned}
0 = \varphi(0) &= \varphi\Big(\frac{a((\|x\|)}{1+a(\|x\|)}x + \frac{1}{1+a(\|x\|)}(-a(\|x\|)x)\Big) \\
&\le \frac{a(\|x\|)}{1+a(\|x\|)}\varphi(x) + \frac{1}{1+a(\|x\|)}\varphi(-a(\|x\|)x) \\
&\le \frac{a(\|x\|)}{1+a(\|x\|)}\varphi(x) + \frac{a(\|x\|)}{1+a(\|x\|)}\varphi(-x)
\end{aligned}$$

$$\leq \frac{a(\|x\|)}{1+a(\|x\|)}\varphi(x) + \frac{1}{1+a(\|x\|)}\varphi(x)$$
$$= \varphi(x)$$

Now suppose that $[x,y],[\hat{x},\hat{y}] \in \partial\varphi$. By the subdifferential inequality, we have:

$$0 \leq \varphi(x) \leq (y,x)$$

Again by the subdifferential inequality and the convexity of φ, we get:

$$\begin{aligned}\varphi(x) - a(\|x\|)\varphi(-\hat{x}) &\leq \varphi(x) - \varphi\big(-a(\|x\|)\hat{x}\big) \\ &\leq \big(y, x + a(\|x\|)\hat{x}\big)\end{aligned}$$

Therefore

$$\frac{1}{a(\|x\|)}\varphi(x) - \varphi(-\hat{x}) \leq \frac{1}{a(\|x\|)}(x,y) + (\hat{x},y) \tag{5.86}$$

and

$$\begin{aligned}\varphi(\hat{x}) - a(\|\hat{x}\|)\varphi(-x) &\leq \varphi(\hat{x}) - \varphi(-a(\|\hat{x}\|)x) \\ &\leq (\hat{y}, \hat{x} + a(\|\hat{x}\|)x)\end{aligned}$$

Hence

$$\frac{1}{a(\|\hat{x}\|)}\varphi(\hat{x}) - \varphi(-x) \leq \frac{1}{a(\|\hat{x}\|)}(\hat{x},\hat{y}) + (x,\hat{y}) \tag{5.87}$$

Adding up (5.86) and (5.87), by the assumption on φ, we get

$$\frac{1}{a(\|x\|)}(x,y) + \frac{1}{a(\|\hat{x}\|)}(\hat{x},\hat{y}) + (\hat{x},y) + (x,\hat{y}) \geq 0$$

Therefore the condition (b) is satisfied by taking $b(\|x\|,\|\hat{x}\|) = \text{Max}\{\frac{1}{a(\|x\|)}, \frac{1}{a(\|\hat{x}\|)}\}$. □

5.8.2 THE CASE $C > 0$

In this subsection, we consider (5.77) with $c > 0$ and f satisfing (5.78). We prove the strong convergence of solutions to (5.77) with $c > 0$, for a general maximal monotone operator A.

Lemma 5.8.10 *Assume u is a solution to* (5.77). *Then there exists $p \in H$ such that:*

$$\Big(\frac{d^2u}{dt^2}(t) + c\frac{du}{dt}(t) - f(t), u(t) - p\Big) \geq 0, \quad \textit{for a.e. } t \in \mathbb{R}^+ \tag{5.88}$$

Proof. Let u be a solution to (5.77), and let $t \in \mathbb{R}^+$ be fixed so that $u(t)$ satisfies (5.77). Then by the monotonicity of A, we get:

$$\Big(u''(t) + cu'(t) - f(t), u(t) - \frac{1}{T}\int_{t_0}^{T} u(s)ds\Big)$$

$$
\begin{aligned}
&\geq \frac{1}{T}\int_{t_0}^{T} \big(u''(s)+cu'(s)-f(s),u(t)-u(s)\big)ds \\
&= \frac{1}{T}\int_{t_0}^{T} \Big[\frac{d}{ds}\big(u'(s),u(t)-u(s)\big)+\|u'(s)\|^2 \\
&\quad -\frac{c}{2}\frac{d}{ds}\|u(t)-u(s)\|^2-\big(f(s),u(t)-u(s)\big)\Big]ds \\
&\geq \frac{1}{T}\Big[\big(u'(T),u(t)-u(T)\big)-\big(u'(t_0),u(t)-u(t_0)\big) \\
&\quad -\frac{c}{2}\|u(t)-u(T)\|^2+\frac{c}{2}\|u(t)-u(t_0)\|^2\Big]-\frac{1}{T}\int_{t_0}^{T} M\|f(s)\|ds
\end{aligned}
$$

where $M := \sup_{s\geq t_0}\|u(t)-u(s)\|$. Integrating the above inequality from $T=\xi$ to $T=T'$ and dividing by T', we get

$$
\begin{aligned}
&\big(u''(t)+cu'(t)-f(t),u(t)-\frac{1}{T'}\int_{\xi}^{T'}\sigma_T dT+\frac{\int_0^{t_0}u(s)ds}{T'}\int_{\xi}^{T'}\frac{dT}{T}\big) \\
&\geq \frac{1}{T'}\int_{\xi}^{T'}\Big[\frac{-1}{2T}\frac{d}{dT}\|u(t)-u(T)\|^2-\frac{1}{T}\big(u'(t_0),u(t)-u(t_0)\big) \\
&\quad -\frac{c}{2T}\|u(t)-u(T)\|^2+\frac{c}{2T}\|u(t)-u(t_0)\|^2-\frac{1}{T}\int_{t_0}^{T}M\|f(s)\|ds\Big]dT
\end{aligned}
$$

There exists a sequence $\{T_n'\}$ such that

$$
\frac{1}{T_n'}\int_{\xi}^{T_n'}\sigma_T dT \rightharpoonup p
$$

as $n\to+\infty$. Replacing T' by $\{T_n'\}$ and letting $n\to+\infty$, after integration by parts we get:

$$
\big(u''(t)+cu'(t)-f(t),u(t)-p\big)\geq 0
$$

where p is a weak cluster point of $\frac{1}{T'}\int_{\xi}^{T'}\sigma_T dT$. □

Theorem 5.8.11 *Let u be a solution to* (5.77). *Then $u(t)\to p$ as $t\to+\infty$, where $p\in A^{-1}(-f_\infty)$.*

Proof. Without loss of generality we assume $f_\infty=0$. Let $h>0$ be fixed. Then from (5.77) and the monotonicity of A, we have

$$
\begin{aligned}
&\big(u''(t+h)-u''(t),u(t+h)-u(t)\big)+c\big(u'(t+h)-u'(t),u(t+h)-u(t)\big) \\
&\geq \big(f(t+h)-f(t),u(t+h)-u(t)\big) \\
&\geq -\|f(t+h)-f(t)\|\|u(t+h)-u(t)\| \\
&\geq -M\|f(t+h)-f(t)\| \\
&\geq -M\|f(t+h)\|-M\|f(t)\|
\end{aligned}
$$

where $M = \sup_{t\geq 0}\|u(t+h)-u(t)\|$. Then

$$\frac{d^2}{dt^2}(\|u(t+h)-u(t)\|^2 + c\int_0^t \|u(s+h)-u(s)\|^2 ds) \geq -2M\|f(t+h)\| - 2M\|f(t)\| \tag{5.89}$$

First we show that

$$\int_0^t \|u(s+h)-u(s)\|^2 ds$$

is bounded from above. From (5.89), we get

$$\begin{aligned}
\|u(s+h)-u(s)\|^2 &\leq (\int_s^{s+h}\|u'(r)\|dr)^2 \leq h\int_s^{s+h}\|u'(r)\|^2 dr \\
&\leq \frac{h}{2}\int_s^{s+h}[\frac{d^2}{dr^2}\|u(r)-p\|^2 + c\frac{d}{dr}\|u(r)-p\|^2 \\
&\quad - (f(r),u(r)-p)]dr \\
&\leq \frac{h}{2}[\frac{d}{ds}\|u(s+h)-p\|^2 - \frac{d}{ds}\|u(s)-p\|^2 \\
&\quad + c\|u(s+h)-p\|^2 - c\|u(s)-p\|^2 + M\int_s^{s+h}\|f(r)\|dr]
\end{aligned}$$

where $M = \sup_{t\geq t_0}\|u(t)-p\|$. Integrating the above inequality on $[0,t]$, we get

$$\begin{aligned}
\int_0^t \|u(s+h)-u(s)\|^2 ds &\leq \frac{h}{2}[\|u(t+h)-p\|^2 - \|u(h)-p\|^2 \\
&\quad - \|u(t)-p\|^2 + \|u(0)-p\|^2 \\
&\quad + c\int_0^t \|u(s+h)-p\|^2 ds - c\int_0^t |u(s)-p|^2 ds \\
&\quad + M\int_0^t ds\int_s^{s+h}\|f(r)\|dr] \\
&\leq \frac{h}{2}[\|u(t+h)-p\|^2 + \|u(0)-p\|^2 \\
&\quad + c\int_t^{t+h}\|u(s)-p\|^2 ds - c\int_0^h \|u(s)-p\|^2 ds \\
&\quad + M(\int_0^t r\|f(r)\|dr + \int_t^{t+h} h\|f(r)\|dr)] \\
&\leq R < +\infty
\end{aligned}$$

Therefore

$$\|u(t+h)-u(t)\|^2 + c\int_0^t \|u(s+h)-u(s)\|^2 ds$$

is bounded from above, hence by (5.89) and Lemma 5.8.2, we have

$$\|u(t+h)-u(t)\|^2 + c\int_0^t \|u(r+h)-u(r)\|^2 dr$$

$$\leq \|u(s+h)-u(s)\|^2 + c\int_0^s \|u(r+h)-u(r)\|^2 dr$$
$$+4M\int_s^{+\infty} r\|f(r)\|dr \tag{5.90}$$

Hence

$$\|u(t+h)-u(t)\|^2 \leq \|u(s+h)-u(s)\|^2 + 4M\int_s^{+\infty} r\|f(r)\|dr \tag{5.91}$$

This implies that u is almost nonexpansive. From (5.90) we also get:

$$c\int_s^t \|u(r+h)-u(r)\|^2 dr \leq \|u(s+h)-u(s)\|^2 - \|u(t+h)-u(t)\|^2$$
$$+4M\int_s^{+\infty} r\|f(r)\|dr$$

Then integrating (5.91) on $[s,t]$, and using the above inequality, we get:

$$c\|u(t+h)-u(t)\|^2(t-s) - 4Mc\int_s^t dr\int_r^{+\infty} r'\|f(r')\|dr' \leq$$
$$\|u(s+h)-u(s)\|^2 + 4M\int_s^{+\infty} r\|f(r)\|dr$$

Then we obtain

$$\|u(t+h)-u(t)\|^2 \leq \frac{\|u(s+h)-u(s)\|^2}{c(t-s)} + \frac{4M}{t-s}\int_s^{+\infty} r\|f(r)\|dr$$
$$+\frac{4M}{c(t-s)}\int_s^{+\infty} r\|f(r)\|dr$$

Hence $u(t)$ is a Cauchy net in H. Therefore $u(t)$ converges strongly as $t\to+\infty$, to some $p\in H$.
Now we prove that $0\in A^{-1}(-f_\infty)$. Suppose that $[x,y]\in A$. By the monotonicity of A and (5.77) we get

$$\big(x-u(t),y\big) \geq \big(x-u(t),u''(t)+cu'(t)-f(t)\big)$$
$$\geq \frac{d}{dt}\big(x-u(t),u'(t)\big) - \frac{c}{2}\frac{d}{dt}\|x-u(t)\|^2 - \big(x-u(t),f(t)\big)$$

Integrating the above inequality on $[t_0,T]$ and dividing by T, we get

$$\frac{T-t_0}{T}(x,y+f_\infty) - \frac{1}{T}\Big(\int_{t_0}^T u(t)dt, y+f_\infty\Big)$$
$$\geq \frac{1}{T}\big(x-u(T),u'(T)\big) - \frac{1}{T}\big(x-u(t_0),u'(t_0)\big)$$
$$-\frac{c}{2T}\|x-u(T)\|^2 + \frac{c}{2T}\|x-u(t_0)\|^2$$

$$-\frac{K}{T}\int_{t_0}^{T}\|f(t)-f_\infty\|dt$$

where $K=\sup_{t\geq t_0}\|u(t)-x\|$. Letting $T\to+\infty$, by Lemma 5.8.10 we obtain $(x-p,y+f_\infty)\geq 0$. Now the maximality of A implies that $0\in A^{-1}(-f_\infty)$. □

Remark 5.8.12 *In the homogeneous case* $(f(t)\equiv 0)$, *by* (5.89)

$$g(t):=\|u(t+h)-u(t)\|^2+c\int_0^t\|u(s+h)-u(s)\|^2ds$$

is convex. Since $g(t)$ *is bounded from above, then* $g(t)$ *is nonincreasing and* $g'(t)\leq 0$. *If* u *is nonconstant, we get:*

$$\begin{aligned}
&\frac{d}{dt}\left(\ln\|u(t+h)-u(t)\|^2\right)\leq -c\\
&\Rightarrow \ln\|u(t+h)-u(t)\|^2-\ln\|u(h)-u(0)\|^2\leq -ct\\
&\Rightarrow \|u(t+h)-u(t)\|^2\leq e^{-ct}\|u(h)-u(0)\|^2\\
&\Rightarrow \|u'(t)\|\leq\|u'(0)\|e^{-\frac{c}{2}t}
\end{aligned}$$

and

$$\begin{aligned}
\|u(t)-p\| &= \lim_{T\to+\infty}\|u(t)-u(T)\|\\
&= \lim_{T\to+\infty}\|\int_t^T u'(s)ds\|\\
&\leq \lim_{T\to+\infty}\int_t^T\|u'(s)\|ds\\
&\leq \|u'(0)\|\lim_{T\to+\infty}\int_t^T e^{-\frac{c}{2}s}ds\\
&= -\frac{2}{c}\|u'(0)\|\lim_{T\to+\infty}(e^{-\frac{c}{2}T}-e^{-\frac{c}{2}t})\\
&= \frac{2}{c}\|u'(0)\|e^{-\frac{c}{2}t}
\end{aligned}$$

Therefore $\|u'(t)\|=O(e^{-\frac{c}{2}t})$ *and* $\|u(t)-p\|=O(e^{-\frac{c}{2}t})$.

REFERENCES

APR1. N. C. Apreutesei, Nonlinear second order evolution equations of monotone type and applications, Pushpa Publishing House, Allahabad, India, 2007.

BAR1. V. Barbu, Nonlinear semigroups and Differential Equations in Banach Spaces, Noordhoff, Leyden, 1976.

BAR2. V. Barbu, Sur un problème aux limites pour une classe d'équations différentielles non linéaires abstraits du deuxième ordre en t, C. R. Acad. Sci. Paris Sér. A–B 274 (1972), A459–A462.

BAR3. V. Barbu, A class of boundary problems for second order abstract differential equations, J. Fac. Sci. Univ. Tokyo Sect. IA Math. 19 (1972), 295–319.

BAI. J. B. Baillon, Un théorème de type ergodique pour les contractions non linéaires dans un espace de Hilbert, C. R. Acad. Sci. Paris 280 (1975), A1511–A1514.

BRE. H. Brézis, "Opérateurs Maximaux Monotones et Semi-groupes de Contractions dans les Espaces de Hilbert", North-Holland Mathematics studies, Vol. 5, North-Holland Publishing Co., Amsterdam-London, (1973).

BRU1. R. E. Bruck, Asymptotic convergence of nonlinear contraction semigroups in Hilbert space, J. Funct. Anal. 18 (1975), 15–26.

BRU2. R. E. Bruck, Periodic forcing of solutions of a boundary value problem for a second order differential equation in Hilbert space, J. Math. Anal. Appl. 76 (1980), 159–173.

DJ1. B. Djafari Rouhani, Ergodic theorems for nonexpansive sequences in Hilbert spaces and related problems, Ph.D. Thesis, Yale University, Part I, pp. 1–76 (1981).

DJ2. B. Djafari Rouhani, Asymptotic behaviour of quasi-autonomous dissipative systems in Hilbert spaces, J. Math. Anal. Appl. 147 (1990), 465–476.

DJ3. B. Djafari Rouhani, Asymptotic behaviour of almost nonexpansive sequences in a Hilbert space, J. Math. Anal. Appl. 151 (1990), 226–235.

DJ4. B. Djafari Rouhani, An ergodic theorem for sequences in a Hilbert space, Nonlinear Anal. Forum, 4 (1999), 33–48.

DJ-KH1. B. Djafari Rouhani and H. Khatibzadeh, Asymptotic behavior of solutions to some homogeneous second-order evolution equations of monotone type, 2007, Article ID 72931, 8 pages.

DJ-KH2. B. Djafari Rouhani and H. Khatibzadeh, Asymptotic behavior of bounded solutions to a class of second order nonhomogeneous evolution equations, Nonlinear Anal. 70 (2009), 4369–4376.

DJ-KH3. B. Djafari Rouhani and H. Khatibzadeh, Asymptotic behavior of bounded solutions to a nonhomogeneous second order evolution equation of monotone type, Nonlinear Anal. 71 (2009), e147–e159.

DJ-KH4. B. Djafari Rouhani and H. Khatibzadeh, Asymptotic behavior of bounded solutions to some second order evolution systems, Rocky Mountain J. Math. 40 (2010), 1289–1311.

DJ-KH5. B. Djafari Rouhani and H. Khatibzadeh, A strong convergence theorem for solutions to a nonhomogeneous second order evolution equation, J. Math. Anal. Appl. 363 (2010), 648–654.

DJ-KH6. B. Djafari Rouhani and H. Khatibzadeh, A note on the strong convergence of solutions to a second order evolution equation, J. Math. Anal. Appl. 401 (2013), 963–966.

DJ-KH7. B. Djafari Rouhani and H. Khatibzadeh, Asymptotic behavior for a general class of homogeneous second order evolution equations in a Hilbert space, Dynam. Systems Appl. 24 (2015), 1–15.

EDE. M. Edelstein, The construction of an asymptotic center with a fixed-point property, Bull. Amer. Math. Soc. 78 (1972), 206–208.

LOR. G. G. Lorentz, A contribution to the theory of divergent sequences, Acta. Math. 80 (1948), 167–190.

MIT. E. Mitidieri, Some remarks on the asymptotic behaviour of the solutions of second order evolution equations, J. Math. Anal. Appl. 107 (1985), 211–221.

VER1. L. Véron, Problèmes d'évolution du second ordre associés à des opérateurs monotones, C. R. Acad. Sci. Paris Sér. A 278 (1974), 1099–1101.

VER2. L. Véron, Equations d'évolution du second ordre associées à des opérateurs maximaux monotones, Proc. Roy. Soc. Edinburgh Sect. A 75 (1975/76), 131–147.

6 Heavy Ball with Friction Dynamical System

6.1 INTRODUCTION

In this brief chapter we consider the following second order evolution equation

$$\begin{cases} x''(t) + \gamma x'(t) + A(x(t)) \ni 0, & a.e.\ t \geq 0 \\ x(0) = x_0, \quad x'(0) = x_1 \end{cases} \tag{6.1}$$

where $\gamma > 0$ is a positive constant, and A is a maximal monotone operator. When $A = \nabla\varphi$, where φ is a convex and differentiable function, this system models the rolling of a heavy ball on the slid of the convex function φ with the friction impact $\gamma > 0$. For general maximal monotone operators, (6.1) does not necessarily have a solution. By assuming the existence of a solution, to the best of our knowledge, the study of the asymptotic behavior of the solution is still an open problem. Only for some special monotone operators, the asymptotic behavior is well-known. In this chapter, we study these special cases where the solution to (6.1) is weakly convergent. We refer the reader to [ALV-ATT1, ALV-ATT2, ATT-ALV1, ATT-GOU-RED] for more details about (6.1) when $A = \nabla\varphi$ and its various variants.

6.2 MINIMIZATION PROPERTIES

The following theorem was proved in [ALV].

Theorem 6.2.1 *Suppose that $\varphi \in C^1(H;\mathbb{R})$ is convex and $\inf\varphi > -\infty$. If $u \in C^2(0,+\infty;H)$ is a solution to (6.1), then $u' \in L^2(0,+\infty;H)$, $u'(t) \to 0$ as $t \to +\infty$, and*

$$\lim_{t\to+\infty} \varphi(u(t)) = \inf\varphi.$$

Moreover, if $\operatorname{Argmin}\varphi \neq \emptyset$, *then there exists* $p \in \operatorname{Argmin}\varphi$ *such that* $u(t) \rightharpoonup p$ *as* $t \to +\infty$.

Proof. Fix $x \in H$, and define the following auxiliary function $\psi(t) := \frac{1}{2}\|u(t) - x\|^2$. Since u is a solution to (6.1), it follows that

$$\psi'' + \gamma\psi' = \langle \nabla\varphi(u), x - u\rangle + \|u'\|^2.$$

Now the convexity of φ yields

$$\psi'' + \gamma\psi' \leq \varphi(x) - \varphi(u) + \|u'\|^2. \tag{6.2}$$

Let:

$$E(t) := \varphi(u(t)) + \frac{1}{2}\|u'(t)\|^2.$$

Then we can rewrite the last inequality as:

$$\psi'' + \gamma\psi' \leq \varphi(x) - E(t) + \frac{3}{2}\|u'\|^2$$

Since $E(t)$ is nonincreasing because $E'(t) \leq 0$, then given $t > 0$, for all $\tau \in [0,t]$ we have:

$$\psi''(\tau) + \gamma\psi'(\tau) \leq \varphi(x) - E(t) + \frac{3}{2}\|u'(\tau)\|^2$$

Multiplying both sides of the above inequality by $e^{\gamma\tau}$, and then integrating from 0 to θ, we get:

$$\psi'(\theta) \leq e^{-\gamma\theta}\psi'(0) + \frac{1}{\gamma}(1-e^{-\gamma\theta})[\varphi(x) - E(t)] + \frac{3}{2}\int_0^\theta e^{-\gamma(\theta-\tau)}\|u'(\tau)\|^2 d\tau$$

Integrating once more on $[0,t]$, we obtain:

$$\psi(t) \leq \psi(0) + \frac{1}{\gamma}(1-e^{-\gamma t})\psi'(0) + \frac{1}{\gamma^2}(\gamma t - 1 + e^{-\gamma t})[\varphi(x) - E(t)] + h(t), \quad (6.3)$$

where

$$h(t) := \frac{3}{2}\int_0^t \int_0^\theta e^{-\gamma(\theta-\tau)}\|u'(\tau)\|^2 d\tau d\theta$$

Since $E(t) \geq \varphi(u(t))$, for $t > \frac{1}{\gamma}$, (6.3) gives:

$$\frac{1}{\gamma^2}(\gamma t - 1 + e^{-\gamma t})\varphi(u(t)) \leq \psi(0) + \frac{1}{\gamma}(1-e^{-\gamma t})\psi'(0) + \frac{1}{\gamma^2}(\gamma t - 1 + e^{-\gamma t})\varphi(x) + h(t)$$

Dividing both sides of the above inequality by $\frac{1}{\gamma^2}(\gamma t - 1 + e^{-\gamma t})$ and letting $t \to +\infty$, we get:

$$\limsup_{t\to\infty} \varphi(u(t)) \leq \varphi(x) + \limsup_{t\to+\infty} \frac{\gamma}{t} h(t) \quad (6.4)$$

Let's show that $h(t)$ remains bounded as $t \to +\infty$. By Fubini's theorem, we have:

$$h(t) = \frac{3}{2}\int_0^t \int_\tau^t e^{-\gamma(\theta-\tau)}\|u'(\tau)\|^2 d\theta d\tau = \frac{3}{2\gamma}\int_0^t \|u'(\tau)\|^2 (1 - e^{-\gamma(t-\tau)}) d\tau.$$

From the equality $E' = -\gamma\|u'\|^2$, it follows that:

$$\frac{1}{2}\|u'\|^2 + \varphi(u) + \gamma\int_0^t \|u'(\tau)\|^2 d\tau = E_0,$$

therefore

$$\int_0^t \|u'(\tau)\|^2 d\tau \leq \frac{E_0 - \inf \varphi}{\gamma} < +\infty.$$

Then $u' \in L^2(0,+\infty;H)$, and

$$\begin{aligned} h(t) &\leq \frac{3}{2\gamma}\int_0^t \|u'(\tau)\|^2 d\tau \\ &\leq \frac{3}{2\gamma}\int_0^\infty \|u'(\tau)\|^2 d\tau < +\infty. \end{aligned}$$

On the other hand, since $E(\cdot)$ is nonincreasing and bounded from below by $\inf \varphi$, it converges as $t \to +\infty$. If $\lim_{t\to+\infty} E(t) > \inf \varphi$, then $\lim_{t\to+\infty} \|u'(t)\| > 0$, because by (6.4), $\lim_{t\to+\infty} \varphi(u(t)) = \inf \varphi$, contradicting the fact that $u' \in L^2$. Therefore, $\lim_{t\to+\infty} E(t) = \inf \varphi$, hence $u'(t) \to 0$ as $t \to +\infty$.

Now we prove the weak convergence of $u(t)$ when $\mathrm{Argmin}\varphi \neq \emptyset$. To this end, we use Opial's lemma [OPI]. Since it follows from (6.3) that $u(t)$ is bounded, then there exists $\hat{x} \in H$ such that $u(t_k) \rightharpoonup \hat{x}$ for some sequence $t_k \to +\infty$. Since φ is convex and continuous, then it is lower semicontinuous. Hence

$$\varphi(\hat{x}) \leq \liminf_{k\to+\infty} \varphi(u(t_k)) = \lim_{t\to+\infty} \varphi(u(t)) = \inf \varphi,$$

which shows that $\hat{x} \in \mathrm{Argmin}\ \varphi$. By Opial's lemma, we only need to show that

$$\forall z \in \mathrm{Argmin}\varphi, \quad \lim_{t\to+\infty} \|u(t) - z\| \text{ exists.}$$

For this, fix $z \in \mathrm{Argmin}\varphi$ and define $v(t) := \frac{1}{2}\|u(t) - z\|^2$. The following lemma provides a sufficient condition on $[v']_+$, the positive part of the derivative, in order to ensure the convergence of v.

Lemma 6.2.2 *Let $\theta \in C^1([0,\infty);\mathbb{R})$ be bounded from below. If $[\theta']_+ \in L^1([0,\infty);\mathbb{R})$, then $\theta(t)$ converges as $t \to +\infty$.*

Proof. Set

$$w(t) := \theta(t) - \int_0^t [\theta'(\tau)]_+ d\tau$$

Since $w(t)$ is bounded from below and $w'(t) \leq 0$, then $w(t)$ converges as $t \to +\infty$, and consequently $\theta(t)$ converges as $t \to +\infty$. □

On account of this result, it is sufficient to prove that $[v']_+$ belongs to $L^1(0,\infty)$. Of course, to obtain information on v' we shall use the fact that $u(t)$ is a solution to (6.1). Due to the optimality of z, it follows from (6.2) that

$$v'' + \gamma v' \leq \|u'\|^2 \tag{6.5}$$

Lemma 6.2.3 *If $\omega \in C^1([0,\infty);\mathbb{R})$ satisfies the differential inequality*

$$\omega'(\tau) + \gamma\omega(\tau) \leq g(\tau), \forall \tau \in [0,\infty), \tag{6.6}$$

with $\gamma > 0$ and $g \in L^1([0,\infty);\mathbb{R})$, then $[\omega]_+ \in L^1([0,\infty);\mathbb{R})$.

Proof. We may assume that $g \geq 0$, for if not, we replace g by $|g|$. Multiplying both sides of (6.6) by $e^{\gamma\tau}$ and integrating on $[0,t]$, we get:

$$\omega(t) \leq e^{-\gamma t}\omega(0) + \int_0^t e^{-\gamma(t-\tau)}g(\tau)d\tau$$

Therefore

$$[\omega(t)]_+ \leq e^{-\gamma t}[\omega(0)]_+ + \int_0^t e^{-\gamma(t-\tau)}g(\tau)d\tau$$

and Fubini's theorem gives $\int_0^\infty \int_0^t e^{-\gamma(t-\tau)}g(\tau)d\tau dt = \frac{1}{\gamma}\int_0^\infty g(\tau)d\tau < +\infty$. □

Recall that we already proved $\|u'\|^2 \in L^1([0,\infty);\mathbb{R})$. Now the proof of the theorem is completed by using Lemma 6.2.3 and (6.5). □

Recently, the asymptotic behavior of (6.1) with nonconstant friction was studied in [MAY, BAL-MAY]. For other recent work and developments on the asymptotic behavior of various versions of (6.1) see [BEN-HAR, CAB-ENG-GAD, CAB-FRA].

REFERENCES

ALV. F. Alvarez, On the minimizing property of a second order dissipative system in Hilbert spaces, SIAM J. Control Optim. 38 (2000), 1102–1119.

ALV-ATT1. F. Alvarez and H. Attouch, Convergence and asymptotic stabilization for some damped hyperbolic equations with non-isolated equilibria, ESAIM Control Optim. Calc. Var. 6 (2001), 539–552.

ALV-ATT2. F. Alvarez and H. Attouch, An inertial proximal method for maximal monotone operators via discretization of a nonlinear oscillator with damping, Set-Valued Anal. 9 (2001), 3–11.

ATT-ALV1. H. Attouch and F. Alvarez, The heavy ball with friction dynamical system for convex constrained minimization problems, Optimization (Namur, 1998), 25–35, Lecture Notes in Econom. and Math. Systems, 481, Springer, Berlin, 2000.

ATT-GOU-RED. H. Attouch, X. Goudou and P. Redont, The heavy ball with friction method, I. The continuous dynamical system: Global exploration of the local minima of a real-valued function by asymptotic analysis of a dissipative dynamical system, Commun. Contemp. Math. 2 (2000), 1–34.

BAL-MAY. M. Balti and R. May, Asymptotic for the perturbed heavy ball system with vanishing damping term, Evol. Equ. Control Theory 6 (2017), 177–186.

BEN-HAR. I. Ben Hassen and A. Haraux, Convergence and decay estimates for a class of second order dissipative equations involving a non-negative potential energy, J. Funct. Anal. 260 (2011), 2933–2963.

CAB-ENG-GAD. A. Cabot, H. Engler and S. Gadat, On the long time behavior of second order differential equations with asymptotically small dissipation, Trans. Amer. Math. Soc. 361 (2009), 5983–6017.

CAB-FRA. A. Cabot and P. Frankel, Asymptotics for some semilinear hyperbolic equations with non-autonomous damping, J. Differential Equations, 252 (2012), 294–322.

MAY. R. May, Long time behavior for a semilinear hyperbolic equation with asymptotically vanishing damping term and convex potential, J. Math. Anal. Appl. 430 (2015), 410–416.

OPI. Z. Opial, Weak convergence of the sequence of successive approximations for nonexpasive mappings, Bull. Amer. Math. Soc. 73 (1967), 591–597.

Part III

Difference Equations of Monotone Type

7 First Order Difference Equations and Proximal Point Algorithm

7.1 INTRODUCTION

Consider the following first order evolution equation of monotone type

$$\begin{cases} -u'(t) \in Au(t) + f(t), \\ u(0) = u_0 \in \overline{D(A)} \end{cases} \tag{7.1}$$

that was studied in Chapter 4. By the backward Euler discretization

$$u'(t) = \frac{u(t) - u(t-h)}{h} + o(h),$$

in (7.1), we get the following difference inclusion

$$\begin{cases} u_{n-1} - u_n \in c_n A u_n + f_n \\ u(0) = u_0 \end{cases} \tag{7.2}$$

Using the resolvent operator, (7.2) can be written as follows:

$$\begin{cases} u_n = J_{c_n}^A (u_{n-1} - f_n) \\ u(0) = u_0. \end{cases} \tag{7.3}$$

In (7.2) and (7.3), we can take $u_0 \in H$, and not necessarily in $\overline{D(A)}$. The existence of the sequence $\{u_n\}$ in (7.2) or equivalently (7.3) follows from the surjectivity of the resolvent operator J_λ^A (see Theorem 3.3.2 of Chapter 3). In this chapter, we prove some results on the boundedness, mean convergence, as well as weak and strong convergence of solutions to (7.2), similar to the results of Chapter 4 for (7.1). The resolvent Equation (7.3) is called the proximal point algorithm, and it was first introduced by Rockafellar [ROC] to approximate a zero of a maximal monotone operator. Rockafellar [ROC] proved that when $A^{-1}(0) \neq \emptyset$, $\liminf_{n \to +\infty} c_n > 0$ and $\sum_{n=1}^{\infty} \|f_n\| < +\infty$, then the sequence $\{u_n\}$ converges weakly to a zero of the maximal monotone operator A. Brézis and Lions [BRE-LIO] and Lions [LIO] continued the study of the convergence of the proximal point algorithm, and proved the weak convergence of the mean and some results on the weak convergence of the sequence $\{u_n\}$. They proved that when $A^{-1}(0) \neq \emptyset$, $\sum_{n=1}^{\infty} c_n^2 = +\infty$ and $\sum_{n=1}^{\infty} \|f_n\| < +\infty$, the sequence $\{u_n\}$ converges weakly to a zero of A. Clearly this condition is weaker than the one used by Rockafellar. Before Rockafellar, the proximal point algorithm was studied by Martinet [MAR] for convex functions. If $f : H \to (-\infty, +\infty]$ is a proper, convex and lower semicontinuous function with $\text{Argmin} f \neq \emptyset$, by Theorem 2.4.3 there exists a sequence $\{u_n\}$ satisfying

$$\begin{cases} u_n = \text{Argmin}_{y \in H}\{f(y) + \frac{1}{2c_n}\|y - u_{n-1}\|^2\}, \ n \geq 1 \\ u_0 \in H. \end{cases} \tag{7.4}$$

Martinet proved that the sequence $\{u_n\}$ given by (7.4) converges weakly to a minimum point of f. It is easy to check that (7.4) is equivalent to (7.3) when the maximal monotone operator A is equal to ∂f. This chapter is devoted to the asymptotic behavior of the sequence $\{u_n\}$ given by (7.3). Section 2 describes sufficient conditions for the boundedness of the sequence $\{u_n\}$. In Section 3, we consider the periodic forcing case. In Section 4, we investigate the weak and strong convergence of the sequence $\{u_n\}$ generated by (7.2) when the error sequence $\{f_n\}$ is summable (i.e. $\sum_{n=1}^{\infty} \|f_n\| < +\infty$). The convergence analysis of the proximal point algorithm with a non-summable error sequence is the subject of Section 5. Finally, in Section 6 we study the rate of convergence. Throughout the chapter we denote $\frac{u_{n-1} - u_n - f_n}{c_n}$ by Au_n.

7.2 BOUNDEDNESS OF SOLUTIONS

It is a well-known result (see also Section 4 of this chapter) that "$\{u_n\}$ is bounded if and only if $A^{-1}(0) \neq \emptyset$ provided that $\sum_{n=1}^{\infty} c_n = +\infty$ and $\sum_{n=1}^{\infty} \|f_n\| < +\infty$". In this very short section, we prove that the coercivity condition on the operator A implies the boundedness of the sequence $\{u_n\}$ without assuming $A^{-1}(0) \neq \emptyset$. The following theorem is also proved by Boikanyo and Morosanu [BOI-MOR] with stronger conditions on $\{f_n\}$ and $\{c_n\}$.

Definition 7.2.1 *The (possibly multivalued) operator A in H is said to be coercive if there exists $y_0 \in H$ such that* $\lim_{\|z\| \to +\infty} \frac{(w, z - y_0)}{\|z - y_0\|} = +\infty, \forall [z, w] \in A$.

Theorem 7.2.2 *Let A be a coercive maximal monotone operator. If the sequence $\{\frac{\|f_n\|}{c_n}\}$ is bounded, then for each $u_0 \in H$, the sequence $\{u_n\}$ generated by (7.3) is bounded.*

Proof. Let $C > 0$ be such that for each $n \geq 1$, $\frac{\|f_n\|}{c_n} < C$. By coerciveness of A, there exist $K > 0$ and $y_0 \in H$ such that for all $[z, w] \in A$, with $\|z\| > K$, $\frac{(w, z - y_0)}{\|z - y_0\|} > C$. Suppose that there exists n such that $\|u_{n+1} - y_0\| > K$. From (7.2), we have

$$u_n - u_{n+1} - f_{n+1} \in c_{n+1} A u_{n+1}. \tag{7.5}$$

Multiplying both sides of (7.5) by $\frac{u_{n+1} - y_0}{\|u_{n+1} - y_0\|}$, we get

$$c_{n+1} C + \|u_{n+1} - y_0\| \leq \|u_n - y_0\| + \|f_{n+1}\|.$$

This implies that

$$\|u_{n+1}-y_0\| \leq \|u_n - y_0\| + c_{n+1}\left(\frac{\|f_{n+1}\|}{c_{n+1}} - C\right) < \|u_n - y_0\|,$$

for each $n \geq 0$ such that $\|u_{n+1}-y_0\| > K$. It follows that for all $n \geq 0$,

$$\|u_{n+1}-y_0\| \leq \max\{\|u_0-y_0\|, K\}.$$

Hence the sequence $\{u_n\}$ is bounded. □

7.3 PERIODIC FORCING

In this section we prove the discrete version of Theorem 4.3.3 for the first order evolution Equation (7.1). But as it is shown in the following, the discrete version holds true for general maximal monotone operators.

Theorem 7.3.1 *Suppose that A is a maximal monotone operator in H. If (7.2) has a solution $\{u_n\}_{n\geq 1}$ satisfying $\liminf_{n\to+\infty}\|s_n\| < +\infty$, where $s_n = \frac{1}{n}\sum_{k=0}^{n-1} u_{kN}$, and $\{f_n\}_{n\geq 1}$ and $\{c_n\}_{n\geq 1}$ are periodic with period N, then there exists an N-periodic solution $\{w_n\}_{n\geq 1}$ of (7.2) such that $u_n - w_n \rightharpoonup 0$ as $n \to +\infty$. Moreover, any two periodic solutions differ by an additive constant.*

Proof. Let $x \in H$, and let $m \geq 0$ be an integer. Since A is maximal monotone, there is a unique solution to:

$$\begin{cases} u_{i-1} - u_i \in c_i A u_i + f_i, & i > m \\ u_m = x. \end{cases} \tag{7.6}$$

For $n \geq m$, we define the operators $Q(n,m): H \to H$ by

$$Q(n,m)x := u_n.$$

Let v_n be a solution of (7.6) with $v_m = y$. The monotonicity of A implies that

$$(u_{n-1} - v_{n-1}, u_n - v_n) - \|u_n - v_n\|^2 \geq 0.$$

Therefore

$$\|u_n - v_n\| \leq \|u_{n-1} - v_{n-1}\|.$$

Hence by definition of $Q(n,m)$, it follows that $Q(n,m)$ is nonexpansive. The nonexpansiveness of $Q(n,m)$ also shows the uniqueness of the solution to (7.6) with $u_m = x$. By the uniqueness of the solution, we have

$$Q(n,m)Q(m,k) = Q(n,k)$$

for $n \geq m \geq k$, and by the periodicity, we also have:

$$Q(n+N, m+N) = Q(n,m)$$

for $n \geq m$. It follows that

$$Q(m+N,m)^n = Q(m+nN,m).$$

In particular, taking $m=0$, it follows that $\{u_{kN}\}_{k\geq 0}$ is a nonexpansive sequence in H. Therefore, by the ergodic theorem proved in [ROU], we deduce that s_n converges weakly in H, and the limit is a fixed point of $Q(N,0)$. This shows the existence of a periodic solution to (7.2), which is therefore bounded. Hence all solutions to (7.2) are bounded. Suppose that $\{u_n\}$ is a bounded solution of (7.2) with $u_0 = x$ and let $\{v_n\}$ be a periodic solution of (7.2) with $v_0 = y$. Let $z_n := u_n - v_n$. From (7.2) we have:

$$z_{n-1} - z_n \in c_n(Au_n - Av_n).$$

Multiplying the above inclusion by z_n, we get

$$(z_{n-1} - z_n, z_n) \geq 0.$$

Hence

$$\|z_n\|^2 \leq (z_{n-1}, z_n) \tag{7.7}$$

which implies that $\|z_n\|$ is non increasing, therefore it has a limit. We also get:

$$(z_{n-1} - z_n, z_n - z_{n-1}) + (z_{n-1} - z_n, z_{n-1}) \geq 0$$

Therefore

$$\|z_n - z_{n-1}\|^2 \leq (z_{n-1} - z_n, z_{n-1})$$

Summing up the above inequality from $n=1$ to m and letting $m \to +\infty$, and using (7.7), we get that $\lim_{n\to+\infty} \|z_n\|$ exists, and:

$$\sum_{n=1}^{+\infty} \|z_n - z_{n-1}\|^2 \leq \sum_{n=1}^{+\infty} (z_{n-1} - z_n, z_{n-1}) \tag{7.8}$$

$$= \sum_{n=1}^{+\infty} (\|z_{n-1}\|^2 - (z_n, z_{n-1})) \tag{7.9}$$

$$\leq \sum_{n=1}^{+\infty} (\|z_{n-1}\|^2 - \|z_n\|^2) = \|z_0\|^2 - \lim_{n\to+\infty} \|z_n\|^2 < +\infty. \tag{7.10}$$

Since $\{v_n\}$ is a periodic solution of (7.6), for each $m \geq 0$, we have

$$u_{m+nN} - u_{m+(n+1)N} = \sum_{i=nN}^{(n+1)N-1} (u_{m+i} - v_{m+i} - (u_{m+i+1} - v_{m+i+1})) \to 0 \tag{7.11}$$

as $n \to +\infty$. Let

$$x_n := Q(m+nN,0)u_0 = u_{m+nN}$$

Then $\{x_n\}_{n\geq 1}$ is a nonexpansive sequence, which is asymptotically regular (i.e. $x_{n+1} - x_n \to 0$, as $n \to +\infty$), by (7.11). It follows from Theorem 5.7.6 of Chapter 4 that:

$$u_{m+nN} \rightharpoonup w_m \tag{7.12}$$

as $n \to +\infty$. Since $\{v_n\}_{n\geq 1}$ is a periodic solution of (7.2), we have

$$\begin{aligned}
&\lim_{n\to+\infty}(u_{m+nN} - u_{m-1+nN} - (v_m - v_{m-1}))\\
=&\lim_{n\to+\infty}(u_{m+nN} - v_{m+nN} - (u_{m-1+nN} - v_{m-1+nN}))\\
=&0
\end{aligned}$$

Therefore, $\lim_{n\to+\infty}(u_{m+nN} - u_{m-1+nN}) = (v_m - v_{m-1})$ exists, and from (7.12), we get:

$$\lim_{n\to+\infty}(u_{m+nN} - u_{m-1+nN}) = w_m - w_{m-1}$$

This implies that $w_m - v_m = w_0 - v_0 =$ constant, showing that any two periodic solutions differ by an additive constant. By (7.2), we have $[u_{m+nN}, \frac{1}{c_m}(u_{m-1+nN} - u_{m+nN} - f_m)] \in A$. Since by Theorem 3.3.4, A is demiclosed, by letting $n \to +\infty$, we get

$$[w_m, \frac{1}{c_m}(w_{m-1} - w_m - f_m)] \in A$$

Therefore $w_m \in D(A)$, and $w_{m-1} - w_m - f_m \in c_m A w_m$. Since w_n is N-periodic

$$\begin{aligned}
u_n - w_n &= u_n - u_{k+mN} + u_{k+mN} - w_{k+mN} + w_{k+mN} - w_n\\
&= u_n - w_n - (u_{k+mN} - w_{k+mN}) + u_{k+mN} - w_k\\
&= z_n - z_{k+mN} + u_{k+mN} - w_k
\end{aligned}$$

where $k + mN \leq n < k + (m+1)N$ and $z_n = u_n - w_n$.
Now from (7.12), we have $u_{k+mN} - w_k \rightharpoonup 0$ as $m \to +\infty$, and from (7.8), we get:

$$\|z_n - z_{k+mN}\| \leq \sum_{i=k+mN}^{i=k+(m+1)N-1} \|z_i - z_{i-1}\| \to 0$$

as $n \to +\infty$. This shows that $u_n - w_n \rightharpoonup 0$, as $n \to +\infty$, and completes the proof of the theorem. □

7.4 CONVERGENCE OF THE PROXIMAL POINT ALGORITHM

As was mentioned in the introduction, the first convergence result for the proximal point algorithm was proved by Rockafellar. Then Brézis and Lions [BRE-LIO] and Lions [LIO] improved his result. All the previous authors studied the convergence of (7.3) by assuming that $A^{-1}(0) \neq \emptyset$. In this chapter, we prove weak and strong convergence results without assuming $A^{-1}(0) \neq \emptyset$. First we recall an elementary lemma without proof.

Lemma 7.4.1 *Suppose that $\{a_n\}_{n\geq 1}$ and $\{b_n\}_{n\geq 1}$ are nonnegative real sequences and $\sum_{n=1}^{+\infty} b_n < +\infty$. If*

$$a_{n+1} \leq a_n + b_n, \text{ for all } n \geq 1,$$

then there exists $\lim_{n\to+\infty} a_n$.

We start with a weak ergodic theorem which was proved by Lions (see [LIO]), by assuming that $A^{-1}(0) \neq \phi$ (see also [MOR], Theorem 3.1, p.139). We denote: $w_n := (\sum_{k=1}^{n} c_k)^{-1} \sum_{k=1}^{n} c_k u_k$.

Theorem 7.4.2 *Assume that $\{u_n\}_{n\geq 1}$ given by (7.2) is bounded, $\sum_{n=1}^{+\infty} c_n = +\infty$ and $\sum_{n=1}^{+\infty} \|f_n\| < +\infty$. Then the sequence $\{w_n\}$ converges weakly as $n \to +\infty$ to an element $p \in A^{-1}(0)$, which is also the asymptotic center of $\{u_n\}$.*

Proof. By the monotonicity of A, we have

$$(Au_n, u_k) + (Au_k, u_n) \leq (Au_n, u_n) + (Au_k, u_k), \text{ for all } k, n \geq 1.$$

Multiplying both sides of the above inequality by $c_k c_n$ and using (7.2), we get

$$(u_{n-1} - u_n - f_n, c_k u_k) + (u_{k-1} - u_k - f_k, c_n u_n)$$

$$\leq c_k(u_{n-1} - u_n - f_n, u_n) + c_n(u_{k-1} - u_k - f_k, u_k).$$

Assume that $w_{m_j} \rightharpoonup p$ as $j \to +\infty$. Summing up both sides of the above inequality from $k = 1$ to $k = m_j$, dividing by $\sum_{k=1}^{m_j} c_k$, and letting $j \to +\infty$, we get

$$(u_{n-1} - u_n - f_n, u_n - p) \geq 0 \tag{7.13}$$

This implies that

$$\|u_n - p\| \leq \|u_{n-1} - p\| + \|f_n\|.$$

By Lemma 7.4.1, there exists $\lim_{n\to+\infty} \|u_n - p\|$. If q is another cluster point of $\{w_n\}_{n\geq 1}$, $\lim_{n\to+\infty} \|u_n - q\|$ exists too. It follows that there exists $\lim_{n\to+\infty} \frac{1}{2}(\|u_n - p\|^2 - \|u_n - q\|^2)$ and hence $\lim_{n\to+\infty}(u_n, p - q)$ exists. Therefore $\lim_{n\to+\infty}(w_n, p - q)$ exists. This implies that $(q, p - q) = (p, p - q)$. Hence $p = q$ and therefore $w_n \rightharpoonup p \in H$ as $n \to +\infty$. Now suppose that $[x, y] \in A$. By the monotonicity of A, we have

$$\begin{aligned}
(y, x - w_n) &= (\sum_{k=1}^{n} c_k)^{-1} \sum_{k=1}^{n} c_k (y, x - u_k) \\
&= (\sum_{k=1}^{n} c_k)^{-1} \sum_{k=1}^{n} c_k [(y - Au_k, x - u_k) + (Au_k, x - u_k)] \\
&\geq (\sum_{k=1}^{n} c_k)^{-1} \sum_{k=1}^{n} (u_{k-1} - u_k - f_k, x - u_k) \\
&\geq (\sum_{k=1}^{n} c_k)^{-1} \sum_{k=1}^{n} (\frac{1}{2}|u_k - x|^2 - \frac{1}{2}|u_{k-1} - x|^2 - |f_k||u_k - x|)
\end{aligned}$$

$$\geq \frac{-(\sum_{k=1}^{n} c_k)^{-1}}{2}|u_0 - x|^2 - (\sum_{k=1}^{n} c_k)^{-1} \sum_{k=1}^{n} |f_k||u_k - x|.$$

Letting $n \to +\infty$, it follows that

$$(y, x - p) \geq 0.$$

Then, by the maximality of A, we have $0 \in A(p)$ as desired. For $x \in H$, we have

$$\|u_n - p\|^2 = \|u_n - x\|^2 - \|x\|^2 + \|p\|^2 - 2(u_n, p - x).$$

Multiplying both sides of this equality by c_n, summing up from $n = 1$ to $n = m$, dividing by $\sum_{n=1}^{m} c_n$, and letting $m \to +\infty$, we obtain

$$\begin{aligned}\lim_{n \to +\infty} \|u_n - p\|^2 &= \limsup_{n \to +\infty} \|u_n - x\|^2 - \|x - p\|^2 \\ &< \limsup_{n \to +\infty} \|u_n - x\|^2, \text{ if } x \neq p.\end{aligned}$$

Hence p is the asymptotic center of the sequence $\{u_n\}$, and the proof of the theorem is now complete. □

Remark 7.4.3 *The proof of Theorem 7.4.2 shows that for $f_n \equiv 0$ and $\{u_n\}$ given by (7.2), the sequence $\{w_n\}$ actually converges weakly as $n \to +\infty$ to some $p \in A^{-1}(0)$, if and only if* $\liminf_{n \to +\infty} \|w_n\| < +\infty$ *(without assuming the boundedness of $\{u_n\}$). Therefore in this case, $\{u_n\}$ is bounded if and only if* $\liminf_{n \to +\infty} \|w_n\| < +\infty$.

Problem 7.4.4 *However, in the nonhomogeneous case where $f_n \neq 0$, we do not know whether in Theorem 7.4.2, the boundedness of $\{u_n\}$ can be replaced by* $\liminf_{n \to +\infty} \|w_n\| < +\infty$.

Now we show the weak convergence of $\{u_n\}$.

Remark 7.4.5 *By Brézis and Lions (see [BRE-LIO], Remark 14, p. 344), the weak (resp. strong) convergence of $\{u_n\}$ given by (7.2) in the homogeneous case where $f_n \equiv 0$ implies the weak (resp. strong) convergence of $\{u_n\}$ in the nonhomogeneous case $f_n \neq 0$, if $\sum_{n=1}^{+\infty} \|f_n\| < +\infty$. Therefore for the study of weak (resp. strong) convergence of $\{u_n\}$, we may assume without loss of generality that $f_n \equiv 0$.*

Theorem 7.4.6 *Assume that $\{u_n\}_{n \geq 1}$ is given by (7.2), $\sum_{n=1}^{+\infty} c_n^2 = +\infty$ and $\sum_{n=1}^{+\infty} \|f_n\| < +\infty$. Then $\{u_n\}$ converges weakly as $n \to +\infty$ to an element $p \in A^{-1}(0)$, which is also the asymptotic center of $\{u_n\}$, if and only if* $\liminf_{n \to +\infty} \|w_n\| < +\infty$.

Proof. Necessity being obvious, let's prove the sufficiency. By Remark 7.4.5, we assume without loss of generality that $f_n \equiv 0$. From (7.2) we have $u_{n-1} - p \in u_n - p + c_n A u_n$ where p is the weak limit of $\{w_n\}$, which exists by Remark 7.4.3, since the assumption on $\{c_n\}_{n \geq 1}$ implies that $\sum_{n=1}^{+\infty} c_n = +\infty$. Squaring this inclusion and taking into account (7.2) and (7.13) we obtain

$$c_n^2\|Au_n\|^2 \leq \|u_{n-1}-p\|^2 - \|u_n - p\|^2.$$

Summing up this inequality from $n = 1$ to m and letting $m \to +\infty$, we get

$$\sum_{n=1}^{+\infty} c_n^2\|Au_n\|^2 \leq \|u_0 - p\|^2 - \lim_{n\to+\infty} \|u_n - p\|^2 < +\infty$$

since by Theorem 7.4.2, $\lim_{n\to+\infty} \|u_n - p\|^2$ exists. By assumption on $\{c_n\}$ we deduce that $\liminf_{n\to+\infty} \|Au_n\|^2 = 0$. On the other hand, since A is monotone, we have

$$(Au_{n-1} - Au_n, u_{n-1} - u_n) \geq 0.$$

From (7.2) when $f_n \equiv 0$, we get

$$(Au_{n-1} - Au_n, c_nAu_n) \geq 0.$$

Hence

$$\|Au_n\|^2 \leq \|Au_{n-1}\|^2.$$

Then $\lim_{n\to+\infty} \|Au_n\| = 0$. Now assume that $u_{n_j} \rightharpoonup q$ as $j \to +\infty$. Then by the monotonicity of A, we have

$$(Au_n - Au_{n_j}, u_n - u_{n_j}) \geq 0.$$

Hence we get

$$(Au_n, u_n - q) \geq 0.$$

From (7.2) when $f_n \equiv 0$, we deduce that

$$\|u_n - q\| \leq \|u_{n-1} - q\|.$$

Thus $\lim_{n\to+\infty} \|u_n - q\|^2$ exists, and by a similar proof as in Theorem 7.4.2, we conclude that: $u_n \rightharpoonup p$ as $n \to +\infty$. □

In our next two theorems, we show the strong convergence of $\{u_n\}$.

Theorem 7.4.7 *Assume that $(I+A)^{-1}$ is compact, $\sum_{n=1}^{+\infty} c_n^2 = +\infty$ and $\sum_{n=1}^{+\infty} \|f_n\| < +\infty$. Then the sequence $\{u_n\}$ given by (7.2) converges strongly as $n \to +\infty$ to an element $p \in A^{-1}(0)$, which is also the asymptotic center of $\{u_n\}$, if and only if $\liminf_{n\to+\infty} \|w_n\| < +\infty$.*

Proof. Necessity being obvious, we prove the sufficiency. Again by Remark 7.4.3, we assume without loss of generality that $f_n \equiv 0$. From the proof of Theorem 7.4.6 it follows that $\lim_{n\to+\infty} \|Au_n\| = 0$, and $u_n \rightharpoonup p$ as $n \to +\infty$. Therefore the sequence $\{u_n + Au_n\}$ is bounded, and hence the compacity of $(I+A)^{-1}$ implies that $\{u_n\}$ has a strongly convergent subsequence $\{u_{n_j}\}$ to p. Now, by the monotonicity of A, we have

$$(Au_n - Au_{n_j}, u_n - u_{n_j}) \geq 0, \ \ \forall\, n \geq 1.$$

Letting $j \to +\infty$, we get

$$(Au_n, u_n - p) \geq 0, \quad \forall\, n \geq 1.$$

Now the proof of Theorem 7.4.6 shows that $\lim_{n\to+\infty} \|u_n - p\|^2$ exists, which implies that $u_n \to p$ as $n \to +\infty$, as desired. □

Theorem 7.4.8 *Assume that A is strongly monotone, $\sum_{n=1}^{+\infty} c_n = +\infty$ and $\sum_{n=1}^{+\infty} \|f_n\| < +\infty$. Then the sequence $\{u_n\}$ given by (7.2) converges strongly as $n \to +\infty$ to an element $p \in A^{-1}(0)$, which is also the asymptotic center of $\{u_n\}$, if and only if $\liminf_{n\to+\infty} \|w_n\| < +\infty$.*

Proof. Necessity being obvious, we prove the sufficiency. Again by Remark 7.4.5, we assume without loss of generality that $f_n \equiv 0$. By the proof of Theorem 7.4.2 (see also Remark 7.4.3), we know that $\{w_n\}$ converges weakly as $n \to +\infty$ to $p \in A^{-1}(0)$, and $\lim_{n\to+\infty} \|u_n - p\|^2$ exists. Since A is strongly monotone and $\{u_n\}$ is given by (7.2), we have

$$\begin{aligned}\alpha c_n \|u_n - p\|^2 &\leq (u_{n-1} - u_n, u_n - p) \\ &\leq \frac{1}{2}\|u_{n-1} - p\|^2 - \frac{1}{2}\|u_n - p\|^2.\end{aligned}$$

Summing up both sides of this inequality from $n = 1$ to m, and letting $m \to +\infty$, we get

$$\sum_{n=1}^{+\infty} c_n \|u_n - p\|^2 < +\infty.$$

Since $\sum_{n=1}^{+\infty} c_n = +\infty$, this implies that:

$$\liminf_{n\to+\infty} \|u_n - p\|^2 = 0,$$

and hence $u_n \to p$ as $n \to +\infty$, as desired. □

7.5 CONVERGENCE WITH NON-SUMMABLE ERRORS

Suppose that u_n is a sequence given by (7.3). By taking $x_{n-1} = u_{n-1} - f_n$ and $e_n = f_{n+1}$ in (7.3), we get

$$x_n = (I + c_n A)^{-1} x_{n-1} + e_n$$

Rockafellar proposed the following open problem: "does the sequence x_n converge weakly to a zero of A if the error sequence $\{e_n\}$ is non-summable?" In this section, we give a partial affirmative answer to this problem and study the convergence of the sequence $\{u_n\}$ in (7.3), without summability assumptions on the error sequence $\{f_n\}$. First we prove an elementary lemma.

Lemma 7.5.1 *Suppose that $\{a_n\}$ and $\{b_n\}$ are positive sequences such that $\sum_{n=1}^{+\infty} b_n = +\infty$ and $\lim_{n\to+\infty} \frac{a_n}{b_n} = 0$, then $\lim_{m\to+\infty} \frac{\sum_{n=1}^{m} a_n}{\sum_{n=1}^{m} b_n} = 0$.*

Proof. Let $\varepsilon > 0$. By the assumption there is $n_0 > 0$ such that for each $n \geq n_0$, $a_n < \varepsilon b_n$. Now

$$\frac{\sum_{n=1}^{m} a_n}{\sum_{n=1}^{m} b_n} < \frac{\sum_{n=1}^{n_0} a_n + \varepsilon \sum_{n=n_0}^{m} b_n}{\sum_{n=1}^{m} b_n}$$
$$< \frac{\sum_{n=1}^{n_0} a_n}{\sum_{n=1}^{m} b_n} + \varepsilon \to \varepsilon$$

as $m \to +\infty$. The result follows now since $\varepsilon > 0$ is arbitrary. □

Theorem 7.5.2 *Let $\{u_n\}$ be a bounded sequence generated by (7.3). If $\frac{\|f_n\|}{c_n} \to 0$ as $n \to +\infty$ and $\sum_{n=1}^{+\infty} c_n = +\infty$, then $A^{-1}(0) \neq \phi$. Moreover, every weak cluster point of w_n belongs to $A^{-1}(0)$.*

Proof. Suppose that $y \in A(x)$. By the monotonicity of A, we have

$$\begin{aligned}
(y, x - w_k) &= (\sum_{n=1}^{k} c_n)^{-1} \sum_{n=1}^{k} c_n (y, x - u_n) \\
&= (\sum_{n=1}^{k} c_n)^{-1} \sum_{n=1}^{k} c_n [(y - Au_n, x - u_n) + (Au_n, x - u_n)] \\
&\geq (\sum_{n=1}^{k} c_n)^{-1} \sum_{n=1}^{k} (u_{n-1} - u_n - f_n, x - u_n) \\
&\geq (\sum_{n=1}^{k} c_n)^{-1} \sum_{n=1}^{k} (\frac{1}{2}\|u_n - x\|^2 - \frac{1}{2}\|u_{n-1} - x\|^2 - \|f_n\|\|u_n - x\|) \\
&\geq \frac{-(\sum_{n=1}^{k} c_n)^{-1}}{2}\|u_0 - x\|^2 - (\sum_{n=1}^{k} c_n)^{-1} \sum_{n=1}^{k} \|f_n\|\|u_n - x\|.
\end{aligned}$$

Since $\{w_k\}$ is bounded, it has a subsequence $\{w_{k_j}\}$ such that $w_{k_j} \rightharpoonup p \in H$ as $j \to +\infty$; substituting k by k_j in the above inequality and letting $j \to +\infty$, by Lemma 7.5.1 and our assumptions, we get

$$(y, x - p) \geq 0.$$

Then, by the maximality of A, we have $p \in A^{-1}(0)$ as desired. □

In the following two theorems, we show that the weak $\omega-$limit set of the bounded sequence $\{u_n\}$ generated by (7.3) is a subset of $A^{-1}(0)$.

Theorem 7.5.3 *Let $\{u_n\}$ be a bounded sequence generated by (7.3) and $\sum_{n=1}^{+\infty} c_n^2 = +\infty$. If $\sum_{n=1}^{\infty} \frac{\|f_n\|^2}{c_n^2} < +\infty$, and $\frac{\|f_n\|}{c_n^2} \to 0$ as $n \to +\infty$, then $\omega_w(u_n) \subset A^{-1}(0)$.*

Proof. By the monotonicity of A and (7.3), we have

$$(Au_{n-1}-Au_n, Au_n+\frac{f_n}{c_n})\geq 0.$$

This implies that

$$\begin{aligned}\|Au_n\|^2 &\leq (Au_{n-1},Au_n)+(Au_{n-1}-Au_n,\frac{f_n}{c_n})\\ &\leq \frac{1}{2}\|Au_{n-1}\|^2+\frac{1}{2}\|Au_n\|^2-\frac{1}{2}\|Au_n-Au_{n-1}\|^2+\frac{1}{2}\|Au_{n-1}-Au_n\|^2\\ &\quad+\frac{1}{2}\frac{\|f_n\|^2}{c_n^2}.\end{aligned}$$

Then

$$\|Au_n\|^2\leq \|Au_{n-1}\|^2+\frac{\|f_n\|^2}{c_n^2}. \tag{7.14}$$

Since $\sum_{n=1}^{\infty}\frac{\|f_n\|^2}{c_n^2}<+\infty$, there exists $\lim_{n\to+\infty}\|Au_n\|=l$. Let $L=\sup_{n\geq 1}\|Au_n\|$. Theorem 7.5.2 shows that $A^{-1}(0)\neq\phi$. Let $p\in A^{-1}(0)$. By the monotonicity of A and (7.3), we get

$$(u_{n-1}-u_n-f_n, u_n-p)\geq 0.$$

This implies that

$$\|u_n-u_{n-1}\|^2\leq \|u_{n-1}-p\|^2-\|u_n-p\|^2+M\|f_n\|,$$

where $M:=2\sup_{n\geq 1}\|u_n-p\|$. Summing up from $n=1$ to k, by (7.3), we get

$$\sum_{n=1}^{k}c_n^2\|Au_n+\frac{f_n}{c_n}\|^2\leq \|u_0-p\|^2-\|u_k-p\|^2+M\sum_{n=1}^{k}\|f_n\|.$$

It follows that

$$\sum_{n=1}^{k}c_n^2(\|Au_n\|^2+\frac{\|f_n\|^2}{c_n^2}-2\|Au_n\|\frac{\|f_n\|}{c_n})\leq \|u_0-p\|^2-\|u_k-p\|^2+M\sum_{n=1}^{k}\|f_n\|.$$

Hence

$$\sum_{n=1}^{k}c_n^2\|Au_n\|^2\leq 2L\sum_{n=1}^{k}c_n\|f_n\|+\|u_0-p\|^2-\|x_k-p\|^2+M\sum_{n=1}^{k}\|f_n\|.$$

By (7.14), we get

$$\|Au_k\|^2\leq \|Au_n\|^2+\sum_{i=n+1}^{k}\frac{\|f_i\|^2}{c_i^2}.$$

Then

$$\|Au_k\|^2 \sum_{n=1}^{k} c_n^2 \leq \sum_{n=1}^{k} c_n^2 \sum_{i=n+1}^{k} \frac{\|f_i\|^2}{c_i^2} + 2L\sum_{n=1}^{k} c_n\|f_n\| + \|u_0 - p\|^2 - \|u_k - p\|^2$$
$$+ M\sum_{n=1}^{k} \|f_n\|.$$

From the assumptions and Lemma 7.5.1, it follows that $Au_k \to 0$ as $k \to +\infty$. If $u_{n_k} \rightharpoonup q$ as $k \to +\infty$, then the demiclosedness of A implies that $q \in A^{-1}(0)$. □

Theorem 7.5.4 *Let $\{u_n\}$ be a bounded sequence generated by (7.3) and $A = \partial\varphi$, where $\varphi : H \to (-\infty, +\infty]$ is a proper, convex and lower semicontinuous function. If $\sum_{n=1}^{\infty} c_n = +\infty$, $\sum_{n=1}^{\infty} \frac{\|f_n\|^2}{c_n} < \infty$ and $\frac{\|f_n\|}{c_n} \to 0$ as $n \to +\infty$, then $\omega_w(u_n) \subset A^{-1}(0)$.*

Proof. By the subdifferential inequality and (7.3), we get

$$\begin{aligned} c_i(\varphi(u_i) - \varphi(u_{i-1})) &\leq c_i(\partial\varphi(u_i), u_i - u_{i-1}) \\ &= (u_{i-1} - u_i - f_i, u_i - u_{i-1}) \\ &\leq -\|u_i - u_{i-1}\|^2 + \frac{1}{2}\|u_i - u_{i-1}\|^2 + \frac{1}{2}\|f_i\|^2 \\ &\leq \frac{1}{2}\|f_i\|^2. \end{aligned} \tag{7.15}$$

Dividing both sides of (7.15) by c_i and summing up from $i = n+1$ to k, we obtain

$$\varphi(u_k) \leq \varphi(u_n) + \frac{1}{2}\sum_{i=n+1}^{k} \frac{\|f_i\|^2}{c_i}. \tag{7.16}$$

By Theorem 7.5.2, we know that $A^{-1}(0) \neq \phi$. Let $p \in A^{-1}(0)$. By the subdifferential inequality and (7.3), we have

$$\begin{aligned} c_n(\varphi(u_n) - \varphi(p)) &\leq (u_{n-1} - u_n - f_n, u_n - p) \\ &\leq (\frac{1}{2}\|u_{n-1} - p\|^2 - \frac{1}{2}\|u_n - p\|^2) + M\|f_n\|, \end{aligned} \tag{7.17}$$

where $M := \sup_{n\geq 1} \|u_n - p\|$. Summing up from $n = 1$ to k, we obtain

$$\sum_{n=1}^{k} c_n(\varphi(u_n) - \varphi(p)) \leq \frac{1}{2}\|u_0 - p\|^2 - \frac{1}{2}\|u_k - p\|^2 + M\sum_{n=1}^{k} \|f_n\|.$$

From (7.16), we get:

$$(\varphi(u_k)-\varphi(p))\sum_{n=1}^{k}c_n \leq \frac{1}{2}\sum_{n=1}^{k}c_n\sum_{i=n+1}^{k}\frac{\|f_i\|^2}{c_i}+\frac{1}{2}\|u_0-p\|^2-\frac{1}{2}\|u_k-p\|^2$$
$$+M\sum_{n=1}^{k}\|f_n\|.$$

From the assumptions and Lemma 7.5.1, it follows that: $\varphi(u_k)-\varphi(p)\to 0$ as $k\to+\infty$. If $u_{k_j}\rightharpoonup q$ as $j\to+\infty$, then $\varphi(q)\leq\liminf_{j\to+\infty}\varphi(u_{k_j})=\varphi(p)$. Hence $q\in A^{-1}(0)$. □

Remark 7.5.5 *Theorems 7.5.3 and 7.5.4 show that if $A^{-1}(0)$ is a singleton (which happens if, for example, A is strictly monotone in Theorem 7.5.3 or φ is strictly convex in Theorem 7.5.4), then $u_n\rightharpoonup p$, where p is the unique element of $A^{-1}(0)$.*

Remark 7.5.6 *Although Theorems 7.5.3 and 7.5.4 do not imply the weak convergence of u_n to $p\in A^{-1}(0)$ unless when A is strictly monotone or φ is strictly convex, they improve the errors in the proximal point algorithm. They show that the error sequence $\{\|f_n\|\}$ can go to infinity, provided that a suitable assumption on c_n holds.*

Lemma 7.5.7 *Assume that $\{y_n\}$ is a positive real sequence satisfying the following inequality:*

$$b_ny_n\leq y_{n-1}-y_n+a_n, \tag{7.18}$$

where $\{b_n\}$ and $\{a_n\}$ are positive sequences; then the following hold:
i) If $\{\frac{a_n}{b_n}\}$ is bounded, then the sequence $\{y_n\}$ is bounded.
ii) If $\lim_{n\to+\infty}\frac{a_n}{b_n}=0$, then there exists $\lim_{n\to+\infty}y_n$.
iii) If $\lim_{n\to+\infty}\frac{a_n}{b_n}=0$ and $\sum_{n=1}^{+\infty}b_n=+\infty$, then $y_n\to 0$ as $n\to+\infty$.

Proof. i) First we prove the boundedness of $\{y_n\}$. There exists B such that $\frac{a_n}{b_n}\leq B$, for all $n\geq 1$. If $y_n>B$, then

$$y_n\leq y_{n-1}+b_n(\frac{a_n}{b_n}-y_n)<y_{n-1}+b_n(\frac{a_n}{b_n}-B)\leq y_{n-1}.$$

It follows that: $y_n\leq max\{y_0,B\}$.
ii) For each $\varepsilon>0$ there is $m_0>0$ such that for each $m\geq m_0$, $\frac{a_m}{b_m}<\varepsilon$. By the above argument for all $k>m\geq m_0$, we have

$$y_k\leq max\{y_{k-1},\frac{a_k}{b_k}\}\leq\cdots\leq max\{y_m,\frac{a_{m+1}}{b_{m+1}},\cdots,\frac{a_k}{b_k}\}\leq max\{y_m,\varepsilon\}\leq y_m+\varepsilon.$$

Then, there exists $\lim_{n\to+\infty}y_n$.
iii) We divide both sides of (7.18) by b_n and take liminf as $n\to+\infty$. It suffices to show that $\liminf_{n\to+\infty}\frac{1}{b_n}(y_{n-1}-y_n)\leq 0$. Suppose to the contrary that

$$\liminf_{n\to+\infty}\frac{1}{b_n}(y_{n-1}-y_n)>\lambda>0.$$

Then, there exists n_0 such that for each $n \geq n_0$, $\frac{1}{b_n}(y_{n-1} - y_n) > \lambda$. Multiplying both sides of this inequality by b_n, summing up from $n = n_0$ to m and letting $m \to \infty$, we get a contradiction. □

Theorem 7.5.8 *Let $\{u_n\}$ be the sequence generated by* (7.3) *and A be a maximal monotone and strongly monotone operator. If $\sum_{n=1}^{\infty} c_n = +\infty$ and $\frac{|f_n|}{c_n} \to 0$ as $n \to +\infty$, then $u_n \to p$ as $n \to +\infty$, where p is the unique element of $A^{-1}(0)$.*

Proof. Theorem 7.2.2 implies the boundedness of $\{u_n\}$. Now, by Theorem 7.5.2, $A^{-1}(0)$ is nonempty. Suppose that p is the single element of $A^{-1}(0)$. By the strong monotonicity of A and (7.3), we get

$$(u_{n-1} - u_n - f_n, u_n - p) \geq \alpha c_n \|u_n - p\|^2.$$

It follows that

$$2\alpha c_n \|u_n - p\|^2 \leq \|u_{n-1} - p\|^2 - \|u_n - p\|^2 + 2M\|f_n\|,$$

where $M := \sup_{n\geq 1} \|u_n - p\|$. The theorem is now proved by using the assumptions and Lemma 7.5.7(iii). □

7.6 RATE OF CONVERGENCE

Güler [GUL] computed the rate of convergence of $\varphi(u_n)$ to $\varphi(p)$ as $n \to +\infty$, where $\{u_n\}$ is generated by (7.3) with $A = \partial\varphi$ and p is a minimum point of φ, provided that $u_n \to p$ as $n \to +\infty$. In this section, we give a simple proof to Güler's result without assuming that $u_n \to p$, which does not always occur as it has been shown by Güler [GUL]. First we prove the following elementary lemma.

Lemma 7.6.1 *Suppose that $\{a_n\}$ and $\{b_n\}$ are two positive real sequences such that $\{a_n\}$ is nonincreasing and convergent to zero, and $\sum_{n=1}^{+\infty} a_n b_n < +\infty$; then $(\sum_{k=1}^{n} b_k) a_n \to 0$ as $n \to +\infty$.*

Proof. By the assumptions on $\{a_n\}$ and $\{b_n\}$, for each $m > k$, we have:

$$a_m(\sum_{n=1}^{m} b_n) \leq a_m(\sum_{n=1}^{k} b_n) + \sum_{n=k+1}^{m} a_n b_n.$$

Taking limsup as $m \to +\infty$, we get:

$$\limsup_{m\to+\infty} a_m(\sum_{n=1}^{m} b_n) \leq \sum_{n=k+1}^{+\infty} a_n b_n.$$

The lemma is now proved by letting $k \to +\infty$. □

Theorem 7.6.2 *Suppose that $\{u_n\}$ is a bounded sequence generated by* (7.3) *with $f_n \equiv 0$ and $A = \partial\varphi$, where φ is a proper, convex and lower semicontinuous function. If $\sum_{n=1}^{+\infty} c_n = +\infty$, then $\varphi(u_n) - \varphi(p) = o((\sum_{i=1}^{n} c_i)^{-1})$, where p is a minimum point of φ.*

Proof. Summing up (7.17) from $n = 1$ to k, we get:

$$\sum_{n=1}^{k} c_n(\varphi(u_n) - \varphi(p)) \leq \frac{1}{2}\|u_0 - p\|^2.$$

By the proof of Theorem 7.5.4, $\varphi(u_n) - \varphi(p)$ is nonincreasing and convergent to 0. Now, the theorem follows by using Lemma 7.6.1. □

REFERENCES

BOI-MOR. O.A. Boikanyo, G. Morosanu, Modified Rockafellar's algorithm, Math. Sci. Res. J. 13 (2009), 101–122.

BRE-LIO. H. Brézis and P. L. Lions, Produits infinis de *ré*solvantes, Israel J. Math. 29 (1978), 329–345.

ROU. B. Djafari Rouhani, Ergodic theorems for nonexpansive sequences in Hilbert spaces and related problems, Ph.D. Thesis, Yale University, Part I (1981), pp. 1–76.

ROU-KHA1. B. Djafari Rouhani and H. Khatibzadeh, On the proximal point algorithm, J. Optim. Theory Appl. 137 (2008), 411–417.

ROU-KHA2. B. Djafari Rouhani and H. Khatibzadeh, Existence and asymptotic behaviour of solutions to first and second-order difference equations with periodic forcing. J. Difference Equ. Appl. 18 (2012), 1593–1606.

GUL. O. Güler, On the convergence of the proximal point algorithm for convex minimization. SIAM J. Control Optim. 29 (1991), 403–419.

KHA. H. Khatibzadeh, Some remarks on the proximal point algorithm. J. Optim. Theory Appl. 153 (2012), 769–778.

LIO. P.L. Lions, Une méthode itérative de résolution d'une inéquation variationnelle, Israel J. Math. 31 (1978), 204–208.

MOR. G. Morosanu, 'Nonlinear Evolution Equations and Applications', Editura Academiei Romane (and D. Reidel publishing Company), Bucharest, 1988.

MAR. B. Martinet, Régularisation d'inéquations variationnelles par approximations successives. Rev. Française Informat. Recherche Opérationnelle 4 (1970), 154–158.

ROC. R.T. Rockafellar, Monotone operators and the proximal point algorithm, SIAM J. Control Optimization 14 (1976), 877–898.

8 Second Order Difference Equations

8.1 INTRODUCTION

Consider the following second order evolution equation of monotone type

$$\begin{cases} p(t)u''(t)+r(t)u'(t)\in Au(t)+f(t), \\ u(0)=u_0, \quad \sup_{t\geq 0}\|u(t)\|<+\infty \end{cases} \tag{8.1}$$

that we studied in Chapter 5. Using the following difference quotients,

$$u'(t)=\frac{u(t)-u(t-h)}{h}+o(h),$$

and

$$u''(t)=\frac{u(t+h)-2u(t)+u(t-h)}{h^2}+o(h)$$

to approximate the first and second derivatives of u in (8.1), we get the following difference inclusion

$$\begin{cases} u_{n+1}-(1+\theta_n)u_n+\theta_n u_{n-1}\in c_nAu_n+f_n \\ u_0=x, \quad \sup_{n\geq 0}\|u_n\|<+\infty, \end{cases} \tag{8.2}$$

where c_n and θ_n are two positive real sequences, f_n is a sequence in H and $x\in H$. In this chapter, we prove some results on the existence, periodicity and asymptotic behavior of solutions to (8.2), similar to the results of Chapter 5 for (8.1). Section 2 is devoted to the study of existence and uniqueness of solutions to (8.2). In Section 3, we consider the periodic forcing case. We prove the existence of a periodic solution if c_n, θ_n and f_n are periodic. Also, we show that each solution asymptotically converges to a periodic solution. In Section 4, the continuous dependence on initial data is studied. In Section 5, ergodic, weak and strong convergence of solutions to (8.2) for the homogeneous case ($f_n\equiv 0$) is investigated. In Section 6, we consider the subdifferential case, that is when the monotone operator A is the subdifferential of a proper, convex and lower semicontinuous function. Section 7 is devoted to the asymptotic behavior of solutions to (8.2) for the non-homogeneous case. Finally in Section 8, we present some applications of these results to convex optimization.

8.2 EXISTENCE AND UNIQUENESS

In this section, we prove the existence and uniqueness of solutions to the following second order difference equations

$$\begin{cases} u_{i+1}^N - (1+\theta_i)u_i^N + \theta_i u_{i-1}^N \in c_i A u_i^N + f_i, & 1 \le i \le N \\ u_0^N = a, \quad u_{N+1}^N = b, \end{cases} \tag{8.3}$$

and

$$\begin{cases} u_{i+1} - (1+\theta_i)u_i + \theta_i u_{i-1} \in c_i A u_i + f_i, & i \ge 1 \\ u_0 = a, \ \sup_{i\ge 1} \|u_i\| < +\infty, \end{cases} \tag{8.4}$$

where $a, b \in H$, $\theta_i, c_i > 0$, $f_i \in H$ for $i \ge 1$ and $A : D(A) \subset H \to H$ is maximal monotone.

Define the sequence a_i as follows.

$$a_0 = 1, \quad a_i = \frac{1}{\theta_1 \cdots \theta_i}, \quad i \ge 1 \tag{8.5}$$

Let $H_{a_i}^N := H \times \cdots \times H$ endowed with the inner product

$$\langle (u_i)_{1\le i\le N}, (v_i)_{1\le i\le N} \rangle = \sum_{i=1}^N a_i (u_i, v_i) \tag{8.6}$$

for each $(u_i)_{1\le i\le N}, (v_i)_{1\le i\le N} \in H^N$. It is obvious that $H_{a_i}^N$ is a Hilbert space and it is algebraically and topologically isomorphic with H^N. Define the operator B by

$$B((u_i)_{1\le i\le N}) = (-u_{i+1} + (1+\theta_i)u_i - \theta_i u_{i-1})_{1\le i\le N} \tag{8.7}$$

$$D(B) := \{(u_i)_{1\le i\le N} \in H^N, \ u_0 = a, \ u_{N+1} = b\} \tag{8.8}$$

Proposition 8.2.1 *The operator B is maximal monotone in $H_{a_i}^N$.*

Proof. (8.7) can be rewritten as:

$$B((u_i)_{1\le i\le N}) = -(\frac{1}{a_i}(\varphi_{i+1} - \varphi_i)_{1\le i\le N}) \tag{8.9}$$

where $\varphi_i = a_{i-1}(u_i - u_{i-1})$. If we take $(u_i)_{1\le i\le N}, (v_i)_{1\le i\le N} \in D(B)$, $\varphi_i = a_{i-1}(u_i - u_{i-1})$, $\psi_i = a_{i-1}(v_i - v_{i-1})$, we have

$$\begin{aligned} &\langle B((u_i)_{1\le i\le N}) - B((v_i)_{1\le i\le N}), (u_i - v_i)_{1\le i\le N} \rangle \\ &= -\sum_{i=1}^N (\varphi_{i+1} - \varphi_i - \psi_{i+1} + \psi_i, u_i - v_i) \\ &= \sum_{i=1}^N a_i \|u_{i+1} - u_i - v_{i+1} + v_i\|^2 - \sum_{i=1}^N (\varphi_i - \psi_i, u_{i+1} - u_i - v_{i+1} + v_i) \end{aligned}$$

$$
\begin{aligned}
&-\sum_{i=1}^{N}(\varphi_{i+1}-\varphi_i-\psi_{i+1}+\psi_i,u_{i+1}-v_{i+1})\\
&=\sum_{i=1}^{N}a_i\|u_{i+1}-u_i-v_{i+1}+v_i\|^2\\
&+\sum_{i=1}^{N}[(\varphi_i-\psi_i,u_i-v_i)-(\varphi_{i+1}-\psi_{i+1},u_{i+1}-v_{i+1})]\\
&=\sum_{i=1}^{N}a_i\|u_{i+1}-u_i-v_{i+1}+v_i\|^2+\|u_1-v_1\|^2\geq 0
\end{aligned}
$$

This proves the monotonicity of B. To prove the maximal monotonicity of B, it is sufficient to show that for each $(h_i)_{1\leq i\leq N}\in H^N$, there exists a sequence $(u_i)_{1\leq i\leq N}\in H^N$ that satisfies

$$
\begin{cases} u_{i+1}-(2+\theta_i)u_i+\theta_i u_{i-1}=h_i, \quad 1\leq i\leq N\\ u_0=a, \quad u_{N+1}=b \end{cases} \tag{8.10}
$$

We are going to look for solutions to (8.10) of the form

$$
u_i=v_i+\alpha_i x+\beta_i y, \quad 1\leq i\leq N \tag{8.11}
$$

where $x,y\in H$ and v_i,α_i and β_i are respective solutions to the following problems:

$$
\begin{cases} v_{i+1}-(2+\theta_i)+\theta_i v_{i-1}=h_i, \quad 1\leq i\leq N\\ v_0=0, \quad v_1=0 \end{cases} \tag{8.12}
$$

$$
\begin{cases} \alpha_{i+1}-(2+\theta_i)\alpha_i+\theta_{i-1}\alpha_{i-1}=0, \quad 1\leq i\leq N\\ \alpha_0=0, \quad \alpha_1=c>0, \end{cases} \tag{8.13}
$$

and

$$
\begin{cases} \beta_{i+1}-(2+\theta_i)\beta_i+\theta_{i-1}\beta_{i-1}=0, \quad 1\leq i\leq N\\ \beta_{N+1}=0, \quad \beta_N=-c \end{cases} \tag{8.14}
$$

We note that u_i given by (8.11) satisfies Equation (8.10) for each $x,y\in H$. In order for the boundary conditions $u_0=a$ and $u_{N+1}=b$ to be satisfied, we choose $x=\frac{(b-v_{N+1})}{\alpha_{N+1}}$ and $y=\frac{a}{\beta_0}$; then the problem (8.10) has a solution and consequently B is maximal monotone in $H^N_{a_i}$. □

Now we prove the existence result for (8.3).

Theorem 8.2.2 *Assume that $A:D(A)\subset H\to H$ is a maximal monotone operator in H such that $0\in D(A)$, and $c_i>0,\theta_i>0$ and $f_i\in H$ for $1\leq i\leq N$ are given sequences. Then for each $a,b\in H$, the finite scheme* (8.3) *has a unique solution $(u_i)_{1\leq i\leq N}\in D(A)^N$.*

Proof. The operator $\mathscr{A}((u_i)_{1\le i\le N}) = (c_1 v_1, \cdots, c_N v_N)$ with $v_i \in Au_i$ for $1 \le i \le N$ is maximal monotone in H^N. Let A_λ be the Yosida approximation of A and $\mathscr{A}_\lambda$ be the Yosida approximation of $\mathscr{A}$. Then

$$\mathscr{A}_\lambda((u_i)_{1\le i\le N}) = (c_1 A_\lambda u_1, \cdots, c_N A_\lambda u_N)$$

Since B is maximal monotone in $H^N_{a_i}$(see Proposition 8.2.1), and $\mathscr{A}_\lambda$ is maximal monotone and everywhere defined, then $B + \mathscr{A}_\lambda$ is maximal monotone in $H^N_{a_i}$(see Theorem 3.3.5); hence $\mathscr{R}(\omega I + B + \mathscr{A}_\lambda) = H^N_{a_i}$ for each $\lambda, \omega > 0$ (see Theorem 3.3.2). Then for a given sequence $(f_i)_{1\le i\le N} \in H^N_{a_i}$, there exists a sequence $(u_i^{\lambda\omega})_{1\le i\le N} \in H^N$, such that:

$$\begin{cases} u_{i+1}^{\lambda\omega} - (1+\theta_i)u_i^{\lambda\omega} + \theta_i u_{i-1}^{\lambda\omega} = c_i A_\lambda u_i^{\lambda\omega} + \omega u_i^{\lambda\omega} + f_i, \quad 1 \le i \le N \\ u_0^{\lambda\omega} = a, \quad u_{N+1}^{\lambda\omega} = b \end{cases} \tag{8.15}$$

We show that $u_i^{\lambda\omega}$ is bounded respect to λ and ω, then we take the limit in (8.15) as $\lambda \to 0$ and $\omega \to 0$.
Multiplying (8.15) by $a_i u_i^{\lambda\omega}$ and summing up from $i = 1$ to $i = N$, we get:

$$\sum_{i=1}^N a_i(u_{i+1}^{\lambda\omega} - u_i^{\lambda\omega}, u_i^{\lambda\omega}) - \sum_{i=1}^N a_i\theta_i(u_i^{\lambda\omega} - u_{i-1}^{\lambda\omega}, u_i^{\lambda\omega})$$

$$= \sum_{i=1}^N c_i a_i(A_\lambda u_i^{\lambda\omega}, u_i^{\lambda\omega}) + \omega\sum_{i=1}^N a_i\|u_i^{\lambda\omega}\|^2 + \sum_{i=1}^N a_i(f_i, u_i^{\lambda\omega}).$$

Without loss of generality, we may assume that $0 \in A0$. Otherwise, we consider $\tilde{A}u_i = Au_i - A^o 0$ instead of A (where $A^o x$ is the least minimum norm element of Ax which was defined in Chapter 3), and $\tilde{f}_i = f_i + c_i A^o 0$ instead of f_i. Since $a_i\theta_i = a_{i-1}$, from the monotonicity of A_λ we conclude that:

$$\omega\sum_{i=1}^N a_i\|u_i^{\lambda\omega}\|^2 \le \sum_{i=1}^N [a_i(u_{i+1}^{\lambda\omega} - u_i^{\omega\lambda}, u_i^{\lambda\omega}) - a_{i-1}(u_i^{\lambda\omega} - u_{i-1}^{\lambda\omega}, u_{i-1}^{\lambda\omega})]$$

$$- \sum_{i=1}^N a_{i-1}\|u_i^{\lambda\omega} - u_{i-1}^{\lambda\omega}\|^2 - \sum_{i=1}^N a_i(f_i, u_i^{\lambda\omega})$$

Therefore

$$\omega\sum_{i=1}^N a_i\|u_i^{\lambda\omega}\|^2 + a_N\|u_N^{\lambda\omega}\|^2 + \sum_{i=1}^N a_{i-1}\|u_i^{\lambda\omega} - u_{i-1}^{\lambda\omega}\|^2$$

$$\le a_N\|b\|\|u_N^{\lambda\omega}\| + \|a\|\|u_1^{\lambda\omega}\| + \|a\|^2 + (\sum_{i=1}^N a_i\|f_i\|^2)^{\frac{1}{2}}(\sum_{i=1}^N a_i\|u_i^{\lambda\omega}\|^2)^{\frac{1}{2}} \tag{8.16}$$

One can see that

$$\|u_k^{\lambda\omega}\| \le \frac{1}{\sqrt{a_k}}(\sum_{i=1}^N a_i\|u_i^{\lambda\omega}\|^2)^{\frac{1}{2}}, \quad 1 \le k \le N \tag{8.17}$$

By using (8.16) and (8.17) one can write

$$\omega\sum_{i=1}^{N} a_i\|u_i^{\lambda\omega}\|^2+\sum_{i=1}^{N} a_{i-1}\|u_i^{\lambda\omega}-u_{i-1}^{\lambda\omega}\|^2\le K_1(\sum_{i=1}^{N} a_i\|u_i^{\lambda\omega}\|^2)^{\frac{1}{2}}+K_2 \tag{8.18}$$

where $K_1,K_2>0$ are constants independent of λ and ω. On the other hand,

$$\|u_k^{\lambda\omega}\|=\sum_{i=1}^{k}(\|u_i^{\lambda\omega}\|-\|u_{i-1}^{\lambda\omega}\|)+\|a\|\le\sum_{i=1}^{k}\|u_i^{\lambda\omega}-u_{i-1}^{\lambda\omega}\|+\|a\|$$

Therefore

$$a_k\|u_k^{\lambda\omega}\|^2\le 2(\sum_{i=1}^{k}\frac{a_k}{a_{i-1}})(\sum_{i=1}^{k} a_{i-1}\|u_i^{\lambda\omega}-u_{i-1}^{\lambda\omega}\|^2)+2\|a\|^2a_k$$

Summing up from $k=1$ to $k=N$, we get:

$$\sum_{k=1}^{N} a_k\|u_k^{\lambda\omega}\|^2\le K_3(\sum_{i=1}^{N} a_{i-1}\|u_i^{\lambda\omega}-u_{i-1}^{\lambda\omega}\|^2)+K_4 \tag{8.19}$$

Combining (8.19) with (8.18) we get

$$\sum_{i=1}^{N} a_{i-1}\|u_i^{\lambda\omega}-u_{i-1}^{\lambda\omega}\|^2\le K_5(\sum_{i=1}^{N} a_{i-1}\|u_i^{\lambda\omega}-u_{i-1}^{\lambda\omega}\|^2)^{\frac{1}{2}}+K_6 \tag{8.20}$$

which implies that

$$\sum_{i=1}^{N} a_{i-1}\|u_i^{\lambda\omega}-u_{i-1}^{\lambda\omega}\|^2\le K_7 \tag{8.21}$$

where $K5,K_6$ and K_7 are constants independent of λ and ω. By (8.19) and (8.17) we conclude that:

$$\sum_{i=1}^{N} a_i\|u_i^{\lambda\omega}\|^2\le K_8,\quad \|u_i^{\lambda\omega}\|\le K_9,\quad 1\le i\le N \tag{8.22}$$

and by (8.15) we get:

$$\|A_\lambda u_i^{\lambda\omega}\|\le K_{10},\quad 1\le i\le N \tag{8.23}$$

We prove the strong convergence of $u_i^{\lambda\omega}$ in $H_{a_i}^N$ as $\lambda\to 0$. To this aim, we subtract (8.15) with μ from (8.15) with λ, then we multiply the difference by $a_i(u_i^{\lambda\omega}-u_i^{\mu\omega})$, and sum up from $i=1$ to $i=N$. We conclude that:

$$\sum_{i=1}^{N} a_i(u_{i+1}^{\lambda\omega}-u_{i+1}^{\mu\omega}-u_i^{\lambda\omega}+u_i^{\mu\omega},u_i^{\lambda\omega}-u_i^{\mu\omega})$$

$$-\sum_{i=1}^{N} a_i\theta_i(u_i^{\lambda\omega}-u_i^{\mu\omega}-u_{i-1}^{\lambda\omega}+u_{i-1}^{\mu\omega},u_i^{\lambda\omega}-u_i^{\mu\omega})$$

$$= \sum_{i=1}^{N} a_i c_i (A_\lambda u_i^{\lambda\omega} - A_\mu u_i^{\mu\omega}, u_i^{\lambda\omega} - u_i^{\mu\omega}) + \omega \sum_{i=1}^{N} a_i \|u_i^{\lambda\omega} - u_i^{\mu\omega}\|^2 \tag{8.24}$$

Suppose that M_1 and M_2 are the left and right hand side of (8.24). By using the initial conditions in (8.15) and the well known relation

$$u_i^{\lambda\omega} = J_\lambda u_i^{\lambda\omega} + \lambda A_\lambda u_i^{\lambda\omega}, \quad 1 \leq i \leq N \tag{8.25}$$

we conclude that:

$$M_1 = -a_N \|u_N^{\lambda\omega} - u_N^{\mu\omega}\|^2 - \sum_{i=1}^{N} a_{i-1} \|u_i^{\lambda\omega} - u_i^{\mu\omega} - u_{i-1}^{\lambda\omega} + u_{i-1}^{\mu\omega}\|^2 \tag{8.26}$$

and

$$\begin{aligned} M_2 = &\sum_{i=1}^{N} a_i c_i (A_\lambda u_i^{\lambda\omega} - A_\mu u_i^{\mu\omega}, J_\lambda u_i^{\lambda\omega} - J_\mu u_i^{\mu\omega}) \\ &+ \sum_{i=1}^{N} a_i c_i (A_\lambda u_i^{\lambda\omega} - A_\mu u_i^{\mu\omega}, \lambda A_\lambda u_i^{\lambda\omega} - \mu A_\mu u_i^{\mu\omega}) \\ &+ \omega \sum_{i=1}^{N} a_i \|u_i^{\lambda\omega} - u_i^{\mu\omega}\|^2. \end{aligned}$$

Since A is monotone and $A_\lambda u_i^{\lambda\omega} \in A(J_\lambda u_i^{\lambda\omega})$ we have:

$$M_2 \geq \sum_{i=1}^{N} a_i c_i (\lambda \|A_\lambda u_i^{\lambda\omega}\|^2 + \mu \|A_\mu u_i^{\mu\omega}\|^2)$$

$$-(\lambda + \mu) \sum_{i=1}^{N} a_i c_i (A_\lambda u_i^{\lambda\omega}, A_\mu u_i^{\mu\omega}) + \omega \sum_{i=1}^{N} a_i \|u_i^{\lambda\omega} - u_i^{\mu\omega}\|^2 \tag{8.27}$$

By using (8.26) and (8.27) in (8.24) and with the aid of (8.23) we obtain:

$$\omega \sum_{i=1}^{N} a_i \|u_i^{\lambda\omega} - u_i^{\mu\omega}\|^2 + \sum_{i=1}^{N} a_{i-1} \|u_i^{\lambda\omega} - u_i^{\mu\omega} - u_{i-1}^{\lambda\omega} + u_{i-1}^{\mu\omega}\|^2 \leq K_{11}(\lambda + \mu) \tag{8.28}$$

Then we conclude that $(u_i^{\lambda\omega})$ and $(u_i^{\lambda\omega} - u_{i-1}^{\lambda\omega})$ are strongly convergent sequences as $\lambda \to 0$. Let $u_i^{\lambda\omega} \to u_i^\omega$ as $\lambda \to 0$. Passing to subsequences if necessary, suppose that $A_\lambda u_i^{\lambda\omega} \rightharpoonup w_i^\omega$ as $\lambda \to 0$ in H. By (8.25), we conclude that $J_\lambda u_i^{\lambda\omega} \to u_i^\omega$ as $\lambda \to 0$ in H. Since A is maximal monotone in H (and therefore strong-weak closed), by taking the limit in the inclusion $A_\lambda u_i^{\lambda\omega} \in A(J_\lambda u_i^{\lambda\omega})$, we get:

$$u_i^\omega \in D(A), \quad w_i^\omega \in A u_i^\omega \tag{8.29}$$

Therefore taking the limit in (8.15) as $\lambda \to 0$, we conclude that:

$$\begin{cases} u_{i+1}^\omega - (1+\theta_i) u_i^\omega + \theta_i u_{i-1}^\omega \in c_i A u_i^\omega + \omega u_i^\omega + f_i, & 1 \leq i \leq N \\ u_0^\omega = a, \quad u_{N+1}^\omega = b \end{cases} \tag{8.30}$$

Since $\|u_i^\omega\| = \lim_{\lambda\to 0}\|u_i^{\lambda\omega}\|$, (8.22) implies that

$$\|u_i^\omega\| \le K_9 \tag{8.31}$$

The next step of the proof consists of showing the strong convergence of $(u_i^\omega - u_{i-1}^\omega)$ as $\omega \to 0$. Considering the Equation (8.30) for $\delta, \omega > 0$, and subtracting them, then multiplying both sides by $a_i(u_i^\omega - u_i^\delta)$ and summing up from $i = 1$ to $i = N$, we get the following equality:

$$\begin{aligned}
&\sum_{i=1}^{N} a_i(u_{i+1}^\omega - u_{i+1}^\delta - u_i^\omega + u_i^\delta, u_i^\omega - u_i^\delta) \\
&- \sum_{i=1}^{N} a_i\theta_i(u_i^\omega - u_i^\delta - u_{i-1}^\omega + u_{i-1}^\delta, u_i^\omega - u_i^\delta) \\
= &\sum_{i=1}^{N} a_i c_i(v_i^\omega - v_i^\delta, u_i^\omega - u_i^\delta) + \sum_{i=1}^{N} a_i(\omega u_i^\omega - \delta u_i^\delta, u_i^\omega - u_i^\delta)
\end{aligned} \tag{8.32}$$

where $v_i^\omega \in Au_i^\omega$ and $v_i^\delta \in Au_i^\delta$. Since the left hand side can be written in the form

$$M_1 = -a_N\|u_N^\omega - u_N^\delta\|^2 - \sum_{i=1}^{N} a_{i-1}\|u_i^\omega - u_i^\delta - u_{i-1}^\omega + u_{i-1}^\delta\|^2$$

then (8.32) implies that

$$\begin{aligned}
\sum_{i=1}^{N} a_{i-1}\|u_i^\omega - u_i^\delta - u_{i-1}^\omega + u_{i-1}^\delta\|^2 &\le -\sum_{i=1}^{N} a_i c_i(v_i^\omega - v_i^\delta, u_i^\omega - u_i^\delta) \\
&- \sum_{i=1}^{N} a_i[\omega\|u_i^\omega\|^2 + \delta\|u_i^\delta\|^2] + (\omega+\delta)\sum_{i=1}^{N} a_i(u_i^\omega, u_i^\delta)
\end{aligned}$$

By the monotonicity of A and the boundedness of u_i^ω which was shown in (8.31), we find that

$$\sum_{i=1}^{N} a_{i-1}\|u_i^\omega - u_i^\delta - u_{i-1}^\omega + u_{i-1}^\delta\|^2 \le K_{12}(\omega+\delta) \tag{8.33}$$

Therefore $(u_i^\omega - u_{i-1}^\omega)_\omega$ is a Cauchy net and hence strongly convergent in H as $\omega \to 0$. On the other hand, passing to a subsequence if necessary, we may assume that u_i^ω is weakly convergent. Suppose that $u_i^\omega \rightharpoonup u_i$ as $\omega \to 0$. By rewriting the Equation (8.30) in the form

$$u_{i+1}^\omega - u_i^\omega - \theta_i(u_i^\omega - u_{i-1}^\omega) \in c_i A u_i^\omega + \omega u_i^\omega + f_i, \quad 1 \le i \le N \tag{8.34}$$

and taking the limit as $\omega \to 0$, we conclude that $u_i \in D(A)$ and u_i is a solution to (8.3). Now we prove the uniqueness. If $(u_i)_{1\le i\le N}$ and $(v_i)_{1\le i\le N}$ are solutions to (8.3) and $w_i = u_i - v_i$, then the monotonicity of A implies that

$$\sum_{i=1}^{N} a_i(w_{i+1} - w_i, w_i) - \sum_{i=1}^{N} a_{i-1}(w_i - w_{i-1}, w_i) \ge 0 \tag{8.35}$$

Since $w_{N+1} = w_0 = 0$, we get

$$\sum_{i=1}^{N} a_{i-1}\|w_i - w_{i-1}\|^2 + a_N\|w_N\|^2 \leq 0 \tag{8.36}$$

which shows the uniqueness of the solution. □

Next we are going to investigate the existence of solutions to (8.4). In each of the two cases $\theta_n \geq 1$ and $0 < \theta_n < 1$, we prove that if (8.4) has a solution for an initial value, then it has a solution for every other initial value. Then we prove the existence of solutions when $A^{-1}(0) \neq \emptyset$. First we prove a lemma.

Lemma 8.2.3 *If the following assumption on* θ_n

$$\sum_{k=1}^{\infty} \frac{1}{h_k} = +\infty, \qquad \text{where } h_k = \sum_{i=1}^{k} \frac{1}{\theta_i \ldots \theta_k} \tag{8.37}$$

is satisfied, then every positive and bounded sequence a_n *satisfying*

$$a_n \leq \frac{1}{1+\theta_n} a_{n+1} + \frac{\theta_n}{1+\theta_n} a_{n-1} \tag{8.38}$$

is nonincreasing.

Proof. (8.38) implies that a_n is nonincreasing or eventually increasing. Suppose to the contrary that a_n is eventually increasing. Then there exists $m > 0$ such that for all $i > m$, $a_i > a_{i-1}$. Form (8.38) we obtain

$$\theta_i \leq \frac{a_{i+1} - a_i}{a_i - a_{i-1}}, \quad \forall i > m$$

then

$$\theta_i \cdots \theta_k \leq \frac{a_{k+1} - a_k}{a_i - a_{i-1}} \qquad \forall i > m$$

and

$$h_k = \sum_{i=1}^{m} \frac{1}{\theta_i \cdots \theta_k} + \sum_{i=m+1}^{k} \frac{1}{\theta_i \cdots \theta_k} \geq \sum_{i=m+1}^{k} \frac{1}{\theta_i \cdots \theta_k} \geq \sum_{i=m+1}^{k} \frac{a_i - a_{i-1}}{a_{k+1} - a_k} = \frac{a_k - a_m}{a_{k+1} - a_k}.$$

Then:

$$\sum_{k=1}^{\infty} \frac{1}{h_k} \leq \sum_{k=1}^{m} \frac{a_{k+1} - a_k}{a_k - a_m} + \sum_{k=m+1}^{\infty} \frac{a_{k+1} - a_k}{a_k - a_m} \leq \sum_{k=1}^{m} \frac{a_{k+1} - a_k}{a_k - a_m} + \frac{1}{\lambda} \sum_{k=m+1}^{\infty} (a_{k+1} - a_k)$$

$$= \sum_{k=1}^{m} \frac{a_{k+1} - a_k}{a_k - a_m} + \frac{1}{\lambda}(l - a_{m+1}) < +\infty$$

where $0 < \lambda = a_{m+1} - a_m \leq a_k - a_m$ for all $k > m$, and $l = \lim_{k \to +\infty} a_k$. This is a contradiction. □

Theorem 8.2.4 *Suppose that $A : D(A) \subset H \to H$ is a maximal monotone operator such that $0 \in D(A)$, and $f_i \in H$, $c_i > 0$ and $\theta_i \in (0,1)$ for all $i \geq 1$ such that (8.37) is satisfied. If the problem (8.4) has a solution for some $u_0 = b \in H$, then for each initial value $u_0 = a \in H$, it has a unique solution. If $(u_i)_{i\geq 1}$ and $(v_i)_{i\geq 1}$ are two solutions to (8.4) with $u_0 = a$ and $v_0 = b$, then $\|u_i - v_i\|$ is nonincreasing and $\|u_i - v_i\| \leq \|a-b\|$.*

Proof. Let $b \in H$ be such that the problem

$$\begin{cases} w_{i+1} - (1+\theta_i)w_i + \theta_i w_{i-1} \in c_i A w_i + f_i, & i \geq 1 \\ w_0 = b, \ \sup_{i\geq 1} \|w_i\| < +\infty \end{cases} \tag{8.39}$$

has a solution $(w_i)_{i\geq 1}$, with $w_i \in D(A)$, $i \geq 1$. If $a \in H$ is an arbitrary element, then we show that problem (8.4) with $u_0 = a$ has a solution $(u_i)_{i\geq 1}$ with $u_i \in D(A)$, $i \geq 1$. By Theorem 8.2.2 the problem

$$\begin{cases} u_{i+1}^N - (1+\theta_i)u_i^N + \theta_i u_{i-1}^N \in c_i A u_i^N + f_i, & 1 \leq i \leq N \\ u_0^N = u_{N+1}^N = a \end{cases} \tag{8.40}$$

has a unique solution $(u_i^N)_{1\leq i\leq N} \in D(A)^N$. We show that the limit $u_i = \lim_{N\to+\infty} u_i^N$ exists uniformly with respect to i belonging to a finite set of natural numbers, and the limit is a solution of (8.4).

Let $y_i = u_i^N - w_i$ for $1 \leq i \leq N$. Since A is monotone, by substituting w_i in (8.4) and subtracting from (8.40), then multiplying the difference by y_i, we conclude that:

$$(y_{i+1} - (1+\theta_i)y_i + \theta_i y_{i-1}, y_i) \geq 0 \tag{8.41}$$

so

$$\|y_i\| \leq \frac{1}{1+\theta_i}\|y_{i+1}\| + \frac{\theta_i}{1+\theta_i}\|y_{i-1}\| \tag{8.42}$$

Since the right hand side of the above inequality is a convex combination, we get:

$$\|u_i^N\| \leq \max\{\|y_0\|, \|y_{N+1}\|\} + \|w_i\| \leq \|a\| + K \tag{8.43}$$

where

$$K = \sup\{\|w_i\|, \ i \geq 0\} \tag{8.44}$$

Inequality (8.41) also implies that

$$(y_{i+1} - y_i, y_i) \geq \theta_i(y_i - y_{i-1}, y_i)$$

therefore

$$\theta_i\|y_i - y_{i-1}\|^2 \leq (y_{i+1} - y_i, y_i) - \theta_i(y_i - y_{i-1}, y_{i-1}), \ \ 1 \leq i \leq N.$$

Multiplying the above inequalities for $1 \leq i \leq N-1$ respectively by $\theta_N \cdots \theta_{i+1}$, and then adding them up from $i = 1$ to $i = N$, we get:

$$\sum_{i=1}^{N} \theta_N \cdots \theta_{i+1}\theta_i\|y_i - y_{i-1}\|^2 \leq (y_{N+1} - y_N, y_N) - \theta_N \cdots \theta_1(y_1 - y_0, y_0) \tag{8.45}$$

For each fixed N_0, (8.44) and (8.45) imply the existence of $M > 0$, independent of N, such that

$$\sum_{i=1}^{N_0} \theta_N \cdots \theta_{i+1}\theta_i \|u_i^N - u_{i-1}^N\|^2 \leq M \tag{8.46}$$

We show that for each N_0, the sequence $(u_i^N)_{1\leq i\leq N_0}$ is a Cauchy sequence with respect to N. Suppose that $N_0 < N_1 < N_2$ are natural numbers and $z_i = u_i^{N_1} - u_i^{N_2},\ 0 \leq i \leq N_1 + 1$. By (8.40)

$$(z_{i+1} - z_i, z_i) \geq \theta_i(z_i - z_{i-1}, z_i),\ \ 1 \leq i \leq N_1$$

which implies the inequality

$$\theta_i\|z_i - z_{i-1}\|^2 \leq (z_{i+1} - z_i, z_i) - \theta_i(z_i - z_{i-1}, z_{i-1}),\ \ 1 \leq i \leq N_1 \tag{8.47}$$

Multiplying (8.47) by $\theta_k\theta_{k-1}\cdots\theta_{i+1}$ (where $k \in \{1,\cdots,N_1\}, 1 \leq i \leq k-1$) and then summing up from $i = 1$ to $i = k$, we get:

$$\sum_{i=1}^{k} \theta_k \cdots \theta_{i+1}\theta_i\|z_i - z_{i-1}\|^2 \leq (z_{k+1} - z_k, z_k),\ \ 1 \leq k \leq N_1 \tag{8.48}$$

because $z_0 = 0$. Therefore

$$\sum_{i=1}^{k} \theta_k \cdots \theta_{i+1}\theta_i\|z_i - z_{i-1}\|^2 \leq \frac{1}{2}(\|z_{k+1}\|^2 - \|z_k\|^2),\ \ 1 \leq k \leq N_1 \tag{8.49}$$

This shows that the sequence $(\|z_k\|)_k$ is nondecreasing. Moreover

$$\|z_k\| = \sum_{i=1}^{k}[\|z_i\| - \|z_{i-1}\|] \leq \sum_{i=1}^{k}\|z_i - z_{i-1}\|$$

Therefore

$$\|z_k\|^2 \leq \Big(\sum_{i=1}^{k}\frac{1}{\theta_k\cdots\theta_{i+1}\theta_i}\Big)\Big(\sum_{i=1}^{k}\theta_k\cdots\theta_{i+1}\theta_i\|z_i - z_{i-1}\|^2\Big),\ \ 1 \leq k \leq N_1 \tag{8.50}$$

Using (8.49) we get:

$$\|z_k\|^2 \leq \frac{1}{2}h_k(\|z_{k+1}\|^2 - \|z_k\|^2),\ \ 1 \leq k \leq N_1 \tag{8.51}$$

Summing up from $k = N_0$ to $k = N_1$, and using (8.43), we get:

$$\sum_{k=N_0}^{N_1}\frac{1}{h_k}\|z_k\|^2 \leq \frac{1}{2}\|z_{N_1+1}\|^2 \leq C$$

where C is a positive constant. Since $(\|z_k\|)$ is nondecreasing, for $i \in \{1, \cdots, N_0\}$ we can write

$$\|z_i\|^2 (\sum_{k=N_0}^{N_1} \frac{1}{h_k}) \leq \sum_{k=N_0}^{N_1} \frac{1}{h_k} \|z_k\|^2 \leq C \tag{8.52}$$

Therefore

$$\|u_i^{N_1} - u_i^{N_2}\|^2 \leq \frac{C}{(\sum_{k=N_0}^{N_1} \frac{1}{h_k})}, \quad 1 \leq i \leq N_0 \tag{8.53}$$

Since by assumption $\sum_{k=1}^{\infty} \frac{1}{h_k} = +\infty$, $\lim_{N_1 \to +\infty} u_i^{N_1} = u_i$ exists uniformly with respect to i belonging to every finite subset of natural numbers. The operator A is maximal monotone in H and therefore demiclosed. By letting $N \to +\infty$ in (8.40), we conclude that u_i satisfies (8.4). To prove the uniqueness, suppose that $(u_i)_{i \geq 1}$ and $(v_i)_{i \geq 1}$ are two solutions of (8.4) with $u_0 = a$ and $v_0 = b$, and let $q_i = u_i - v_i$. Subtracting the corresponding equations for u_i and v_i, and multiplying by q_i, by the monotonicity of A we get:

$$\|q_i\| \leq \frac{1}{1+\theta_i} \|q_{i+1}\| + \frac{\theta_i}{1+\theta_i} \|q_{i-1}\|, \quad i \geq 1 \tag{8.54}$$

By Lemma 8.2.3, $\|q_i\|$ is nonincreasing and therefore $\|u_i - v_i\|$ is nonincreasing. Therefore $\|u_i - v_i\| \leq \|a - b\|$. Choosing $a = b$, we get the uniqueness and the proof is complete. □

Theorem 8.2.5 *Suppose that* $0 < \theta_n < 1$ *and* $\sum_{k=1}^{\infty} \frac{1}{h_k} = +\infty$ *where* $h_k = \sum_{i=1}^{k} \frac{1}{\theta_k \cdots \theta_i}$. *If* $\{f_n\}$ *is a sequence in* H *such that* $\sum_{n=1}^{+\infty} n \frac{\|f_n\|}{\theta_1 \cdots \theta_n} < +\infty$ *and* $A^{-1}(0) \neq \phi$, *then Equation (8.4) has a unique solution.*

Proof. Suppose that $p \in A^{-1}(0)$ or $0 \in A(p)$. By Theorem 8.2.4, it is sufficient to show that the problem (8.4) has a solution for $u_0 = p \in A^{-1}(0)$. Then it will have a unique solution for all $x \in H$. Therefore it suffices to prove the existence of a solution to the following problem:

$$\begin{cases} u_{n+1} - (1+\theta_n)u_n + \theta_n u_{n-1} \in c_n A u_n + f_n \\ u_0 = p \in A^{-1}(0), \qquad \sup_{n \geq 1} \|u_n\| < +\infty \end{cases} \tag{8.55}$$

We define $B(x) = A(x+p)$, and $v_n = u_n - p$. Then (8.55) is equivalent to the following problem:

$$\begin{cases} v_{n+1} - (1+\theta_n)v_n + \theta_n v_{n-1} \in c_n B v_n + f_n \\ v_0 = 0, \qquad \sup_{n \geq 1} \|v_n\| < +\infty \end{cases} \tag{8.56}$$

By Theorem 8.2.2, the following equation has a solution:

$$\begin{cases} v_{n+1}^N - (1+\theta_n)v_n^N + \theta_n v_{n-1}^N \in c_n B v_n^N + f_n, \qquad 1 \leq n \leq N \\ v_0^N = v_{N+1}^N = 0 \end{cases} \tag{8.57}$$

Multiplying both sides of this equation by v_n^N, and using the fact that B is monotone and $0 \in B(0)$, we get:

$$(v_{n+1}^N, v_n^N) - (1+\theta_n)\|v_n^N\|^2 + \theta_n(v_{n-1}^N, v_n^N) \geq (f_n, v_n^N)$$

$$\Rightarrow (1+\theta_n)\|v_n^N\| \leq \|v_{n+1}^N\| + \theta_n\|v_{n-1}^N\| + \|f_n\|$$

$$\Rightarrow a_n(\|v_n^N\| - \|v_{n+1}^N\|) - a_{n-1}(\|v_{n-1}^N\| - \|v_n^N\|) \leq a_{n-1}\frac{\|f_n\|}{\theta_n}$$

where $a_n = \frac{1}{\theta_1 \cdots \theta_n}$. Summing up from $n = r$ to $n = N$, we get:

$$\|v_r^N\| - \|v_{r-1}^N\| \leq \sum_{n=r}^{N} \frac{\|f_n\|}{\theta_1 \cdots \theta_n}$$

Summing up both sides of the above inequality from $r = 1$ to $r = k$, we get

$$\|v_k^N\| - \|v_0^N\| \leq \sum_{r=1}^{r=k}\sum_{n=r}^{N} \frac{\|f_n\|}{\theta_1 \cdots \theta_n}$$

$$\Rightarrow \|v_k^N\| \leq \sum_{r=1}^{r=N}\sum_{n=r}^{N} \frac{\|f_n\|}{\theta_1 \ldots \theta_n} < \sum_{n=1}^{\infty} n\frac{\|f_n\|}{\theta_1 \cdots \theta_n} < +\infty.$$

Now let $N_0 < N_1 < N_2$, and set $z_n = v_n^{N_1} - v_n^{N_2}, \;\; 0 \leq n \leq N_1 + 1$. On the other hand, by (8.57), we get:

$$(z_{n+1} - (1+\theta_n)z_n + \theta_n z_{n-1}, z_n) \geq 0$$

$$\Rightarrow (z_{n+1} - z_n, z_n) - \theta_n(z_n - z_{n-1}, z_{n-1}) \geq \theta_n\|z_n - z_{n-1}\|^2$$

Multiplying the above inequality by $\theta_{n+1} \cdots \theta_k$ where $k \in \{1, \cdots, N\}$ and then summing up from $n = 1$ to $n = k$, we get:

$$\sum_{n=1}^{k} \theta_n \cdots \theta_k\|z_n - z_{n-1}\|^2 \leq (z_{k+1} - z_k, z_k) \leq \frac{1}{2}(\|z_{k+1}\|^2 - \|z_k\|^2) \tag{8.58}$$

In addition

$$\|z_k\| = \sum_{i=1}^{k}(\|z_i\| - \|z_{i-1}\|) \leq \sum_{i=1}^{k}\|z_i - z_{i-1}\|$$

hence

$$\|z_k\|^2 \leq (\sum_{i=1}^{k} \frac{1}{\theta_i \cdots \theta_k})(\sum_{i=1}^{k} \theta_i \cdots \theta_k\|z_i - z_{i-1}\|^2)$$

By (8.58)

$$\|z_k\|^2 \leq \frac{1}{2}h_k(\|z_{k+1}\|^2 - \|z_k\|^2)$$

Summing up from $k = N_0$ to $k = N_1$, we get

$$\sum_{k=N_0}^{N_1} \frac{1}{h_k}\|z_k\|^2 \leq \frac{1}{2}\|z_{N_1+1}\|^2 \leq C$$

where C is a positive constant. Since $\|z_k\|$ is nondecreasing, for $1 \leq i \leq N_0$ we have:

$$\|z_i\|^2\left(\sum_{k=N_0}^{N_1} \frac{1}{h_k}\right) \leq \sum_{k=N_0}^{N_1} \frac{1}{h_k}\|z_k\|^2 \leq C$$

then

$$\|u_i^{N_1} - u_i^{N_2}\|^2 \leq \frac{C}{(\sum_{k=N_0}^{N_1} \frac{1}{h_k})}, \quad 1 \leq i \leq N_0$$

Hence, $v_i = \lim_{n \to +\infty} v_i^n$ exists. The result now follows from the fact that A is demi-closed. □

Theorem 8.2.6 *Let the sequences $c_i > 0$, $\theta_i \geq 1$ and $f_i \in H$ be given. Suppose that $A : D(A) \subset H \to H$ is a maximal monotone operator with $0 \in D(A)$. If* (8.4) *has a solution for an initial value $u_0 = b \in H$, then it has a unique solution for every initial value $u_0 = a \in H$. If $(u_i)_{i \geq 1}$ and $(v_i)_{i \geq 1}$ are two solutions with $u_0 = a$ and $v_0 = b$, then the sequence $(\|u_i - v_i\|)$ is nonincreasing and $\|u_i - v_i\| \leq \|a - b\|$.*

Proof. Suppose that $b \in H$ is given such that the difference inclusion

$$\begin{cases} w_{i+1} - (1+\theta_i)w_i + \theta_i w_{i-1} \in c_i A w_i + f_i, & i \geq 1 \\ w_0 = b, \quad \sup_{i \geq 1} \|w_i\| = K < +\infty \end{cases} \tag{8.59}$$

admits a solution $(w_i)_{i \geq 1}$. Let $a \in H$ be arbitrarily chosen. We prove that (8.4) has a unique solution with initial condition $u_0 = a$. By Theorem 8.2.2, there exists a unique solution $(u_i^N)_{1 \leq i \leq N} \in D(A)^N$ for the auxiliary problem

$$\begin{cases} u_{i+1}^N - (1+\theta_i)u_i^N + \theta_i u_{i-1}^N \in c_i A u_i^N + f_i, & 1 \leq i \leq N \\ u_0^N = u_{N+1}^N = a \end{cases} \tag{8.60}$$

We show that $\lim_{N \to +\infty} u_i^N = u_i$ exists uniformly on every finite set of natural numbers, and that $(u_i)_{i \geq 1}$ is a solution to (8.4). Similar to the proof of Theorem 8.2.4 we find that

$$\|u_i^N\| \leq \|a\| + 2K, \quad \forall N \geq 1, \; 1 \leq i \leq N. \tag{8.61}$$

If $N_0 < N_1 < N_2$ are natural numbers and $z_i = u_i^{N_1} - u_i^{N_2}$, $0 \leq i \leq N_1 + 1$, then subtracting the corresponding equations for $u_i^{N_1}$ and $u_i^{N_2}$, and multiplying the difference by z_i, we get:

$$(z_{i+1} - (1+\theta_i)z_i + \theta_i z_{i-1}, z_i) \geq 0, \quad 1 \leq i \leq N_1 \tag{8.62}$$

which implies that

$$(z_{i+1} - z_i, z_i) - \theta_i(z_i - z_{i-1}, z_{i-1}) \geq \theta_i(\|z_i\| - \|z_{i-1}\|)^2, \quad 1 \leq i \leq N_1$$

Multiplying by $\theta_{i+1}\cdots\theta_k$, where $1\le i\le k-1$ and $1\le k\le N_1$ and summing up from $i=1$ to $i=k$, we get:

$$\sum_{i=1}^{k}\theta_i\theta_{i+1}\cdots\theta_k(\|z_i\|-\|z_{i-1}\|)^2\le\frac{1}{2}(\|z_{k+1}\|^2-\|z_k\|^2),\quad 1\le k\le N_1 \tag{8.63}$$

Now we sum from $k=N_0$ to $k=N_1$. Taking into account (8.61) and $\theta_i\ge 1,\ \forall i\ge 1$, this implies that

$$(N_1-N_0+1)\sum_{i=1}^{N_0}(\|z_i\|-\|z_{i-1}\|)^2\le C \tag{8.64}$$

On the other hand, we see that

$$\begin{aligned}(N_1-N_0+1)\sum_{i=1}^{N_0}(\|z_i\|-\|z_{i-1}\|)^2 &\ge \frac{N_1-N_0+1}{N_0}\Big(\sum_{i=1}^{N_0}(\|z_i\|-\|z_{i-1}\|)\Big)^2\\ &=\frac{N_1-N_0+1}{N_0}\|z_{N_0}\|^2\end{aligned} \tag{8.65}$$

By (8.64) and (8.65) we conclude that

$$\|z_{N_0}\|^2\le\frac{CN_0}{N_1-N_0+1} \tag{8.66}$$

The inequality (8.63) also implies that $(\|z_k\|)$ is a nondecreasing sequence, and so for $1\le i\le N_0$ we have

$$\|u_i^{N_1}-u_i^{N_2}\|\le\frac{CN_0}{N_1-N_0+1}$$

Therefore $\lim_{N\to+\infty}u_i^N=u_i$, exists uniformly with respect to i belonging to every finite set of natural numbers. By taking the limit as $N\to+\infty$ in (8.60), we conclude that $(u_i)_{i\ge 1}$ is a solution to (8.4). The proof of the last part of the theorem is similar to the proof of Theorem 8.2.4. □

Theorem 8.2.7 *Suppose that $\theta_n\ge 1$ and $\{f_n\}$ is a sequence in H such that $\sum_{n=1}^{+\infty}n\frac{\|f_n\|}{\theta_n}<+\infty$ and $A^{-1}(0)\ne\phi$, then Equation* (8.4) *has a unique solution.*

Proof. Suppose that $p\in A^{-1}(0)$ or $0\in A(p)$. By Theorem 8.2.6, it suffices to show that the problem (8.4) has a solution for $u_0=p\in A^{-1}(0)$. Then it will have a unique solution for every initial value $x\in H$. Therefore it is sufficient to prove the existence of a solution for the following difference inclusion:

$$\begin{cases}u_{n+1}-(1+\theta_n)u_n+\theta_nu_{n-1}\in c_nAu_n+f_n\\ u_0=p\in A^{-1}(0),\qquad \sup_{n\ge1}\|u_n\|<+\infty\end{cases} \tag{8.67}$$

We define $B(x)=A(x+p)$, and $v_n=u_n-p$. Then Equation (8.67) is equivalent to the following equation:

$$\begin{cases}v_{n+1}-(1+\theta_n)v_n+\theta_nv_{n-1}\in c_nBv_n+f_n\\ v_0=0,\qquad \sup_{n\ge1}\|v_n\|<+\infty\end{cases} \tag{8.68}$$

By Theorem 8.2.2, the following equation has a solution:

$$\begin{cases} v_{n+1}^N - (1+\theta_n)v_n^N + \theta_n v_{n-1}^N \in c_n B v_n^N + f_n, & 1 \le n \le N \\ v_0^N = v_{N+1}^N = 0 \end{cases} \tag{8.69}$$

We multiply both sides of the above equation by v_n^N. Since B is monotone and $0 \in B(0)$, we get:

$$(v_{n+1}^N, v_n^N) - (1+\theta_n)\|v_n^N\|^2 + \theta_n(v_{n-1}^N, v_n^N) \ge (f_n, v_n^N)$$

$$\Rightarrow (1+\theta_n)\|v_n^N\| \le \|v_{n+1}^N\| + \theta_n\|v_{n-1}^N\| + \|f_n\|$$

$$\Rightarrow a_n(\|v_n^N\| - \|v_{n+1}^N\|) - a_{n-1}(\|v_{n-1}^N\| - \|v_n^N\|) \le a_{n-1}\frac{\|f_n\|}{\theta_n}$$

where $a_n = \frac{1}{\theta_1 \cdots \theta_n}$. Summing up from $n = r$ to $n = N$, we get:

$$a_{r-1}(\|v_r^N\| - \|v_{r-1}^N\|) \le \sum_{n=r}^{N} a_{n-1}\frac{\|f_n\|}{\theta_n} \le a_{r-1}\sum_{n=r}^{N}\frac{\|f_n\|}{\theta_n}$$

Summing up both sides of the above inequality from $r = 1$ to $r = k$, we get:

$$\|v_k^N\| - \|v_0^N\| \le \sum_{r=1}^{r=k}\sum_{n=r}^{N}\frac{\|f_n\|}{\theta_n}$$

$$\Rightarrow \|v_k^N\| \le \sum_{r=1}^{r=N}\sum_{n=r}^{N}\frac{\|f_n\|}{\theta_n} = \sum_{n=1}^{N} n\frac{\|f_n\|}{\theta_n} < \sum_{n=1}^{\infty} n\frac{\|f_n\|}{\theta_n} < +\infty.$$

Now let $N_0 < N_1 < N_2$, and set $z_n = v_n^{N_1} - v_n^{N_2}$, $0 \le n \le N_1 + 1$. On the other hand by (8.69) we get:

$$(z_{n+1} - (1+\theta_n)z_n + \theta_n z_{n-1}, z_n) \ge 0$$

$$\Rightarrow a_n(z_{n+1} - z_n, z_n) - a_{n-1}(z_n - z_{n-1}, z_{n-1}) \ge a_{n-1}\|z_n - z_{n-1}\|^2$$

Summing up from $n = 1$ to $n = k$, we get:

$$\sum_{n=1}^{k} a_{n-1}\|z_n - z_{n-1}\|^2 \le a_k(z_{k+1} - z_k, z_k) \le \frac{a_k}{2}(\|z_{k+1}\|^2 - \|z_k\|^2)$$

Since the sequence $\{a_k\}$ is nonincreasing, we deduce that:

$$\sum_{n=1}^{k}\|z_n - z_{n-1}\|^2 \le \frac{1}{2}(\|z_{k+1}\|^2 - \|z_k\|^2) \tag{8.70}$$

Summing up (8.70) from $k = N_0$ to N_1, we get:

$$(N_1 - N_0 + 1)\sum_{n=1}^{N_0}\|z_n - z_{n-1}\|^2 \le \frac{1}{2}\|z_{N_1+1}\|^2 \le 2(\sum_{n=1}^{\infty} n\frac{\|f_n\|}{\theta_n})^2.$$

On the other hand

$$\begin{aligned}
\frac{N_1-N_0+1}{N_0}\|z_{N_0}\|^2 &= \frac{N_1-N_0+1}{N_0}[\sum_{i=1}^{N_0}(\|z_i\|-\|z_{i-1}\|)]^2 \\
&\leq \frac{N_1-N_0+1}{N_0}N_0\sum_{i=1}^{N_0}\|z_i-z_{i-1}\|^2 \\
&= (N_1-N_0+1)\sum_{i=1}^{N_0}\|z_i-z_{i-1}\|^2 \\
&\leq 2(\sum_{n=1}^{\infty} n\frac{\|f_n\|}{\theta_n})^2.
\end{aligned}$$

Therefore

$$\|z_{N_0}\|^2 \leq \frac{2N_0}{N_1-N_0+1}(\sum_{n=1}^{\infty} n\frac{\|f_n\|}{\theta_n})^2.$$

Since by (8.70), $\|z_n\|$ is nondecreasing, this implies that $v_i=\lim_{n\to+\infty}v_i^n$ exists uniformly with respect to i belonging to every finite set of natural numbers. Then the result follows because A is demiclosed. Now we prove the uniqueness. If u_n and v_n are two solutions to (8.4) with $u_0=v_0=x$, then by the monotonicity of A and Lemma 8.2.3, $\|u_n-v_n\|$ is nonincreasing. This implies uniqueness. □

8.3 PERIODIC FORCING

In this section, we assume that c_n, θ_n and f_n in the difference inclusion (8.4) are periodic, and we prove the existence of a periodic solution, as well as the weak convergence of any bounded solution to a periodic solution. This section contains a theorem from [ROU-KHA4]. Also in [APR3, POF-REI] the reader can find some special cases of some theorems similar to the following theorem.

Theorem 8.3.1 *Suppose that $A: D(A)\subset H\to H$ is a maximal monotone operator with $0\in D(A)$, and c_n, θ_n and f_n are periodic with period $N>0$ such that either* (8.37) *is satisfied or $\theta_n\geq 1$. If* (8.4) *has a solution u_n for an initial value, then it has a periodic solution w_n with period N, and $u_n-w_n\rightharpoonup 0$ as $n\to+\infty$. Moreover, any two periodic solutions differ by an additive constant.*

Proof. Let $x\in H$, and let $m\geq 0$ be an integer. Since A is maximal monotone, there is a unique solution to:

$$\begin{cases} u_{i+1}-(1+\theta_i)u_i+\theta_i u_{i-1}\in c_iAu_i+f_i, & i>m \\ u_m=x. \end{cases}$$

For $n\geq m$, we define the operators $Q(n,m): H\to H$ by

$$Q(n,m)x := u_n.$$

Let v_n be a solution to (8.4) with $v_m = y$. The monotonicity of A implies that:

$$(u_{n+1} - v_{n+1} - (1+\theta_n)(u_n - v_n) + \theta_n(u_{n-1} - v_{n-1}), u_n - v_n) \geq 0.$$

Therefore

$$(1+\theta_n)\|u_n - v_n\| \leq \|u_{n+1} - v_{n+1}\| + \theta_n\|u_{n-1} - v_{n-1}\|.$$

By Lemma 8.2.3 and (8.37), we get:

$$\|u_n - v_n\| \leq \|u_{n-1} - v_{n-1}\|.$$

Hence from the definition of $Q(n,m)$, it follows that $Q(n,m)$ is nonexpansive. The Nonexpansiveness of $Q(n,m)$ also shows the uniqueness of the solution to (8.4) with $u_m = x$. By the uniqueness of the solution, we have:

$$Q(n,m)Q(m,k) = Q(n,k)$$

for $n \geq m \geq k$, and by the periodicity, we also have:

$$Q(n+N, m+N) = Q(n,m)$$

for $n \geq m$. It follows that

$$Q(m+N,m)^n = Q(m+nN,m).$$

In particular, taking $m = 0$, it follows that $\{u_{kN}\}_{k\geq 0}$ is a nonexpansive sequence in H. Therefore, by the ergodic theorem proved in [DJA1], we deduce that the sequence $s_n = \frac{1}{n}\sum_{k=0}^{n-1} u_{kN}$ converges weakly in H, and the limit is a fixed point of $Q(N,0)$. This shows the existence of a periodic solution to (8.4), which is therefore bounded. Hence all solutions to (8.4) are bounded. Suppose that u_n and v_n are respectively a bounded and a periodic solution to (8.4). Set $z_n = u_n - v_n$. Then by (8.4) and the monotonicity of A, we get

$$\left(z_{n+1} - (1+\theta_n)z_n + \theta_n z_{n-1}, z_n\right) \geq 0$$

This implies that

$$\|z_{n+1} - z_n\|^2 \leq (z_{n+1} - z_n, z_{n+1}) - \theta_n(z_n - z_{n-1}, z_n)$$

$$= \frac{1}{2}\|z_{n+1} - z_n\|^2 + \frac{1}{2}\|z_{n+1}\|^2 - \frac{1}{2}\|z_n\|^2 - \frac{\theta_n}{2}\|z_n - z_{n-1}\|^2 - \frac{\theta_n}{2}\|z_n\|^2 + \frac{\theta_n}{2}\|z_{n-1}\|^2$$

Since by Lemma 8.2.3, $\|z_n\|$ is nonincreasing, we get:

$$\|z_n - z_{n-1}\|^2 \leq \|z_{n-1}\|^2 - \|z_n\|^2$$

Summing up from $n = 1$ to m and then letting $m \to +\infty$, we get $\sum_{n=1}^{+\infty} \|z_{n+1} - z_n\|^2 < +\infty$. Since $\{v_n\}$ is a periodic solution of (8.4), for each $m \geq 0$, we have

$$u_{m+nN} - u_{m+(n+1)N} = \sum_{i=nN}^{(n+1)N-1} (u_{m+i} - v_{m+i} - (u_{m+i+1} - v_{m+i+1})) \to 0 \qquad (8.71)$$

as $n \to +\infty$. Let

$$x_n := Q(m+nN,0)u_0 = u_{m+nN}$$

Then $\{x_n\}_{n\geq 1}$ is a nonexpansive sequence, which is asymptotically regular by (8.71). It follows from [DJA2, DJA3] that:

$$u_{m+nN} \rightharpoonup w_m \tag{8.72}$$

as $n \to +\infty$. Since $\{v_n\}_{n\geq 1}$ is a periodic solution of (8.4), we have

$$\lim_{n\to+\infty}(u_{m+nN} - u_{m-1+nN} - (v_m - v_{m-1}))$$

$$= \lim_{n\to+\infty}(u_{m+nN} - v_{m+nN} - (u_{m-1+nN} - v_{m-1+nN})) = 0$$

Therefore, $\lim_{n\to+\infty}(u_{m+nN} - u_{m-1+nN}) = (v_m - v_{m-1})$ exists, and from (8.72), we get:

$$\lim_{n\to+\infty}(u_{m+nN} - u_{m-1+nN}) = w_m - w_{m-1}$$

This implies that $w_m - v_m = w_0 - v_0 =$ constant, showing that any two periodic solutions differ by an additive constant, as we show below that w_n is a periodic solution to (8.4). By (8.4) we have

$$[u_{m+nN}, \frac{1}{c_m}(u_{m+1+nN} - (1+\theta_{m+nN})u_{m+nN} + \theta_{m+nN}u_{m-1+nN} - f_m)] \in A.$$

Since A is demiclosed, by letting $n \to +\infty$, we get:

$$[w_m, \frac{1}{c_m}(w_{m+1} - (1+\theta_m)w_m + \theta_m w_{m-1} - f_m)] \in A$$

Therefore $w_m \in D(A)$, and $w_{m+1} - (1+\theta_m)w_m + \theta_m w_{m-1} - f_m \in c_m A w_m$, which shows that w_n is a periodic solution to (8.4). Since w_n is N-periodic, we have:

$$\begin{aligned}
u_n - w_n &= u_n - u_{k+mN} + u_{k+mN} - w_{k+mN} + w_{k+mN} - w_n \\
&= u_n - w_n - (u_{k+mN} - w_{k+mN}) + u_{k+mN} - w_k \\
&= z_n - z_{k+mN} + u_{k+mN} - w_k
\end{aligned}$$

where $k + mN \leq n < k + (m+1)N$ and $z_n = u_n - w_n$. Now from (8.72), we have $u_{k+mN} - w_k \rightharpoonup 0$ as $m \to +\infty$, and from $\|z_n - z_{n-1}\| \to 0$, we get:

$$\|z_n - z_{k+mN}\| \leq \sum_{i=k+mN}^{i=k+(m+1)N-1} \|z_i - z_{i-1}\| \to 0$$

as $n \to +\infty$. This shows that $u_n - w_n \rightharpoonup 0$, as $n \to +\infty$, and completes the proof of the theorem. □

8.4 CONTINUOUS DEPENDENCE ON INITIAL CONDITIONS

Consider the following second order difference equation:

$$\begin{cases} u_{i+1} - (1+\theta_i)u_i + \theta_i u_{i-1} \in c_i A u_i, & i \geq 1 \\ u_0 = a, \quad \sup_{i\geq 1} \|u_i\| < +\infty, \end{cases} \tag{8.73}$$

where $a \in H$ and A is a maximal monotone operator in H, for which we already proved the existence and uniqueness of the solution. Our aim in this section is to show that the function that associates to the initial data $\{a, A\}$ the solution $(u_i)_{i\geq 1}$ to (8.73) is continuous. The results of this section are from [APR-APR1] and [APR-APR2] to which we refer for some of the proofs. First we recall a similar theorem for (8.3) from [APR-APR1], to which we refer for its proof.

Theorem 8.4.1 *Suppose that $(u_i^N)_{1\leq i\leq N}$ and $(u_i^{nN})_{1\leq i\leq N}$ are respectively solutions to the bilocal problems*

$$\begin{cases} u_{i+1}^N - (1+\theta_i)u_i^N + \theta_i u_{i-1}^N \in c_i A u_i^N + f_i, & 1 \leq i \leq N \\ u_0^N = a, \quad u_{N+1}^N = b \end{cases} \tag{8.74}$$

and

$$\begin{cases} u_{i+1}^{nN} - (1+\theta_i)u_i^{nN} + \theta_i u_{i-1}^{nN} \in c_i A^n u_i^{nN} + f_i^n, & 1 \leq i \leq N \\ u_0^{nN} = a_n, \quad u_{N+1}^{nN} = b_n \end{cases} \tag{8.75}$$

where A and A^n are maximal monotone operators in H with $0 \in D(A) \cap D(A^n)$, $a, b, a_n, b_n, f_i, f_i^n \in H$, $c_i > 0, \theta_i > 0, 1 \leq i \leq N$. If $a_n \to a$, $b_n \to b$, $f_i^n \to f_i$ and A^n converges to A in the sense of resolvent, (that is $(I+\lambda A^n)^{-1}x \to (I+\lambda A)^{-1}x, \forall \lambda > 0, \forall x \in H$), then

$$\lim_{n\to+\infty} u_i^{nN} = u_i^N, \quad 1 \leq i \leq N$$

In this section, we prove a similar result for (8.4). Let $u^n = (u_i^n)_{i\geq 1}$ be the solution to the following difference inclusion:

$$\begin{cases} u_{i+1}^n - (1+\theta_i)u_i^n + \theta_i u_{i-1}^n \in c_i A^n u_i^n, & i \geq 1 \\ u_0^n = a^n, \quad \sup_{i\geq 1} \|u_i^n\| = C_n < +\infty, & n \geq 1 \end{cases} \tag{8.76}$$

where $A^n : D(A^n) \subset H \to H$ is a sequence of maximal monotone operators in H and $a^n \in H$.

Theorem 8.4.2 *Suppose that $A : D(A) \subset H \to H$ and $A^n : D(A^n) \subset H \to H$ are maximal monotone operators in H such that $0 \in D(A) \cap D(A^n)$ and $0 \in A0 \cap A^n 0$. Also suppose that $a, a^n \in H$, $c_i > 0$, and $\theta_i > 0$, $i \geq 1$ is a sequence such that*

$$0 < \theta_i < 1, \quad \sum_{i=1}^{\infty} \frac{1}{h_i} = +\infty, \quad \textit{where } h_i = \sum_{j=1}^{i} \frac{1}{\theta_j \cdots \theta_i} \tag{8.77}$$

or

$$\theta_i \geq 1. \tag{8.78}$$

Assume that $u = (u_i)_{i\geq 1}$ *and* $u^n = (u_i^n)_{i\geq 1}$ *are respectively solutions to* (8.73) *and* (8.76). *If* $a^n \to a$ *in* H *and* $A^n \to A$ *in the sense of resolvent, then* $u_i^n \to u_i$ *as* $n \to +\infty$, *uniformly on every finite subset of integers.*

Proof. We are going to estimate the solutions of problems (8.73) and (8.76) by the solutions to the following problems on the finite set $\{1,\cdots,N\}$.

$$\begin{cases} u_{i+1}^N - (1+\theta_i)u_i^N + \theta_i u_{i-1}^N \in c_i A u_i^N, & 1 \leq i \leq N \\ u_0^N = u_{N+1}^N = a, \end{cases} \tag{8.79}$$

and

$$\begin{cases} u_{i+1}^{nN} - (1+\theta_i)u_i^{nN} + \theta_i u_{i-1}^{nN} \in c_i A^n u_i^{nN}, & 1 \leq i \leq N \\ u_0^{nN} = u_{N+1}^{nN} = a^n, \quad n \geq 1 \end{cases} \tag{8.80}$$

By Theorem 8.2.2, we know that these problems have unique solutions

$$u^N = (u_i^N)_{1\leq i\leq N} \in D(A)^N$$

$$u^{nN} = (u_i^{nN})_{1\leq i\leq N} \in D(A^n)^N.$$

Moreover, we can prove the following auxiliary result. □

Lemma 8.4.3 *the sequence* C_n *in* (8.76) *is bounded in* $\mathbb{R}$.

Proof. Since $0 \in D(A^n)$ and $0 \in A^n 0$ for each $n \geq 1$, $u_i^n \equiv 0$ is the unique solution of problem (8.76) with $u_0^n = 0$. Then Theorems 8.2.4 and 8.2.6 imply that $\|u_i^n\| \leq \|a^n\|$, $\forall n, i \geq 1$ where $u^n = (u_i^n)_{i\geq 1}$ is the solution to (8.76) with initial data $u_0^n = a^n$. Therefore $C_n = \sup_{i\geq 1} \|u_i^n\|$ is bounded in $\mathbb{R}$. □

By Theorems 8.2.6 and 8.2.4, $u_i^N \to u_i$ and $u_i^{nN} \to u_i^n$ as $N \to +\infty$, uniformly for i belonging to every finite subset of the integers. The following lemma is in Lemma 2.2 of [APR-APR2].

Lemma 8.4.4 *With the same assumptions as in Theorem 8.4.1,* $\lim_{N\to+\infty} u_i^{nN} = u_i^n$ *exists for each* $n \geq 1$, *uniformly on every finite subset of the integers. Moreover, for each integer* $N_0 \geq 1$ *and* $1 \leq i \leq N_0$, *if* θ_i *satisfies either* (8.77) *or* (8.78), *then the following respective estimates hold:*

$$\|u_i^{nN} - u_i^n\| \leq \frac{2\|a^n\|}{\sqrt{\sum_{k=N_0}^{N} \frac{1}{h_k}}} \tag{8.81}$$

or

$$\|u_i^{nN} - u_i^n\| \leq \frac{2N_0\|a^n\|}{(N - N_0 + 1)} \tag{8.82}$$

Also, similar estimates hold for the respective solutions u *and* u^N *to problems* (8.73) *and* (8.74).

Proof. (for Theorem 1.4.2). Suppose that N is a fixed integer. We have

$$\|u_i^n - u_i\| \leq \|u_i^n - u_i^{nN}\| + \|u_i^{nN} - u_i^N\| + \|u_i^N - u_i\|, \quad 1 \leq i \leq N,\ n \geq 1$$

Suppose that $1 \leq N_0 \leq N$. Then applying Lemma 8.4.4 to both problems in (8.79), we get:

$$\|u_i^n - u_i\| \leq \frac{2\|a^n\| + 2\|a\|}{\sqrt{\sum_{k=N_0}^{N} \frac{1}{h_k}}} + \|u_i^{nN} - u_i^N\|, \quad 1 \leq i \leq N_0 \tag{8.83}$$

if we assume (8.77) for θ_n, and we get:

$$\|u_i^n - u_i\| \leq \frac{2N_0(\|a^n\| + \|a\|)}{N - N_0 + 1} + \|u_i^{nN} - u_i^N\|, \quad 1 \leq i \leq N_0 \tag{8.84}$$

if θ_n satisfies (8.78). For N fixed, by Theorem 3.1 of [APR-APR1], we have $u_i^{nN} \to u_i^N$ in H as $n \to +\infty$, for $1 \leq i \leq N$. By taking limsup as $n \to +\infty$ in (8.83) and (8.84) we obtain either

$$\limsup_{n \to +\infty} \|u_i^n - u_i\| \leq \frac{4\|a\|}{\sqrt{\sum_{k=N_0}^{N} \frac{1}{h_k}}}$$

or

$$\limsup_{n \to +\infty} \|u_i^n - u_i\| \leq \frac{4N_0\|a\|}{N - N_0 + 1}, \quad 1 \leq i \leq N_0$$

respectively, depending on whether condition (8.77) or (8.78) is satisfied by θ_n. In both cases, the result follows by letting $N \to +\infty$. □

8.5 ASYMPTOTIC BEHAVIOR FOR THE HOMOGENEOUS CASE

In this section, we concentrate on the homogeneous case when $f_n \equiv 0$. We study the convergence of solutions to (8.4) and their weighted averages to a zero of the maximal monotone operator A, with some suitable assumptions on the parameters c_n and θ_n. Throughout this section, we denote by $M = \sup_{n \geq 0} \|u_n\|$, $a_n = (\theta_1 \cdots \theta_n)^{-1}$, with $a_0 = 1$, and Au_n, the element $\frac{u_{n+1} - (1+\theta_n)u_n + \theta_n u_{n-1}}{c_n}$ in H.

8.5.1 WEAK ERGODIC CONVERGENCE

In this subsection, we study the asymptotic behavior of the weighted averages of solutions to (8.4). We introduce three types of weighted averages, and prove the weak or strong convergence of each of them to a zero of the maximal monotone operator A. Moreover, we show that the existence of a solution to (8.4) implies that $A^{-1}(0) \neq \emptyset$, provided that the appropriate conditions on the parameters c_n and θ_n hold.

We denote $w_n := (\sum_{i=1}^n c_i)^{-1}(\sum_{i=1}^n c_i u_i)$, $z_n := (\sum_{i=1}^n a_i c_i)^{-1} \sum_{i=1}^n a_i c_i u_i$, $\sigma_n := (\sum_{i=1}^n \sum_{k=i}^\infty a_k c_k)^{-1}(\sum_{i=1}^n \sum_{k=i}^\infty a_k c_k u_k)$.

We will need the following elementary lemmas in order to prove our main results.

Lemma 8.5.1 *Let $\{x_i\}$ and $\{y_i\}$ be two sequences of real numbers. Then,*

$$\sum_{i=1}^{k}(\Delta x_i)y_i = x_{k+1}y_{k+1} - x_1y_1 - \sum_{i=1}^{k} x_{i+1}(\Delta y_i),$$

where $\Delta x_i = x_{i+1} - x_i$.

The following lemma was proved in Chapter 7 (see Lemma 1.6.1). We recall it for the convenience of the reader.

Lemma 8.5.2 *Let $\{a_n\}$ and $\{b_n\}$ be two sequences of real positive numbers. If $\{a_n\}$ is nonincreasing and convergent to zero and $\sum_{n=1}^{+\infty} a_nb_n < +\infty$, then $(\sum_{k=1}^{n} b_k)a_n \to 0$ as $n \to +\infty$.*

Lemma 8.5.3 *Let $\{a_n\}$ be a sequence of positive numbers with $\sum_{n=1}^{+\infty} a_n^{-1} = +\infty$. If $\{b_n\}$ is a bounded sequence, then $\liminf_{n\to+\infty} a_n(b_{n+1} - b_n) \le 0$, and $\liminf_{n\to+\infty} a_n(b_n - b_{n+1}) \le 0$.*

Proof. It suffices to show that $\liminf_{n\to+\infty} a_n(b_{n+1} - b_n) \le 0$. The second inequality is proved in a similar way. Suppose to the contrary that there is an integer $n_0 > 0$ and $\lambda > 0$ such that for each integer $n \ge n_0$, we have $a_n(b_{n+1} - b_n) > \lambda > 0$. Then dividing by a_n and summing up from $n = n_0$ to ∞, we get a contradiction. □

Lemma 8.5.4 *Let $\{u_n\}$ be a solution to* (8.4). *Then $a_{n-1}\|u_n - u_{n-1}\|$ is either nonincreasing or eventually increasing.*

Proof. From the monotonicity of A, we have

$$(Au_{i+1} - Au_i, u_{i+1} - u_i) \ge 0, \qquad \forall i \ge 1.$$

By (8.4), we get

$$\frac{1}{c_{i+1}}(u_{i+2} - u_{i+1}, u_{i+1} - u_i) - \frac{\theta_{i+1}}{c_{i+1}}\|u_{i+1} - u_i\|^2 - \frac{1}{c_i}\|u_{i+1} - u_i\|^2$$
$$+\frac{\theta_i}{c_i}(u_i - u_{i-1}, u_{i+1} - u_i) \ge 0, \qquad \forall i \ge 1.$$

It follows that

$$\frac{1}{c_i}\|u_{i+1} - u_i\| - \frac{\theta_i}{c_i}\|u_i - u_{i-1}\| \le \frac{1}{c_{i+1}}\|u_{i+2} - u_{i+1}\| - \frac{\theta_{i+1}}{c_{i+1}}\|u_{i+1} - u_i\|, \qquad (8.85)$$

for all $i \ge 1$. If $\{a_{i-1}\|u_i - u_{i-1}\|\}$ is not nonincreasing, then there exists $j \ge 1$ such that $a_j\|u_{j+1} - u_j\| > a_{j-1}\|u_j - u_{j-1}\|$. Then (8.85) and the identity $a_n\theta_n = a_{n-1}$ imply that the sequence $\{a_{i-1}\|u_i - u_{i-1}\|\}_{i\ge j}$ is increasing. □

Lemma 8.5.5 *Suppose that u_i is a solution to* (8.4) *and $p \in A^{-1}(0)$. Then $\|u_n - p\|$ is nonincreasing or eventually increasing. Moreover, if $\sum_{n=1}^{+\infty} \theta_1 \cdots \theta_n = +\infty$, then $\|u_n - p\|$ and $a_{n-1}\|u_n - u_{n-1}\|$ are nonincreasing and $a_{n-1}\|u_n - u_{n-1}\|$ converges to zero as $n \to +\infty$.*

Proof. From the monotonicity of A and (8.4), we get

$$\big(u_{i+1}-(1+\theta_i)u_i+\theta_i u_{i-1}, u_i-p\big)\geq 0. \tag{8.86}$$

This implies that

$$\|u_{i+1}-p\|^2-\|u_i-p\|^2+\theta_i(\|u_{i-1}-p\|^2-\|u_i-p\|^2)\geq \|u_{i+1}-u_i\|^2+\theta_i\|u_i-u_{i-1}\|^2\geq 0. \tag{8.87}$$

If $\|u_i-p\|$ is not nonincreasing, there is $j>0$ such that $\|u_j-p\|<\|u_{j+1}-p\|$. Then by (8.87), the sequence $\{\|u_i-p\|\}_{i\geq j+1}$ is increasing.
For the second part of the lemma, multiplying (8.87) by a_i, we get

$$a_{i-1}\|u_i-u_{i-1}\|^2\leq a_i(\|u_{i+1}-p\|^2-\|u_i-p\|^2)-a_{i-1}(\|u_i-p\|^2-\|u_{i-1}-p\|^2).$$

Summing up from $i=k$ to m, we get:

$$\sum_{i=k}^{m}a_{i-1}\|u_i-u_{i-1}\|^2\leq a_m(\|u_{m+1}-p\|^2-\|u_m-p\|^2)-a_{k-1}(\|u_k-p\|^2-\|u_{k-1}-p\|^2).$$

Taking liminf when $m\to+\infty$, by our assumption and Lemma 8.5.3,

$$\liminf_{m\to+\infty}a_m(\|u_{m+1}-p\|^2-\|u_m-p\|^2)\leq 0,$$

then we get:

$$\sum_{i=k}^{\infty}a_{i-1}\|u_i-u_{i-1}\|^2\leq a_{k-1}(\|u_{k-1}-p\|^2-\|u_k-p\|^2). \tag{8.88}$$

(8.88) implies that $\{\|u_k-p\|^2\}$ is nonincreasing and $\sum_{i=1}^{\infty}a_{i-1}^{-1}a_{i-1}^2\|u_i-u_{i-1}\|^2<+\infty$. The assumption on $\{\theta_i\}$ implies that $\liminf_{i\to+\infty}a_{i-1}\|u_i-u_{i-1}\|=0$. By Lemma 8.5.4, $a_{i-1}\|u_i-u_{i-1}\|$ is nonincreasing and therefore $\lim_{i\to+\infty}a_{i-1}\|u_i-u_{i-1}\|=0$. □

Theorem 8.5.6 *Let $\{u_n\}_{n\geq 1}$ be a solution to (8.4). Assume that $\sum_{n=1}^{+\infty}c_n=+\infty$, $\sum_{n=1}^{\infty}\theta_1\cdots\theta_n=+\infty$ and $\frac{\sum_{k=1}^{n}|\theta_k-\theta_{k-1}|}{\sum_{k=1}^{n}c_k}\to 0$ as $n\to+\infty$. Then $A^{-1}(0)\neq\emptyset$ and $w_n\rightharpoonup p\in A^{-1}(0)$.*
In particular, the latter condition holds if either one of the following two conditions is satisfied:
1) $\sum_{n=1}^{\infty}|\theta_n-\theta_{n-1}|<+\infty$
2) $\lim_{n\to+\infty}\frac{|\theta_n-\theta_{n-1}|}{c_n}=0.$

Proof. By the monotonicity of A and (8.4), we have for all $k,n\geq 1$,

$$\big(u_{n+1}-(1+\theta_n)u_n+\theta_n u_{n-1}, c_k u_k\big)+\big(u_{k+1}-(1+\theta_k)u_k+\theta_k u_{k-1}, c_n u_n\big)$$

$$\leq c_k\big(u_{n+1}-(1+\theta_n)u_n+\theta_n u_{n-1}, u_n\big)+c_n\big(u_{k+1}-(1+\theta_k)u_k+\theta_k u_{k-1}, u_k\big). \tag{8.89}$$

Summing up the above inequality from $k=1$ to m, and using the technique of Lemma 8.5.1, we get:

$$\begin{aligned}
&\big(u_{n+1}-(1+\theta_n)u_n+\theta_n u_{n-1},\sum_{k=1}^{m}c_k u_k\big)\\
&\le \sum_{k=1}^{m}c_k\big(u_{n+1}-(1+\theta_n)u_n+\theta_n u_{n-1},u_n\big)-\big(u_{m+1}-u_1,c_n u_n\big)\\
&+\theta_m(u_m,c_n u_n)-\theta_0(u_0,c_n u_n)\\
&-\sum_{k=1}^{m}(\theta_k-\theta_{k-1})(u_{k-1},c_n u_n)+\frac{c_n}{2}[\|u_{m+1}\|^2-\|u_1\|^2-\theta_m\|u_m\|^2\\
&+\theta_0\|u_0\|^2+\sum_{k=1}^{m}(\theta_k-\theta_{k-1})\|u_{k-1}\|^2].
\end{aligned}$$

Suppose that $w_{m_j}\rightharpoonup p$. Then dividing both sides by $\sum_{k=1}^{m}c_k$, substituting m by m_j and letting $j\to+\infty$ in the above inequality, we get (8.86). Now the proof of Lemma 8.5.5 implies that $\|u_n-p\|$ is nonincreasing. If q is another cluster point of w_n, then $\|u_n-q\|$ is nonincreasing too. Therefore $\lim_{n\to+\infty}(u_n,p-q)$ exists. This implies that $\lim_{n\to+\infty}(w_n,p-q)$ exists. Then $(p,p-q)=(q,p-q)$, which implies that $p=q$. Therefore $w_n\rightharpoonup p$. Now we prove that $p\in A^{-1}(0)$. Let $[x,y]\in A$. By the monotonicity of A, we have:

$$(y,u_k-x)\le(Au_k,u_k-x).$$

Multiplying the above inequality by c_k and summing up from $k=1$ to m, by applying Lemma 8.5.1, we get:

$$\begin{aligned}
\sum_{k=1}^{m}c_k(y,u_k-x)&\le\frac{1}{2}[(\|u_{m+1}-x\|^2-\|u_1-x\|^2)\\
&\quad-\sum_{k=1}^{m}\theta_k(\|u_k-x\|^2-\|u_{k-1}-x\|^2)]\\
&\le\frac{1}{2}(\|u_{m+1}-x\|^2-\|u_1-x\|^2)-\frac{\theta_m}{2}\|u_m-x\|^2\\
&\quad+\frac{\theta_1}{2}\|u_0-x\|^2+\frac{1}{2}\sum_{k=1}^{m-1}(\theta_{k+1}-\theta_k)\|u_k-x\|^2\\
&\le\frac{1}{2}\|u_{m+1}-x\|^2+\frac{\theta_1}{2}\|u_0-x\|^2+\frac{1}{2}\sum_{k=1}^{m-1}(\theta_{k+1}-\theta_k)\|u_k-x\|^2.
\end{aligned}$$

Dividing both sides of the above inequality by $\sum_{k=1}^{m}c_k$, and letting $m\to+\infty$, we get

$$(y,p-x)\le 0,$$

for all $[x,y]\in A$. Since A is maximal monotone, we conclude that $p\in A^{-1}(0)$. □

Theorem 8.5.7 *Assume that* $\{u_n\}_{n\geq 1}$ *is a solution to* (8.4) *and either one of the following two conditions holds:*
1) $\theta_n \geq 1$
2) $0 < \theta_n < 1$, $\sum_{n=1}^{+\infty} \theta_1 \cdots \theta_n = +\infty$ *and* $\frac{a_n}{\sum_{k=1}^{n} a_k c_k} \to 0$ *as* $n \to +\infty$.
If $\sum_{n=1}^{+\infty} a_n c_n = +\infty$, *then* $A^{-1}(0) \neq \emptyset$. *Moreover,* z_n *converges weakly as* $n \to +\infty$ *to some* $p \in A^{-1}(0)$.

Proof. Multiplying both sides of (8.89) by a_k and summing up from $k = 1$ to m, we get:

$$\begin{aligned}
&\big(u_{n+1} - (1+\theta_n)u_n + \theta_n u_{n-1}, \sum_{k=1}^{m} a_k c_k u_k\big) \\
&+ \big(a_m(u_{m+1} - u_m) - a_0(u_1 - u_0), c_n u_n\big) \\
&\leq (\sum_{k=1}^{m} a_k c_k)\big(u_{n+1} - (1+\theta_n)u_n + \theta_n u_{n-1}, u_n\big) \\
&+ \frac{1}{2} c_n [a_m(\|u_{m+1}\|^2 - \|u_m\|^2) - a_0(\|u_1\|^2 - \|u_0\|^2)].
\end{aligned}$$

Dividing both sides by $\sum_{k=1}^{m} a_k c_k$, we get:

$$\begin{aligned}
&\big(u_{n+1} - (1+\theta_n)u_n + u_{n-1}, z_m\big) \\
&\leq (\sum_{k=1}^{m} a_k c_k)^{-1} a_0(u_0 - u_1, c_n u_n) \\
&(\sum_{k=1}^{m} a_k c_k)^{-1} a_m(u_m - u_{m+1}, c_n u_n) \\
&+ \big(u_{n+1} - (1+\theta_n)u_n + \theta_n u_{n-1}, u_n\big) \\
&+ \frac{1}{2}(\sum_{k=1}^{m} a_k c_k)^{-1} c_n a_0(\|u_0\|^2 - \|u_1\|^2) \\
&+ \frac{1}{2}(\sum_{k=1}^{m} a_k c_k)^{-1} c_n a_m(\|u_{m+1}\|^2 - \|u_m\|^2)
\end{aligned} \tag{8.90}$$

Suppose that $z_{m_j} \rightharpoonup p$. Substituting m by m_j in the above inequality, and letting $j \to +\infty$, we get (8.86). By the proof of Lemma 8.5.5, this implies that: $\|u_n - p\| \leq \|u_{n-1} - p\|$. If q is another cluster point of z_n, then there exists $\lim_{n\to+\infty} \|u_n - q\|$. This implies that there exists $\lim_{n\to+\infty}(z_n, p-q)$, then $(p, p-q) = (q, p-q)$, hence $p = q$. Therefore $z_n \rightharpoonup p$. Now let $[x,y] \in A$. By the monotonicity of A, we get

$$(y, u_k - x) \leq (Au_k, u_k - x).$$

Multiplying by $a_k c_k$ and summing up from $k = 1$ to m, we get

$$\sum_{k=1}^{m} a_k c_k (y, u_k - x)$$

$$\begin{aligned}
&\leq \sum_{k=1}^{m} a_k(u_{k+1}-u_k,u_k-x)-\sum_{k=1}^{m} a_k\theta_k(u_k-u_{k-1},u_k-x)\\
&\leq \frac{1}{2}\sum_{k=1}^{m} a_k(\|u_{k+1}-x\|^2-\|u_k-x\|^2)-\frac{1}{2}\sum_{k=1}^{m} a_{k-1}(\|u_k-x\|^2-\|u_{k-1}-x\|^2)\\
&=\frac{1}{2}a_m\|u_{m+1}-x\|^2-\frac{1}{2}a_1\|u_1-x\|^2+\frac{1}{2}\sum_{k=1}^{m-1}(a_k-a_{k+1})\|u_{k+1}-x\|^2\\
&\quad -\frac{1}{2}\|u_0-x\|^2+\frac{1}{2}a_m\|u_m-x\|^2+\frac{1}{2}\sum_{k=1}^{m}(a_{k-1}-a_k)\|u_k-x\|^2. \qquad (8.91)
\end{aligned}$$

If $\theta_n \geq 1$, (8.91) implies that

$$\sum_{k=1}^{m} a_k c_k(y,u_k-x) \leq \frac{3}{2}(M+\|x\|)^2.$$

Dividing by $\sum_{k=1}^{m} a_k c_k$ and letting $m \to +\infty$, we get $(y,p-x) \leq 0$. If $0<\theta_n<1$ we have

$$\sum_{k=1}^{m} a_k c_k(y,u_k-x) \leq (M+\|x\|)^2 a_m.$$

Dividing by $\sum_{k=1}^{m} a_k c_k$ and letting $m \to +\infty$, we get $(y,p-x) \leq 0$. By the maximality of A, we have $p \in A^{-1}(0)$, which is therefore nonempty. □

Theorem 8.5.8 *Assume that $\{u_n\}_{n\geq 1}$ is a solution to* (8.4) *with either $\theta_n \geq 1$, or $0<\theta_n<1$, $\sum_{n=1}^{+\infty}\theta_1\cdots\theta_n=+\infty$ and $\frac{a_n}{\sum_{i=1}^{n}\sum_{k=i}^{+\infty}a_kc_k}\to 0$ as $n\to+\infty$. If $\sum_{n=1}^{+\infty} na_nc_n=+\infty$, then $A^{-1}(0)\neq\emptyset$. Moreover, if $\sum_{n=1}^{+\infty}a_nc_n<+\infty$ and $\sum_{n=1}^{+\infty}na_nc_n=+\infty$, then σ_n converges weakly as $n\to+\infty$ to some $p\in A^{-1}(0)$.*

Proof. Multiplying both sides of (8.89) by a_k and summing up from $k=i$ to m, we get

$$\begin{aligned}
&\big(u_{n+1}-(1+\theta_n)u_n+\theta_n u_{n-1},\sum_{k=i}^{m} a_k c_k u_k\big)\\
&\quad+\big(a_m(u_{m+1}-u_m)-a_{i-1}(u_i-u_{i-1}),c_n u_n\big)\\
&\leq \big(\sum_{k=i}^{m} a_k c_k\big)\big(u_{n+1}-(1+\theta_n)u_n+\theta_n u_{n-1},u_n\big)\\
&\quad+\frac{1}{2}c_n\big[a_m(\|u_{m+1}\|^2-\|u_m\|^2)-a_{i-1}(\|u_i\|^2-\|u_{i-1}\|^2)\big].
\end{aligned}$$

Taking lim inf when $m\to+\infty$, and using the first inequality in Lemma 8.5.3 for the right hand side and the second for the left hand side, we get:

$$\big(u_{n+1}-(1+\theta_n)u_n+\theta_n u_{n-1},\sum_{k=i}^{+\infty} a_k c_k u_k\big)$$

$$\begin{aligned}&\leq a_{i-1}(u_i-u_{i-1},c_nu_n)\\&\quad+(\sum_{k=i}^{\infty}a_kc_k)\big(u_{n+1}-(1+\theta_n)u_n+\theta_nu_{n-1},u_n\big)\\&\quad+\frac{1}{2}c_na_{i-1}(\|u_{i-1}\|^2-\|u_i\|^2).\end{aligned}\tag{8.92}$$

From Lemma 8.5.1, if $\theta_n \geq 1$, we get:

$$\begin{aligned}&\sum_{i=1}^{m}a_{i-1}[(u_i,c_nu_n)-(u_{i-1},c_nu_n)]\\&=a_m(u_m,c_nu_n)-a_0(u_0,c_nu_n)+\sum_{i=1}^{m}(a_{i-1}-a_i)(u_i,c_nu_n)\\&\leq -a_0(u_0,c_nu_n)+c_nM^2a_m+c_nM^2(a_0-a_m)\\&\leq 2c_nM^2<+\infty,\end{aligned}\tag{8.93}$$

and if $0<\theta_n<1$, we have:

$$\begin{aligned}\sum_{i=1}^{m}a_{i-1}[(u_i,c_nu_n)-(u_{i-1},c_nu_n)]&\leq c_nM^2a_0+c_nM^2a_m+c_nM^2(a_m-a_0)\\&=2c_nM^2a_m\end{aligned}\tag{8.94}$$

If $\theta_n \geq 1$, by Lemma 8.5.1, we get:

$$\begin{aligned}\sum_{i=1}^{m}a_{i-1}(\|u_{i-1}\|^2-\|u_i\|^2)&=a_0\|u_0\|^2-a_m\|u_m\|^2+\sum_{i=1}^{m}(a_i-a_{i-1})\|u_i\|^2\\&\leq\|u_0\|^2<+\infty,\end{aligned}\tag{8.95}$$

and if $0<\theta_n<1$, we have:

$$\begin{aligned}\sum_{i=1}^{m}a_{i-1}(\|u_{i-1}\|^2-\|u_i\|^2)&\leq\|u_0\|^2-a_m\|u_m\|^2+M^2(a_m-a_0)\\&\leq M^2a_m.\end{aligned}\tag{8.96}$$

Summing up (8.92) from $i=1$ to m, and then dividing by $\sum_{i=1}^{m}\sum_{k=i}^{\infty}a_kc_k$, if $\theta_n \geq 1$ by (8.93) and (8.95), we get

$$\begin{aligned}&\big(u_{n+1}-(1+\theta_n)u_n+\theta_nu_{n-1},(\sum_{i=1}^{m}\sum_{k=i}^{\infty}a_kc_k)^{-1}\sum_{i=1}^{m}\sum_{k=i}^{+\infty}a_kc_ku_k\big)\\&\leq\big(u_{n+1}-(1+\theta_n)u_n+\theta_nu_{n-1},u_n\big)\\&\quad+\frac{1}{2}c_n(\sum_{i=1}^{m}\sum_{k=i}^{\infty}a_kc_k)^{-1}(\|u_0\|^2+4M^2),\end{aligned}$$

and if $0 < \theta_n < 1$, by (8.94) and (8.96), we get

$$\begin{aligned}
&\big(u_{n+1} - (1+\theta_n)u_n + \theta_n u_{n-1}, (\sum_{i=1}^{m}\sum_{k=i}^{\infty} a_k c_k)^{-1} \sum_{i=1}^{m}\sum_{k=i}^{+\infty} a_k c_k u_k\big) \\
&\leq \big(u_{n+1} - (1+\theta_n)u_n + \theta_n u_{n-1}, u_n\big) \\
&\quad + \frac{1}{2} c_n (\sum_{i=1}^{m}\sum_{k=i}^{\infty} a_k c_k)^{-1} (M^2 a_m + 4M^2 a_m) \\
&= \frac{5}{2} c_n (\sum_{i=1}^{m}\sum_{k=i}^{\infty} a_k c_k)^{-1} M^2 a_m + \big(u_{n+1} - (1+\theta_n)u_n + \theta_n u_{n-1}, u_n\big).
\end{aligned}$$

Assume $\sigma_{m_j} \rightharpoonup p$. Substituting m by m_j and letting $j \to +\infty$, we get (8.86). By the proof of Lemma 8.5.5, $\|u_n - p\|$ is nonincreasing. If q is another cluster point of σ_n, then there exists $\lim_{n\to+\infty} \|u_n - q\|$. This implies that there exists $\lim_{n\to+\infty} (\sigma_n, p - q)$, then $(p, p-q) = (q, p-q)$, hence $p = q$. Therefore $\sigma_n \rightharpoonup p$ as $n \to +\infty$. Now we prove that $p \in A^{-1}(0)$. Let $[x,y] \in A$. By the monotonicity of A, we have:

$$(y, u_k - x) \leq (Au_k, u_k - x).$$

Multiplying by $a_k c_k$ and then summing up from $k = i$ to m, and using the identity $a_k \theta_k = a_{k-1}$, we get:

$$\begin{aligned}
&\sum_{k=i}^{m} a_k c_k (y, u_k - x) \\
&\leq \frac{1}{2} \sum_{k=i}^{m} a_k \big(\|u_{k+1} - x\|^2 - \|u_k - x\|^2\big) - \frac{1}{2} \sum_{k=i}^{m} a_k \theta_k \big(\|u_k - x\|^2 - \|u_{k-1} - x\|^2\big) \\
&= \frac{1}{2} \big[a_m \big(\|u_{m+1} - x\|^2 - \|u_m - x\|^2\big) - a_{i-1} \big(\|u_i - x\|^2 - \|u_{i-1} - x\|^2\big)\big].
\end{aligned}$$

Since $\sum_{k=1}^{\infty} a_k c_k < +\infty$, then taking liminf when $m \to +\infty$, and using Lemma 8.5.3, we get:

$$\sum_{k=i}^{\infty} a_k c_k (y, u_k - x) \leq \frac{1}{2} a_{i-1} \big(\|u_{i-1} - x\|^2 - \|u_i - x\|^2\big).$$

Summing up from $i = 1$ to m, and dividing by $\sum_{i=1}^{m} \sum_{k=i}^{\infty} a_k c_k$, we get:

$$\begin{aligned}
&(y, \sigma_m - x) \\
&\leq \frac{1}{2} (\sum_{i=1}^{m}\sum_{k=i}^{\infty} a_k c_k)^{-1} \sum_{i=1}^{m} a_{i-1} \big(\|u_{i-1} - x\|^2 - \|u_i - x\|^2\big) \\
&= \frac{1}{2} (\sum_{i=1}^{m}\sum_{k=i}^{\infty} a_k c_k)^{-1} \big[a_0 \|u_0 - x\|^2 - a_m \|u_m - x\|^2 + \sum_{i=1}^{m} (a_i - a_{i-1}) \|u_i - x\|^2\big].
\end{aligned}$$

If $\theta_n \geq 1$, then letting $m \to +\infty$, we get $(y, x - p) \geq 0$, for all $[x,y] \in A$. Otherwise (i.e. if $0 < \theta_n < 1$),

$$(y, \sigma_m - x)$$

$$\leq \frac{1}{2}(\sum_{i=1}^{m}\sum_{k=i}^{\infty}a_kc_k)^{-1}\left[\|u_0-x\|^2+(M+\|x\|)^2\sum_{i=1}^{m}(a_i-a_{i-1})\right]$$
$$=\frac{1}{2}(\sum_{i=1}^{m}\sum_{k=i}^{\infty}a_kc_k)^{-1}\left[\|u_0-x\|^2+(M+\|x\|)^2(a_m-a_0)\right].$$

Now letting $m \to +\infty$, we get again $(y,x-p) \geq 0$, for all $[x,y] \in A$. Since A is maximal monotone, we obtain $p \in A^{-1}(0)$. □

8.5.2 STRONG ERGODIC CONVERGENCE

In this short subsection, we show that when the maximal monotone operator A is odd, the weighted average z_n converges strongly to a zero of A.

Proposition 8.5.9 *Suppose that u_n is a solution to* (8.4) *and* $A^{-1}(0) \neq \emptyset$.
1) *If* $0 < \theta_n < 1$ *and* $\sum_{n=1}^{+\infty}\theta_1\cdots\theta_n = +\infty$, *then* $\lim_{n\to+\infty} n\|u_n-u_{n-1}\| = 0$.
2) *If* $\theta_n \geq 1$, *then* $\lim na_n\|u_n-u_{n-1}\| = 0$.

Proof. 1) Let $p \in A^{-1}(0)$. From the monotonicity of A and (8.4), we get (8.86). By Lemma 8.5.5, this implies that $\|u_n-p\|$ is nonincreasing. Now we have:

$$\|u_{i+1}-p\|^2-\|u_i-p\|^2+\theta_i(\|u_{i-1}-p\|^2-\|u_i-p\|^2) \geq \|u_{i+1}-u_i\|^2+\theta_i\|u_i-u_{i-1}\|^2 \geq 0. \tag{8.97}$$

Multiplying (8.97) by a_i, we get

$$a_{i-1}\|u_i-u_{i-1}\|^2 \leq a_i(\|u_{i+1}-p\|^2-\|u_i-p\|^2)-a_{i-1}(\|u_i-p\|^2-\|u_{i-1}-p\|^2).$$

Summing up from $i=k$ to m, we get:

$$\sum_{i=k}^{m}a_{i-1}\|u_i-u_{i-1}\|^2 \leq a_m(\|u_{m+1}-p\|^2-\|u_m-p\|^2)-a_{k-1}(\|u_k-p\|^2-\|u_{k-1}-p\|^2).$$

Taking liminf when $m \to +\infty$, by our assumption and Lemma 8.5.3, we have:

$$\liminf_{m\to+\infty} a_m(\|u_{m+1}-p\|^2-\|u_m-p\|^2) \leq 0.$$

Therefore we get:

$$\sum_{i=k}^{\infty}a_{i-1}\|u_i-u_{i-1}\|^2 \leq a_{k-1}(\|u_{k-1}-p\|^2-\|u_k-p\|^2). \tag{8.98}$$

Since $\{a_i\}$ is increasing, we have

$$\sum_{i=k}^{\infty}\|u_i-u_{i-1}\|^2 \leq \|u_{k-1}-p\|^2-\|u_k-p\|^2.$$

Summing up from $k=1$ to ∞, we get:

$$\sum_{k=1}^{\infty}k\|u_k-u_{k-1}\|^2 \leq \|u_0-p\|^2-l(p)^2 \leq \|u_0-p\|^2 < +\infty,$$

where $l(p) = \lim_{k\to+\infty} \|u_k - p\|$. Therefore $\lim_{k\to+\infty} \|u_k - u_{k-1}\|^2 = 0$. Since by Lemma 8.5.5, $a_{k-1}\|u_k - u_{k-1}\|$ is nonincreasing and a_k is increasing then $\|u_k - u_{k-1}\|$ is nonincreasing. Now, by Lemma 8.5.2 $(\sum_{k=1}^{n} k)\|u_n - u_{n-1}\|^2 \to 0$, then $n\|u_n - u_{n-1}\| \to 0$ as $n \to +\infty$.
2) Summing up (8.98) from $k = 1$ to ∞, we get

$$\sum_{k=1}^{\infty}\sum_{i=k}^{\infty} a_{i-1}\|u_i - u_{i-1}\|^2 \le \sum_{k=1}^{\infty} a_{k-1}(\|u_{k-1} - p\|^2 - \|u_k - p\|^2).$$

Therefore since $a_k \le 1$, we have

$$\begin{aligned}\sum_{k=1}^{\infty} ka_{k-1}\|u_k - u_{k-1}\|^2 &\le \sum_{k=1}^{\infty}(\|u_{k-1} - p\|^2 - \|u_k - p\|^2)\\ &\le \|u_0 - p\|^2 - \lim_{k\to+\infty}\|u_k - p\|^2\\ &\le \|u_0 - p\|^2 < +\infty\end{aligned}$$

Hence $\sum_{k=1}^{\infty} \frac{k}{a_k} a_k^2 \|u_{k+1} - u_k\|^2 = \sum_{k=1}^{\infty} ka_k\|u_{k+1} - u_k\|^2 < +\infty$, because by Lemma 8.5.5, $a_k\|u_{k+1} - u_k\|$ is nonincreasing. Since it is also convergent to zero, by Lemma 8.5.2, we conclude that:

$$(\sum_{k=1}^{n} k)a_n^2\|u_{n+1} - u_n\|^2 \le (\sum_{k=1}^{n} \frac{k}{a_k})a_n^2\|u_{n+1} - u_n\|^2 \to 0$$

as $n \to +\infty$. Therefore $na_n\|u_{n+1} - u_n\| \to 0$ as $n \to +\infty$. □

Theorem 8.5.10 *Suppose that $\{u_n\}$ is a solution to* (8.4) *with $\theta_n \ge 1$ and the monotone operator A is odd. If $\sum_{n=1}^{\infty} a_n c_n = +\infty$, then $z_n \to p \in A^{-1}(0)$ as $n \to +\infty$.*

Proof. Since A is monotone and odd,

$$4\big(u_n, A(u_n)\big) = \big(u_n - (-u_n), A(u_n) - A(-u_n)\big) \ge 0.$$

Therefore (8.86) is satisfied with $p = 0$. Hence $\|u_n\|$ is nonincreasing. Since A is monotone and odd, we have:

$$\pm[(u_i, Au_j) + (u_j, Au_i)] \le (u_i, Au_i) + (u_j, Au_j).$$

Multiplying both sides by $a_i a_j c_i c_j$ and using (8.4), we get:

$$\begin{aligned}&\pm\big[(a_i(u_{i+1} - u_i) - a_{i-1}(u_i - u_{i-1}), a_j c_j u_j)\\ &+ (a_j(u_{j+1} - u_j) - a_{j-1}(u_j - u_{j-1}), a_i c_i u_i)\big]\\ \le\ & \frac{a_j c_j}{2}\{a_i(\|u_{i+1}\|^2 - \|u_i\|^2) - a_{i-1}(\|u_i\|^2 - \|u_{i-1}\|^2)\}\\ &+ \frac{a_i c_i}{2}\{a_j(\|u_{j+1}\|^2 - \|u_j\|^2) - a_{j-1}(\|u_j\|^2 - \|u_{j-1}\|^2)\}.\end{aligned} \tag{8.99}$$

Summing up from $i, j = k$ to n, we get

$$\begin{aligned}
&\pm 2\big[(a_n(u_{n+1}-u_n) - a_{k-1}(u_k - u_{k-1}), \sum_{i=k}^{n} a_i c_i u_i)\big] \\
&\leq (\sum_{i=k}^{n} a_i c_i)\{a_n(\|u_{n+1}\|^2 - \|u_n\|^2) - a_{k-1}(\|u_k\|^2 - \|u_{k-1}\|^2)\}. \qquad (8.100)
\end{aligned}$$

For any fixed k, we have

$$\lim_{n\to+\infty} \big[z_n - (\sum_{i=k}^{n} a_i c_i)^{-1} \sum_{i=k}^{n} a_i c_i u_i\big] = 0. \qquad (8.101)$$

By (8.100) and (8.101), since by Proposition 8.5.9, $\lim_{n\to+\infty} a_n\|u_{n+1}-u_n\| = 0$, we get

$$\limsup_{n\to+\infty} |(u_k - u_{k-1}, w_n)| \leq \frac{1}{2}(\|u_{k-1}\|^2 - \|u_k\|^2). \qquad (8.102)$$

On the other hand, for any $k < n$

$$\begin{aligned}
|(u_k - u_n, z_n)| &= |(\sum_{i=k+1}^{n} (u_i - u_{i-1}), z_n)| \\
&\leq \sum_{i=k+1}^{n} |(u_i - u_{i-1}, z_n)|.
\end{aligned}$$

By taking limsup, we get

$$\begin{aligned}
\limsup_{n\to+\infty} |(u_k - u_n, z_n)| &\leq \limsup_{n\to+\infty} \sum_{i=k+1}^{n} |(u_i - u_{i-1}, z_n)| \\
&\leq \sum_{i=k+1}^{+\infty} \limsup_{n\to+\infty} |(u_i - u_{i-1}, z_n)| \\
&\leq \sum_{i=k+1}^{+\infty} (\frac{1}{2}\|u_{i-1}\|^2 - \frac{1}{2}\|u_i\|^2) \\
&= \frac{1}{2}\|u_k\|^2 - \frac{1}{2}l^2,
\end{aligned}$$

where $l = \lim_{n\to+\infty}\|u_n\|$. Therefore $\lim_{k\to+\infty}\limsup_{n\to+\infty}|(u_k - u_n, z_n)| = 0$. The rest of the proof is exactly similar to the proof of Lemma 1 in [BRU1] (see also Theorem 3.3, pp. 147–149 in [MOR]). Finally, $p \in A^{-1}(0)$ by the first part of Theorem 8.5.7. □

Theorem 8.5.11 *Suppose that $\{u_n\}$ is a solution to (8.4) with $0 < \theta_n < 1$ and $\liminf_{n\to+\infty}\theta_1\cdots\theta_n > 0$. If the monotone operator A is odd and $\sum_{n=1}^{\infty} a_n c_n = +\infty$, then $z_n \to p \in A^{-1}(0)$ as $n \to +\infty$.*

Proof. The assumption on θ_n implies that a_n is bounded and $\sum_{n=1}^{+\infty}\theta_1\cdots\theta_n = +\infty$. Therefore by Proposition 8.5.9, we have $a_n\|u_{n+1}-u_n\| \to 0$. Now the result follows by a proof similar to that of Theorem 8.5.10. Moreover, $p \in A^{-1}(0)$ by the second part of Theorem 8.5.7. □

8.5.3 WEAK CONVERGENCE OF SOLUTIONS

In this section, we study the weak convergence of u_n to a zero of A.

Theorem 8.5.12 *Suppose that u_n is a solution to (8.4) with $\theta_n \geq 1$ and $A^{-1}(0) \neq \emptyset$. If $\liminf_{n\to+\infty} \sum_{i=1}^{kn} a_{n+i}c_{n+i} > 0$, for some positive integer k, then $u_n \rightharpoonup p \in A^{-1}(0)$.*

Proof. By Proposition 8.5.9, $\lim_{n\to+\infty} na_n\|u_{n+1} - u_n\| = 0$. Assume that $u_{n_j} \rightharpoonup p$. Without loss of generality, assume that $\sum_{i=1}^{kn_j} c_{n_j+i} \geq \frac{\alpha}{2} > 0$, $\forall j \geq 1$. Then, $a_n|u_{n+1} - u_n| \leq \frac{\varepsilon_n}{n}$, where $\lim_{n\to+\infty} \varepsilon_n = 0$. By the monotonicity of A, we have:

$$(Au_k - Au_{n_j+i}, u_k - u_{n_j+i}) \geq 0.$$

Multiplying by $c_k a_{n_j+i}c_{n_j+i}$, and summing up from $i = 1$ to kn_j and dividing by $(\sum_{i=1}^{kn_j} a_{n_j+i}c_{n_j+i})^{-1}$, by (8.4) we get:

$$\begin{aligned}
&\big(u_{m+1} - (1+\theta_m)u_m + \theta_m u_{m-1}, (\sum_{i=1}^{kn_j} a_{n_j+i}c_{n_j+i})^{-1} \sum_{i=1}^{kn_j} a_{n_j+i}c_{n_j+i}u_{n_j+i}\big) + \\
&(\sum_{i=1}^{kn_j} a_{n_j+i}c_{n_j+i})^{-1} \sum_{i=1}^{kn_j} \big(a_{n_j+i}(u_{n_j+i+1} - u_{n_j+i}) - a_{n_j+i-1}(u_{n_j+i} - u_{n_j+i-1}), c_m u_m\big) \\
&\leq (u_{m+1} - (1+\theta_m)u_m + \theta_m u_{m-1}, u_m) + c_m(\sum_{i=1}^{kn_j} a_{n_j+i}c_{n_j+i})^{-1} \\
&\big[\sum_{i=1}^{kn_j} \big(a_{n_j+i}(u_{n_j+i+1} - u_{n_j+i}) - a_{n_j+i-1}(u_{n_j+i} - u_{n_j+i-1}), u_{n_j+i}\big)\big].
\end{aligned} \tag{8.103}$$

On the other hand

$$\begin{aligned}
&(\sum_{i=1}^{kn_j} a_{n_j+i}c_{n_j+i})^{-1} \sum_{i=1}^{kn_j} \big(a_{n_j+i}(u_{n_j+i+1} - u_{n_j+i}) - a_{n_j+i-1}(u_{n_j+i} - u_{n_j+i-1}), c_m u_m\big) \\
&= (\sum_{i=1}^{kn_j} a_{n_j+i}c_{n_j+i})^{-1} \big(a_{(k+1)n_j}(u_{(k+1)n_j+1} - u_{(k+1)n_j}) - a_{n_j}(u_{n_j+1} - u_{n_j}), c_m u_m\big) \\
&\leq c_m M(\sum_{i=1}^{kn_j} a_{n_j+i}c_{n_j+i})^{-1} \big[a_{(k+1)n_j}\|u_{(k+1)n_j+1} - u_{(k+1)n_j}\| + a_{n_j}\|u_{n_j+1} - u_{n_j}\|\big] \\
&\leq \frac{2}{\alpha}\Big(\frac{\varepsilon_{(k+1)n_j}}{(k+1)n_j} + \frac{\varepsilon_{n_j}}{n_j}\Big) \to 0,
\end{aligned} \tag{8.104}$$

as $j \to +\infty$. Also,

$$(\sum_{i=1}^{kn_j} a_{n_j+i}c_{n_j+i})^{-1} \sum_{i=1}^{kn_j} \big(a_{n_j+i}(u_{n_j+i+1} - u_{n_j+i}) - a_{n_j+i-1}(u_{n_j+i} - u_{n_j+i-1}), u_{n_j+i}\big)$$

$$= (\sum_{i=1}^{kn_j} a_{n_j+i}c_{n_j+i})^{-1}[\sum_{i=1}^{kn_j} (a_{n_j+i}(u_{n_j+i+1}-u_{n_j+i}) - a_{n_j+i-1}(u_{n_j+i}-u_{n_j+i-1}), u_{n_j+i}-u_{n_j})$$
$$+ \sum_{i=1}^{kn_j} (a_{n_j+i}(u_{n_j+i+1}-u_{n_j+i}) - a_{n_j+i-1}(u_{n_j+i}-u_{n_j+i-1}), u_{n_j})].$$

But

$$(\sum_{i=1}^{kn_j} a_{n_j+i}c_{n_j+i})^{-1}\sum_{i=1}^{kn_j} (a_{n_j+i}(u_{n_j+i+1}-u_{n_j+i}) - a_{n_j+i-1}(u_{n_j+i}-u_{n_j+i-1}), u_{n_j}) \to 0$$

as $j \to +\infty$, in a similar way as in (8.104). And,

$$\begin{aligned}
&(\sum_{i=1}^{kn_j} a_{n_j+i}c_{n_j+i})^{-1}\sum_{i=1}^{kn_j} (a_{n_j+i}(u_{n_j+i+1}-u_{n_j+i}) - a_{n_j+i-1}(u_{n_j+i}-u_{n_j+i-1}), u_{n_j+i}-u_{n_j})\\
&\le \frac{1}{2}(\sum_{i=1}^{kn_j} a_{n_j+i}c_{n_j+i})^{-1}\sum_{i=1}^{kn_j} [a_{n_j+i}(\|u_{n_j+i+1}-u_{n_j}\|^2 - \|u_{n_j+i}-u_{n_j}\|^2)\\
&-a_{n_j+i-1}(\|u_{n_j+i}-u_{n_j}\|^2 - \|u_{n_j+i-1}-u_{n_j}\|^2)]\\
&\le \frac{1}{2}(\sum_{i=1}^{kn_j} a_{n_j+i}c_{n_j+i})^{-1} a_{(k+1)n_j}(\|u_{(k+1)n_j+1}-u_{n_j}\|^2 - \|u_{(k+1)n_j}-u_{n_j}\|^2)\\
&\le \frac{1}{2}(\sum_{i=1}^{kn_j} a_{n_j+i}c_{n_j+i})^{-1} a_{(k+1)n_j}\|u_{(k+1)n_j+1}-u_{(k+1)n_j}\|\|u_{(k+1)n_j+1}+u_{(k+1)n_j}-2u_{n_j}\|\\
&\le 2M(\sum_{i=1}^{kn_j} a_{n_j+i}c_{n_j+i})^{-1} a_{(k+1)n_j}\|u_{(k+1)n_j+1}-u_{(k+1)n_j}\|\\
&\le 4M\frac{\varepsilon_{(k+1)n_j}}{\alpha(k+1)n_j} \to 0, \qquad (8.105)
\end{aligned}$$

as $j \to +\infty$. Then letting $j \to +\infty$ in (8.103), we get:

$$\big(u_{m+1} - (1+\theta_m)u_m + \theta_m u_{m-1}, u_m - p\big) \ge 0.$$

Therefore (8.86) holds, and by the proof of Lemma 8.5.5, $\|u_n - p\|$ is nonincreasing. Now by a similar proof as in Theorem 8.5.6, it follows that $u_n \rightharpoonup p$ as $n \to +\infty$. Finally, we need to show that $p \in A^{-1}(0)$. By Part 2 of Proposition 8.5.9, $na_n\|u_n - u_{n-1}\| \to 0$ as $n \to +\infty$. Then:

$$\begin{aligned}
na_nc_n\|Au_n\| &= na_n\|u_{n+1} - (1+\theta_n)u_n + \theta_n u_{n-1}\|\\
&\le na_n\|u_{n+1} - u_n\| + na_{n-1}\|u_n - u_{n-1}\| \to 0
\end{aligned}$$

as $n \to +\infty$. Therefore for each $\varepsilon > 0$ there exists $n_0 > 0$ such that for every $n \ge n_0$, $na_nc_n\|Au_n\| \le \varepsilon$. Then $(n_j+i)a_{n_j+i}c_{n_j+i}\|Au_{n_j+i}\| \le \varepsilon$ for each $j \ge j_0$. This implies that

$$\sum_{i=1}^{kn_j} c_{n_j+i}a_{n_j+i}\|Au_{n_j+i}\| \le \sum_{i=1}^{kn_j} \frac{\varepsilon}{n_j+i}$$

$$
\begin{aligned}
&\le \sum_{i=1}^{kn_j} \frac{\varepsilon}{n_j} \\
&= \varepsilon k, \quad \forall j \ge j_0.
\end{aligned}
$$

Therefore $\liminf_{n\to+\infty} \|Au_n\| = 0$. Now, the demiclosedness of A implies that $p \in A^{-1}(0)$. □

Theorem 8.5.13 *Suppose that u_n is a solution of* (8.4) *with* $0 < \theta_n < 1$ *and* $A^{-1}(0) \neq \emptyset$. *If* $\sum_{n=1}^{+\infty} \theta_1 \cdots \theta_n = +\infty$ *and* $\liminf_{n\to+\infty} \sum_{i=1}^{kn} c_{n+i} > 0$, *for some positive integer k, then* $u_n \rightharpoonup p \in A^{-1}(0)$.

Proof. By Proposition 8.5.9, $\lim_{n\to+\infty} n\|u_{n+1} - u_n\| = 0$. Assume that $u_{n_j} \rightharpoonup p$. Without loss of generality, assume that $\sum_{i=1}^{kn_j} c_{n_j+i} \ge \frac{\alpha}{2} > 0$, $\forall j \ge 1$. Given $\varepsilon > 0$, there is an integer n_0 such that $\|u_{n+1} - u_n\| \le \frac{\varepsilon}{n}$, $\forall n \ge n_0$. Then we have:

$$
\begin{aligned}
&(\sum_{i=1}^{kn_j} c_{n_j+i})^{-1} \sum_{i=1}^{kn_j} c_{n_j+i} \|u_{n_j+i} - u_{n_j}\| \\
&\le (\sum_{i=1}^{kn_j} c_{n_j+i})^{-1} \sum_{i=1}^{kn_j} c_{n_j+i} \sum_{l=1}^{i} \|u_{n_j+l} - u_{n_j+l-1}\| \\
&\le \varepsilon (\sum_{i=1}^{kn_j} c_{n_j+i})^{-1} \sum_{i=1}^{kn_j} c_{n_j+i} \sum_{l=1}^{i} \frac{1}{n_j + l} \\
&\le \varepsilon (\sum_{i=1}^{kn_j} c_{n_j+i})^{-1} \sum_{i=1}^{kn_j} c_{n_j+i} \int_{n_j}^{n_j+i} \frac{dx}{x} \\
&\le \varepsilon \int_{n_j}^{(k+1)n_j} \frac{dx}{x} \\
&= \varepsilon \ln(k+1), \quad \forall j \ge j_0.
\end{aligned}
\tag{8.106}
$$

Also

$$
\begin{aligned}
(\sum_{i=1}^{kn_j} c_{n_j+i})^{-1} \sum_{i=1}^{kn_j} \|u_{n_j+i+1} - u_{n_j+i}\| &\le \frac{2\varepsilon}{\alpha} \sum_{i=1}^{kn_j} \frac{1}{n_j + i + 1} \\
&\le \frac{2\varepsilon}{\alpha} \sum_{i=1}^{kn_j} \frac{1}{n_j} \\
&= \frac{2k\varepsilon}{\alpha}.
\end{aligned}
\tag{8.107}
$$

Similarly

$$
(\sum_{i=1}^{kn_j} c_{n_j+i})^{-1} \sum_{i=1}^{kn_j} \|u_{n_j+i} - u_{n_j+i-1}\| \le \frac{2k\varepsilon}{\alpha}, \quad \forall j \ge j_0. \tag{8.108}
$$

By the monotonicity of A, we have:

$$\begin{aligned}
&(\sum_{i=1}^{kn_j} c_{n_j+i})^{-1}\sum_{i=1}^{kn_j}\big(u_{n_j+i+1}-(1+\theta_{n_j+i})u_{n_j+i}+\theta_{n_j+i}u_{n_j+i-1}, c_m u_m\big)\\
&+(\sum_{i=1}^{kn_j} c_{n_j+i})^{-1}\big(u_{m+1}-(1+\theta_m)u_m+\theta_m u_{m-1}, \sum_{i=1}^{kn_j} c_{n_j+i}u_{n_j+i}\big)\\
&\le(\sum_{i=1}^{kn_j} c_{n_j+i})^{-1}c_m\sum_{i=1}^{kn_j}\big(u_{n_j+i+1}-(1+\theta_{n_j+i})u_{n_j+i}+\theta_{n_j+i}u_{n_j+i-1}, u_{n_j+i}\big)\\
&+\big(u_{m+1}-(1+\theta_m)u_m+\theta_m u_{m-1}, u_m\big).
\end{aligned}$$

Letting $j\to+\infty$, by (8.104), (8.105) and (8.106), we get

$$(u_{m+1}-(1+\theta_m)u_m+\theta_m u_{m-1}, u_m-p).\ge 0 \tag{8.109}$$

By the proof of Lemma 8.5.5, the sequence $\|u_m-p\|$ is nonincreasing. Then a similar proof as in Theorem 8.5.6 shows that $u_n\rightharpoonup p$ as $n\to+\infty$. Now we show that $p\in A^{-1}(0)$. We know that $n\|u_{n+1}-u_n\|\to 0$ as $n\to+\infty$. Therefore

$$\begin{aligned}
nc_n\|Au_n\| &= n\|u_{n+1}-(1+\theta_n)u_n+\theta_n u_{n-1}\|\\
&\le n\|u_{n+1}-u_n\|+n\theta_n\|u_n-u_{n-1}\|\to 0
\end{aligned}$$

as $n\to+\infty$, because $0<\theta_n<1$. Therefore for each $\varepsilon>0$ there exists $n_0>0$ such that for every $n\ge n_0$, $nc_n\|Au_n\|\le\varepsilon$. Then $(n_j+i)c_{n_j+i}\|Au_{n_j+i}\|\le\varepsilon$ for each $j\ge j_0$. This implies that

$$\sum_{i=1}^{kn_j} c_{n_j+i}\|Au_{n_j+i}\|\le\sum_{i=1}^{kn_j}\frac{\varepsilon}{n_j+i}\le\sum_{i=1}^{kn_j}\frac{\varepsilon}{n_j}=\varepsilon k,\ \ \forall j\ge j_0.$$

Therefore $\liminf_{n\to+\infty}\|Au_n\|=0$. Now the demiclosedness of A implies that $p\in A^{-1}(0)$. □

Corollary 8.5.14 *The conclusions of Theorems 8.5.12 and 8.5.13 hold without assuming that $A^{-1}(0)\neq\emptyset$, if we assume the additional condition* $\lim_{n\to+\infty}\frac{\sum_{k=1}^{n}|\theta_k-\theta_{k-1}|}{\sum_{k=1}^{n}c_k}=0.$

Proof. Since the condition on $\{c_n\}$ in Theorems 8.5.12 and 8.5.13 implies that $\sum_{n=1}^{\infty}c_n=+\infty$, then we know from Theorem 8.5.6 that $A^{-1}(0)\neq\emptyset$. □

Example 8.5.15 *Let* $c_n=\begin{cases}1, & \text{if } n \text{ is even}\\ \frac{1}{n^2}, & \text{if } n \text{ is odd.}\end{cases}$ *Then $\{c_n\}$ satisfies the condition in Theorems 8.5.12 and 8.5.13 with $k=1$, and either $\theta_n\equiv 1$ or $0<\theta_n<1$.*

8.5.4 STRONG CONVERGENCE OF SOLUTIONS

In this section, with additional assumptions on the maximal monotone operator A, we study the strong convergence of the sequence u_n to a zero of A.

Theorem 8.5.16 *Assume that A is strongly monotone, and let $\{u_n\}_{n\geq 1}$ be a solution to* (8.4),
1) *If $\sum_{n=1}^{\infty} na_nc_n = +\infty$, and $\theta_n \geq 1$, then u_n converges strongly as $n \to +\infty$ to some $p \in A^{-1}(0)$ and $\|u_n - p\| = o((\sum_{k=1}^{n} ka_kc_k)^{-\frac{1}{2}})$.*
2) *If $\sum_{n=1}^{\infty} nc_n = +\infty$, $0 < \theta_n < 1$, $\sum_{n=1}^{\infty} \theta_1 \cdots \theta_n = +\infty$ and $\frac{a_m}{\sum_{k=1}^{m} a_kc_k} \to 0$ as $n \to +\infty$, then u_n converges strongly as $n \to +\infty$ to some $p \in A^{-1}(0)$ and $\|u_n - p\| = o((\sum_{k=1}^{n} kc_k)^{-\frac{1}{2}})$.*

Proof. In both cases (1) and (2), we have $\sum_{n=1}^{\infty} na_nc_n = +\infty$. If $\sum_{n=1}^{\infty} a_nc_n = +\infty$, then by the strong monotonicity of A, and Theorem 8.5.7 with $p = \text{weak} - \lim_{n\to+\infty} z_n$, we get:

$$(Au_n, u_n - p) = \liminf_{k\to+\infty}(Au_n, u_n - z_k) \geq \liminf_{k\to+\infty} \alpha\|u_n - z_k\|^2 \geq \alpha\|u_n - p\|^2.$$

Similarly, if $\sum_{n=1}^{\infty} a_nc_n < +\infty$, since $\frac{a_m}{\sum_{k=1}^{m} a_kc_k} \to 0$, we have $\frac{a_m}{\sum_{i=1}^{m}\sum_{k=i}^{+\infty} a_kc_k} \to 0$, then with $p = \text{weak} - \lim_{n\to+\infty} \sigma_n$, by Theorem 8.5.8, we get:

$$(Au_n, u_n - p) = \liminf_{k\to+\infty}(Au_n, u_n - \sigma_k) \geq \liminf_{k\to+\infty} \alpha\|u_n - \sigma_k\|^2 \geq \alpha\|u_n - p\|^2.$$

Multiplying both sides of the above inequality by a_nc_n and using (8.4), we get:
$\alpha a_nc_n\|u_n - p\|^2 \leq a_n(\|u_{n+1} - p\|^2 - \|u_n - p\|^2) + a_{n-1}(\|u_{n-1} - p\|^2 - \|u_n - p\|^2)$.
If $\theta_n \geq 1$, then

$$\sum_{n=1}^{+\infty} na_nc_n\|u_n - p\|^2 < +\infty. \tag{8.110}$$

Hence $\liminf_{n\to+\infty} \|u_n - p\|^2 = 0$. Since $\|u_n - p\|$ is nonincreasing, then $u_n \to p$ as $n \to +\infty$. On the other hand, by (8.110) and Lemma 8.5.2, since $\|u_n - p\|$ is nonincreasing and converges to zero, we get $\|u_n - p\| = o((\sum_{k=1}^{n} ka_kc_k)^{-\frac{1}{2}})$. Otherwise, in the second case $0 < \theta_n < 1$, since a_n is increasing, then

$$\alpha c_n\|u_n - p\|^2 \leq (\|u_{n+1} - p\|^2 - \|u_n - p\|^2) - (\|u_n - p\|^2 - \|u_{n-1} - p\|^2).$$

Therefore

$$\sum_{n=1}^{+\infty} nc_n\|u_n - p\|^2 < +\infty$$

and by Lemma 8.5.2, $\|u_n - p\| = o((\sum_{k=1}^{n} kc_k)^{\frac{-1}{2}})$. □

Theorem 8.5.17 *Assume that u_n is a solution to* (8.4), *and A satisfies the following condition (weaker than the strong monotonicity of A):*

$$\|u - v\| \geq \alpha\|x - y\|, \quad \forall u \in Ax, \quad \forall v \in Ay$$

If either one of the following conditions holds:
1) $\theta_n \geq 1$ *and* $\limsup_{n\to+\infty} na_nc_n > 0$,
2) $0 < \theta_n < 1$, $\sum_{n=1}^{+\infty} \theta_1 \cdots \theta_n = +\infty$ *and* $\limsup_{n\to+\infty} nc_n > 0$,
then $A^{-1}(0) \neq \emptyset$ *and* $u_n \to p \in A^{-1}(0)$ *as* $n \to +\infty$.

Proof. In both cases $\sum_{n=1}^{\infty} na_nc_n = +\infty$. Theorem 8.5.8 implies that $A^{-1}(0) \neq \emptyset$. Let $p = \text{weak} - \lim_{n\to+\infty} z_n$ or σ_n, as in Theorems 8.5.7 and 8.5.8. It follows from (8.4) that:

$$\begin{aligned} \alpha na_nc_n\|u_n - p\| &\leq na_nc_n\|Au_n\| \\ &= na_n\|u_{n+1} - (1+\theta_n)u_n + \theta_n u_{n-1}\| \\ &\leq na_n\|u_{n+1} - u_n\| + na_{n-1}\|u_n - u_{n-1}\| \end{aligned} \tag{8.111}$$

By Part 2 of Proposition 8.5.9 for $\theta_n \geq 1$, $na_nc_n\|u_n - p\| \to 0$ as $n \to +\infty$. Therefore $u_n \to p$ as $n \to +\infty$. Otherwise if $0 < \theta_n < 1$, from (8.111), we have

$$\alpha nc_n\|u_n - p\| \leq n\|u_{n+1} - u_n\| + n\|u_n - u_{n-1}\|.$$

The result follows now from Part 1 of Proposition 8.5.9. □

Theorem 8.5.18 *Assume that* $(I+A)^{-1}$ *is compact, and let* $\{u_n\}_{n\geq 1}$ *be a solution to* (8.4). *If either*
1) $\theta_n \geq 1$, *and* $\sum_{n=1}^{\infty} na_n^2c_n^2 = +\infty$, *or*
2) $0 < \theta_n < 1$, $\sum_{n=1}^{\infty} \theta_1 \cdots \theta_n = +\infty$, $\frac{a_m}{\sum_{k=1}^{m} a_kc_k} \to 0$ *as* $m \to \infty$ *and* $\sum_{n=1}^{\infty} nc_n^2 = +\infty$,
then u_n *converges strongly as* $n \to +\infty$ *to some* $p \in A^{-1}(0)$.

Proof. 1) If $\sum_{n=1}^{\infty} a_nc_n = +\infty$, then by Theorem 8.5.7, $A^{-1}(0) \neq \emptyset$. Otherwise $\sum_{n=1}^{\infty} na_nc_n = +\infty$ and by Theorem 8.5.8, we have $A^{-1}(0) \neq \emptyset$. By (8.4), we have

$$c_nAu_n + (1+\theta_n)u_n = u_{n+1} + \theta_n u_{n-1}.$$

Subtracting $(1+\theta_n)p$ from both sides of the above equation, where $p \in A^{-1}(0)$, and then multiplying by a_n, we get

$$c_na_nAu_n + (a_n + a_{n-1})(u_n - p) = a_n(u_{n+1} - p) + a_{n-1}(u_{n-1} - p).$$

Squaring both sides and using the monotonicity of A, we get:

$$\begin{aligned} c_n^2a_n^2\|Au_n\|^2 &\leq (a_n^2 + a_na_{n-1})(\|u_{n+1} - p\|^2 - \|u_n - p\|^2) \\ &\quad + (a_{n-1}^2 + a_na_{n-1})(\|u_{n-1} - p\|^2 - \|u_n - p\|^2). \end{aligned} \tag{8.112}$$

Since $\theta_n \geq 1$ (note that in this case $a_n \leq 1$ and $a_n \leq a_{n-1}$) and $\|u_n - p\|$ is nonincreasing, we have

$$c_n^2a_n^2\|Au_n\|^2 \leq 2a_n^2(\|u_{n+1} - p\|^2 - \|u_n - p\|^2) - 2a_{n-1}^2(\|u_n - p\|^2 - \|u_{n-1} - p\|^2).$$

Summing up from $n = k$ to infinity, we get

$$\sum_{n=k}^{+\infty} c_n^2 a_n^2 \|Au_n\|^2 \le \lim_{n\to+\infty} 2a_n^2(\|u_{n+1}-p\|^2 - \|u_n - p\|^2) - 2a_{k-1}^2(\|u_k - p\|^2 - \|u_{k-1}-p\|^2)$$
$$= 2a_{k-1}^2(\|u_{k-1}-p\|^2 - \|u_k - p\|^2).$$

Now summing up from $k = 1$ to $+\infty$, we obtain

$$\sum_{k=1}^{+\infty}\sum_{n=k}^{+\infty} c_n^2 a_n^2 \|Au_n\|^2 \le 2\sum_{k=1}^{\infty} a_{k-1}^2(\|u_{k-1}-p\|^2 - \|u_k - p\|^2)$$
$$\le 2\sum_{k=1}^{\infty}(\|u_{k-1}-p\|^2 - \|u_k - p\|^2) < +\infty.$$

This implies that

$$\sum_{n=1}^{\infty} n a_n^2 c_n^2 \|Au_n\|^2 < +\infty.$$

Therefore $\liminf_{n\to+\infty}\|Au_n\| = 0$. Let u_{n_j} be a subsequence of $\{u_n\}$ such that $\lim_{j\to+\infty}\|Au_{n_j}\| = 0$. Then $\{u_{n_j} + Au_{n_j}\}$ is bounded. Since $(I+A)^{-1}$ is compact, there exists a subsequence of $\{u_{n_j}\}$, denoted again by $\{u_{n_j}\}$, such that $u_{n_j} \to q$. By the monotonicity of A, we have $(Au_n - Au_{n_j}, u_n - u_{n_j}) \ge 0$. Letting $j \to +\infty$, we get $(Au_n, u_n - q) \ge 0$. This implies that $\|u_n - q\|$ is nonincreasing and hence $u_n \to q = p$.
2) The assumption $\sum_{n=1}^{\infty} nc_n^2 = +\infty$ implies that $\sum_{n=1}^{\infty} na_n^2 c_n^2 = +\infty$. A similar argument as in Part 1 implies that $A^{-1}(0) \neq \emptyset$. Let $p \in A^{-1}(0)$. Now the assumption $0 < \theta_n < 1$ (note that in this case $a_{n-1} < a_n$ and $a_n > 1$), together with (8.112) implies that

$$c_n^2 a_n^2 \|Au_n\|^2 \le (a_n + a_{n-1})\{a_n(\|u_{n+1}-p\|^2 - \|u_n - p\|^2)$$
$$- a_{n-1}(\|u_n - p\|^2 - \|u_{n-1}-p\|^2)\}$$
$$\le 2a_n^2(\|u_{n+1}-p\|^2 - \|u_n - p\|^2) + 2a_n a_{n-1}(\|u_{n-1}-p\|^2 - \|u_n - p\|^2)$$
$$\le 2a_n^2(\|u_{n+1}-p\|^2 - \|u_n - p\|^2) + 2a_n^2(\|u_{n-1}-p\|^2 - \|u_n - p\|^2).$$

Then

$$c_n^2\|Au_n\|^2 \le 2(\|u_{n+1}-p\|^2 - \|u_n - p\|^2) + 2(\|u_{n-1}-p\|^2 - \|u_n - p\|^2).$$

This implies that $\sum_{n=1}^{+\infty} nc_n^2\|Au_n\|^2 < +\infty$, and the rest of the proof is similar to Part 1 of the proof. □

In the following theorem we prove the strong convergence of solutions without additional assumptions on the monotone operator but we assume $A^{-1}(0) \neq \emptyset$.

Theorem 8.5.19 *Assume that u_n is a solution to* (8.4), $A^{-1}(0) \neq \emptyset$, *and* $\sum_{n=1}^{\infty} \theta_1 \ldots \theta_n < \infty$. *If c_n satisfies either one of the following assumptions*
1) $\liminf_{n\to+\infty} \sqrt{n}c_n > 0$,
2) $\limsup_{n\to+\infty} c_n > 0$,
then $u_n \to p$ as $n \to +\infty$, where $p \in A^{-1}(0)$; moreover, $\|u_n - p\| = O(\sum_{k=n}^{\infty} \theta_1 \ldots \theta_k)$.

Proof. If $\|u_i - p\|$ is nonincreasing, from (8.87), we get

$$\|u_i - u_{i-1}\|^2 \leq \|u_{i-1} - p\|^2 - \|u_i - p\|^2. \tag{8.113}$$

Otherwise, by Lemma 8.5.5, $\|u_i - p\|$ is eventually increasing and by (8.87), we obtain

$$\|u_{i+1} - u_i\|^2 \leq \|u_{i+1} - p\|^2 - \|u_i - p\|^2, \tag{8.114}$$

for large i. Summing up (10.14) and (8.114) from $i = 1$ to ∞, we get

$$\sum_{i=1}^{\infty} \|u_{i+1} - u_i\|^2 < +\infty. \tag{8.115}$$

On the other hand, multiplying the inequalities (10.14) and (8.114) by i and taking liminf, by Lemma 8.5.3, we get:

$$\liminf_{i \to +\infty} i\|u_{i+1} - u_i\|^2 = 0. \tag{8.116}$$

Summing up (8.85) from $n = k$ to $n = m-1$, we obtain

$$\frac{1}{c_k}\|u_{k+1} - u_k\| - \frac{1}{c_m}\|u_{m+1} - u_m\| \leq \frac{\theta_k}{c_k}\|u_k - u_{k-1}\| - \frac{\theta_m}{c_m}\|u_m - u_{m-1}\|. \tag{8.117}$$

Taking liminf as $m \to +\infty$, if $\liminf_{m \to +\infty} \sqrt{m}c_m > 0$ then, by (8.116), we get

$$\|u_{k+1} - u_k\| \leq \theta_k \|u_k - u_{k-1}\|. \tag{8.118}$$

If $\limsup c_m > 0$, then there exists a subsequence $c_{m_j} > c_0 > 0$. Substituting m by m_j in (8.117) and letting $j \to +\infty$, by (9.1.2), we get again (8.118). For each $n > m$, (8.118) implies

$$\|u_n - u_m\| = \|\sum_{k=m}^{n-1}(u_{k+1} - u_k)\| \leq \sum_{k=m}^{n-1}\|u_{k+1} - u_k\| \leq \|u_1 - u_0\| \sum_{k=m}^{n-1} \theta_1 \ldots \theta_k. \tag{8.119}$$

It follows that u_n is Cauchy, and therefore $u_n \to p \in H$. Now we prove that $p \in A^{-1}(0)$. Suppose that $[x, y] \in A$. By the monotonicity of A and (8.4), we have:

$$\begin{aligned}
(x - u_n, y) &= (x - u_n, y - Au_n) + (x - u_n, Au_n) \\
&\geq \frac{1}{c_n}(x - u_n, c_n Au_n) \\
&= \frac{1}{c_n}(x - u_n, u_{n+1} - (1 + \theta_n)u_n + \theta_n u_{n-1}) \\
&= \frac{1}{c_n}(x - u_n, u_{n+1} - u_n) - \frac{\theta_n}{c_n}(x - u_n, u_n - u_{n-1}) \\
&\geq (\frac{-\sqrt{n}}{\sqrt{n}c_n}\|u_{n+1} - u_n\| - \frac{\sqrt{n}}{\sqrt{n}c_n}\|u_{n-1} - u_n\|)(M + \|x\|),
\end{aligned}$$

where $M := \sup_{n\geq 0}\|u_n\|$. If $\liminf_{n\to+\infty}\sqrt{n}c_n > 0$, from the above inequality and by (8.116), we get $(x-p,y)\geq 0$. If $\limsup_{n\to+\infty} c_n > 0$, there is a subsequence $c_{n_j}\geq c_0 > 0$; substituting n_j for n in the above inequality, letting $j\to+\infty$, then using (9.1.2) we get $(x-p,y)\geq 0$. Since A is maximal monotone then $p\in A^{-1}(0)$. Now by letting $n\to+\infty$ in (8.119), we get

$$\|u_m - p\| \leq \|u_0 - u_1\| \sum_{k=m}^{\infty} \theta_1\cdots\theta_k.$$

This implies that $\|u_m - p\| = O(\sum_{k=m}^{\infty}\theta_1\cdots\theta_k)$. □

8.6 SUBDIFFERENTIAL CASE

In this section, we consider an important special case where the maximal monotone operator A is the subdifferential of a proper, convex and lower semicontinuous function φ. We study the weak convergence of the sequence given by (8.4) to a minimum point of φ, as well as the rate of convergence of $\varphi(u_n)$ to the minimum value of φ, which is important in optimization problems. We also prove the strong convergence of u_n with additional assumptions on φ. In this section we also concentrate on the homogeneous case i.e. $f_n\equiv 0$.

Theorem 8.6.1 *Suppose that u_n is a solution to* (8.4) *with $A=\partial\varphi$, where $\varphi : H\to(-\infty,+\infty]$ is a proper, convex and lower semicontinuous function and $A^{-1}(0)\neq\emptyset$. If $\theta_n\geq 1$ and $\sum_{n=1}^{+\infty} nc_na_n = +\infty$, then $u_n\rightharpoonup p\in A^{-1}(0)$, which is a minimum point of φ; moreover, $\varphi(u_n)-\varphi(p) = o((\sum_{i=1}^{n} ic_ia_i)^{-1})$.*

Proof. By the subdifferential inequality, Lemma 8.5.5 and (8.4), we have

$$\begin{aligned}
\varphi(u_i)-\varphi(u_{i-1}) &\leq (\partial\varphi(u_i), u_i-u_{i-1})\\
&= \frac{1}{c_i}(u_{i+1}-(1+\theta_i)u_i+\theta_i u_{i-1}, u_i-u_{i-1})\\
&\leq \frac{1}{c_ia_i}(a_i\|u_{i+1}-u_i\| - a_{i-1}\|u_i-u_{i-1}\|)\|u_i-u_{i-1}\|\\
&\leq 0.
\end{aligned} \tag{8.120}$$

Let $p\in A^{-1}(0)$. Using again the subdifferential inequality and (8.4), we get:

$$\begin{aligned}
a_ic_i(\varphi(u_i)-\varphi(p)) &\leq a_i(u_{i+1}-(1+\theta_i)u_i+\theta_iu_{i-1}, u_i-p)\\
&\leq \frac{1}{2}\big[a_i(\|u_{i+1}-p\|^2-\|u_i-p\|^2)\\
&\quad -a_{i-1}(\|u_i-p\|^2-\|u_{i-1}-p\|^2)\big].
\end{aligned}$$

Summing up from $i=k$ to m, taking liminf when $m\to+\infty$, and then again summing up from $k=1$ to $+\infty$, by Lemmas 8.5.3 and 8.5.5, since $a_i\leq 1$, we obtain:

$$\sum_{i=1}^{+\infty} ic_ia_i(\varphi(u_i)-\varphi(p)) \leq \frac{1}{2}\|x-p\|^2 < +\infty. \tag{8.121}$$

From the assumption in the theorem, it follows that $\liminf_{i\to+\infty}(\varphi(u_i)-\varphi(p))=0$. Since by (8.120), $\varphi(u_i)$ is nonincreasing, then $\lim_{i\to\infty}\varphi(u_i)=\varphi(p)$. If $u_{i_j}\rightharpoonup q$ as $j\to+\infty$, then $\liminf_{j\to+\infty}\varphi(u_{i_j})\geq\varphi(q)$. Hence $\varphi(p)=\lim_{i\to+\infty}\varphi(u_i)\geq\varphi(q)$. This implies that $q\in A^{-1}(0)$. Now the weak convergence of u_n follows from Opial's Lemma [OPI]. The rate of convergence of $\varphi(u_n)$ to $\varphi(p)$ follows directly from (8.121) and Lemma 8.5.2. □

Theorem 8.6.2 *Suppose that u_n is a solution to* (8.4) *with $A=\partial\varphi$, where $\varphi: H\to(-\infty,+\infty]$ is a proper, convex and lower semicontinuous function and $A^{-1}(0)\neq\emptyset$. If $0<\theta_n<1$, $\sum_{n=1}^{\infty}\theta_1\cdots\theta_n=+\infty$ and $\sum_{n=1}^{+\infty}nc_n=+\infty$, then $u_n\rightharpoonup p\in A^{-1}(0)$, which is a minimum point of φ; moreover, $\varphi(u_n)-\varphi(p)=o((\sum_{i=1}^{n}ic_i)^{-1})$.*

Proof. By (8.120) and Lemma 8.5.5, $\{\varphi(u_i)\}$ is nonincreasing. Let $p\in A^{-1}(0)$. Using again the subdifferential inequality and (8.4), we get:

$$\begin{aligned}a_ic_i(\varphi(u_i)-\varphi(p))&\leq a_i\left(u_{i+1}-(1+\theta_i)u_i+\theta_iu_{i-1},u_i-p\right)\\&\leq\frac{1}{2}\big[a_i(\|u_{i+1}-p\|^2-\|u_i-p\|^2)\\&\quad-a_{i-1}(\|u_i-p\|^2-\|u_{i-1}-p\|^2)\big].\end{aligned}$$

Summing up from $i=k$ to m, and taking liminf when $m\to+\infty$, by Lemma 8.5.3, we get

$$\begin{aligned}a_{k-1}\sum_{i=k}^{\infty}c_i(\varphi(u_i)-\varphi(p))&\leq\sum_{i=k}^{+\infty}a_ic_i(\varphi(u_i)-\varphi(p))\\&\leq\frac{1}{2}a_{k-1}(\|u_{k-1}-p\|^2-\|u_k-p\|^2)\end{aligned}$$

(where in the first inequality we used the assumption that $0<\theta_i<1$ which implies that a_i is increasing). Summing up again from $k=1$ to $+\infty$, we obtain

$$\sum_{i=1}^{+\infty}ic_i(\varphi(u_i)-\varphi(p))<+\infty. \tag{8.122}$$

The assumption in the theorem implies that $\liminf_{i\to+\infty}(\varphi(u_i)-\varphi(p))=0$. Since by (8.120), $\varphi(u_i)$ is nonincreasing then $\lim_{i\to\infty}\varphi(u_i)=\varphi(p)$. If $u_{i_j}\rightharpoonup q$ as $j\to+\infty$, then $\liminf_{j\to+\infty}\varphi(u_{i_j})\geq\varphi(q)$. Hence $\varphi(p)=\lim_{i\to+\infty}\varphi(u_i)\geq\varphi(q)$. This implies that $q\in A^{-1}(0)$. The rest of the proof is similar to that of Theorem 8.6.1. □

Theorem 8.6.3 *Assume that u_n is a solution to* (8.4) *with $A=\partial\varphi$, where $\varphi: H\to(-\infty,+\infty]$ is a proper, convex and lower semicontinuous function on H satisfying the following conditions:*

$$D(\varphi)=-D(\varphi),$$

$$\varphi(x)-\varphi(0)\geq a(\|x\|)(\varphi(-x)-\varphi(0)),\ \forall x\in D(\varphi), \tag{8.123}$$

where $a : \mathbb{R}^+ \to (0,1)$ *is a continuous function. If either one of the following two assumptions is satisfied*
a) $\theta_n \geq 1$ *and* $\sum_{n=1}^{+\infty} nc_n a_n = +\infty$,
b) $0 < \theta_n < 1$, $\sum_{n=1}^{\infty} \theta_1 \cdots \theta_n = +\infty$ *and* $\sum_{n=1}^{+\infty} nc_n = +\infty$,
then u_n *converges strongly as* $n \to +\infty$ *to some* $p \in A^{-1}(0)$, *which is a minimum point of* φ.

Proof. Without loss of generality, we may assume that $\varphi(0) = 0$. By the proof of Theorem 4.10.2 in Chapter 4, we have $\varphi(0) \leq \varphi(x)$, $\forall x \in D(\varphi)$. Then 0 is a minimum point of φ, and $0 \in (\partial\varphi)^{-1}(0)$. Hence by Lemma 8.5.5, $\|u_n\|$ is nonincreasing. By the subdifferential inequality and (8.4), we obtain

$$\begin{aligned} c_i(\varphi(u_i) - \varphi(0)) &\leq c_i(\partial\varphi(u_i), u_i) \\ &= \left(u_{i+1} - (1+\theta_i)u_i + \theta_i u_{i-1}, u_i\right) \\ &= (u_{i+1}, u_i) - (1+\theta_i)\|u_i\|^2 + \theta_i(u_{i-1}, u_i) \\ &\leq \frac{1}{2}[\|u_{i+1}\|^2 - \|u_i\|^2 + \theta_i(\|u_{i-1}\|^2 - \|u_i\|^2)]. \end{aligned}$$

Now (a) implies (8.121) and (b) implies (8.122) with $p = 0$. By (8.120), and Lemma 8.5.5, $\{\varphi(u_n)\}$ is nonincreasing. On the other hand, by (8.121) and (8.122) and our hypothesis on c_i, $\liminf_{i\to+\infty} \varphi(u_i) = 0$. Therefore $\lim_{i\to+\infty} \varphi(u_i) = 0$. For $m \leq n$,we define

$$g_m := (1 + a(\|u_n\|))(\|u_m\|^2 - \|u_n\|^2) - a(\|u_n\|)\|u_n - u_m\|^2.$$

We obtain

$$\begin{aligned} g_m - g_{m-1} &= (1 + a(\|u_n\|))(\|u_m\|^2 - \|u_{m-1}\|^2) \\ &\quad - a(\|u_n\|)(\|u_m - u_n\|^2 - \|u_{m-1} - u_n\|^2) \\ &= (1 + a(\|u_n\|))(u_m - u_{m-1}, u_m + u_{m-1}) \\ &\quad - a(\|u_n\|)(u_m - u_{m-1}, u_m + u_{m-1} - 2u_n) \\ &= \|u_m\|^2 - \|u_{m-1}\|^2 + 2a(\|u_n\|)(u_m - u_{m-1}, u_n) \end{aligned}$$

Similarly, for $m < n$, we have:

$$g_m - g_{m+1} = \|u_m\|^2 - \|u_{m+1}\|^2 + 2a(\|u_n\|)(u_m - u_{m+1}, u_n).$$

On the other hand, since φ is convex, by (8.4) and the assumption on φ, for $m \leq n$, we get

$$\begin{aligned} \varphi(u_m) &\geq \varphi(u_n) \\ &\geq a(\|u_n\|)\varphi(-u_n) \\ &\geq \varphi(-a(\|u_n\|)u_n) \\ &\geq \varphi(u_m) + \left(\partial\varphi(u_m), -a(\|u_n\|)u_n - u_m\right) \\ &= \varphi(u_m) + \frac{1}{c_m}\left((1+\theta_m)u_m - u_{m+1} - \theta_m u_{m-1}, a(\|u_n\|)u_n + u_m\right). \end{aligned}$$

This implies that

$$\theta_m(u_m - u_{m-1}, a(\|u_n\|)u_n + u_m) + (u_m - u_{m+1}, a(\|u_n\|)u_n + u_m) \le 0. \tag{8.124}$$

From (8.124), we get:

$$\begin{aligned}
(1+\theta_m)g_m - g_{m+1} - \theta_m g_{m-1} &= \theta_m(g_m - g_{m-1}) + g_m - g_{m+1} \\
&= \theta_m\|u_m\|^2 - \theta_m\|u_{m-1}\|^2 \\
&\quad + 2\theta_m a(\|u_n\|)(u_m - u_{m-1}, u_n) \\
&\quad + \|u_m\|^2 - \|u_{m+1}\|^2 + 2a(\|u_n\|)(u_m - u_{m+1}, u_n) \\
&\le \theta_m\|u_m\|^2 - \theta_m\|u_{m-1}\|^2 + \|u_m\|^2 - \|u_{m+1}\|^2 \\
&\quad - 2\theta_m(u_m - u_{m-1}, u_m) - 2(u_m - u_{m+1}, u_m) \\
&= -\theta_m\|u_m\|^2 - \|u_m\|^2 - \theta_m\|u_{m-1}\|^2 - \|u_{m+1}\|^2 \\
&\quad + 2\theta_m(u_{m-1}, u_m) + 2(u_{m+1}, u_m) \\
&= -\theta_m\big(\|u_m\|^2 + \|u_{m-1}\|^2 - 2(u_{m-1}, u_m)\big) \\
&\quad - \big(\|u_m\|^2 + \|u_{m+1}\|^2 - 2(u_{m+1}, u_m)\big) \le 0.
\end{aligned}$$

Hence

$$\theta_m(g_m - g_{m-1}) \le g_{m+1} - g_m. \tag{8.125}$$

It is obvious that $g_n = 0$. We prove that $g_{n-1} \ge 0$. Since $0 < a(\|u_n\|) < 1$, and $\{u_n\}$ is nonincreasing, we get

$$\begin{aligned}
g_{n-1} &= (1 + a(\|u_n\|))(\|u_{n-1}\|^2 - |u_n\|^2) - a(\|u_n\|)\|u_{n-1} - u_n\|^2 \\
&\ge \|u_{n-1}\|^2 - \|u_n\|^2 - \|u_{n-1} - u_n\|^2 \\
&= \|u_{n-1}\|^2 - \|u_n\|^2 - \|u_{n-1}\|^2 - \|u_n\|^2 + 2(u_{n-1}, u_n) \\
&= 2(u_n, u_{n-1} - u_n) \\
&= 2\big(u_n, \frac{c_n}{\theta_n}\partial\varphi(u_n) + \frac{1}{\theta_n}(u_n - u_{n+1})\big) \\
&= \frac{2}{\theta_n}(u_n, u_n - u_{n+1}) + 2\frac{c_n}{\theta_n}(\partial\varphi(u_n), u_n) \\
&\ge \frac{2}{\theta_n}\|u_n\|^2 - \frac{2}{\theta_n}(u_n, u_{n+1}) \\
&\ge \frac{2}{\theta_n}\|u_n\|^2 - \frac{1}{\theta_n}\|u_n\|^2 - \frac{1}{\theta_n}\|u_{n+1}\|^2 \\
&= \frac{1}{\theta_n}(\|u_n\|^2 - \|u_{n+1}\|^2) \ge 0
\end{aligned}$$

It follows from (8.125) that $g_m \ge 0$ for all $m \le n$. This implies that

$$\|u_m - u_n\|^2 \le \frac{1 + a(\|u_n\|)}{a(\|u_n\|)}(\|u_m\|^2 - \|u_n\|^2)$$

$$< \frac{2}{a(\|u_n\|)}(\|u_m\|^2 - \|u_n\|^2).$$

If $\|u_n\| \to 0$, then $u_n \to 0$ and this yields the theorem. Otherwise, if $\|u_n\| \to r > 0$, from the continuity of a, we have

$$\lim_{n\to+\infty} a(\|u_n\|) = a(\lim_{n\to+\infty} \|u_n\|) = a(r) > 0.$$

Therefore $\{u_n\}$ is a Cauchy sequence in H, hence $u_n \to p \in A^{-1}(0)$ as $n \to +\infty$. □

8.7 ASYMPTOTIC BEHAVIOR FOR THE NON-HOMOGENEOUS CASE

8.7.1 MEAN ERGODIC CONVERGENCE

In this section, we study the weak convergence of the weighted average $v_n := (\sum_{k=1}^n \frac{c_k}{\theta_k})^{-1}(\sum_{k=1}^n \frac{c_k}{\theta_k} u_k)$, to a zero of the monotone operator A for the nonhomogeneous case. Throughout this section, we denote the element $\frac{1}{c_n}(u_{n+1} - (1+\theta_n)u_n + \theta_n u_{n-1} - f_n)$ in H by Au_n. We consider the following assumptions on θ_n, c_n and f_n: θ_n is a sequence of positive real numbers satisfying one of the following assumptions:
T_1) $\theta_n \geq 1$ and nonincreasing.
T_2) θ_n is a monotone sequence, and $\theta := \lim_{n\to+\infty} \theta_n \neq 0, 1$, and $+\infty$.
T_3) θ_n is a monotone sequence with $0 < \theta_n < 1$, and $\sum_{n=1}^{+\infty}(1-\theta_n) < +\infty$.
c_n is a positive sequence satisfying one of the following assumptions:
C_1) $\sum_{n=1}^{\infty} \frac{c_n}{\theta_n} = +\infty$.
C_2) $\liminf_{n\to+\infty} c_n > 0$.
Finally f_n is a sequence in H satisfying one of the following assumptions:
F_1)$\sum_{n=1}^{\infty} \|f_n\| < +\infty$.
F_2) $\sum_{n=1}^{\infty} n\frac{\|f_n\|}{\theta_n} < +\infty$.

Theorem 8.7.1 *Assume that $\{u_n\}_{n\geq 1}$ is a solution to* (8.4). *If θ_n, c_n and $\{f_n\}_{n\geq 1}$ satisfy the assumptions* (T_1), (C_1) *and* (F_2), *respectively, then v_n converges weakly as $n \to +\infty$ to some $p \in A^{-1}(0)$, which is also the asymptotic center of $\{u_n\}_{n\geq 1}$.*

Proof. For all $k, n \geq 0$, by the monotonicity of A, we have

$$(Au_n, u_k) + (Au_k, u_n) \leq (Au_n, u_n) + (Au_k, u_k).$$

Multiplying both sides by $c_n c_k$ and using by (8.4), we get

$$\begin{aligned}&(u_{n+1} - (1+\theta_n)u_n + \theta_n u_{n-1} - f_n, c_k u_k)\\&+ (u_{k+1} - (1+\theta_k)u_k + \theta_k u_{k-1} - f_k, c_n u_n)\\&\leq c_k(u_{n+1} - (1+\theta_n)u_n + \theta_n u_{n-1} - f_n, u_n)\end{aligned}$$

$$+c_n(u_{k+1}-(1+\theta_k)u_k+\theta_k u_{k-1}-f_k,u_k). \tag{8.126}$$

Dividing both sides of the above inequality by θ_n and rearranging the terms, we obtain

$$\begin{aligned}
&(u_{n-1}-u_n,c_ku_k)+(u_{k+1}-u_k,\frac{c_n}{\theta_n}u_n)+\theta_k(u_{k-1}-u_k,\frac{c_n}{\theta_n}u_n)\\
&\le \frac{c_k}{\theta_n}(u_{n+1}-u_n,u_n)-\frac{c_k}{\theta_n}(u_{n+1}-u_n,u_k)+(\frac{f_n}{\theta_n},c_ku_k)+(f_k,\frac{c_n}{\theta_n}u_n)\\
&-c_k(\frac{f_n}{\theta_n},u_n)-\frac{c_n}{\theta_n}(f_k,u_k)+c_k(u_{n-1}-u_n,u_n)+\frac{c_n}{\theta_n}(u_{k+1}-u_k,u_k)\\
&+\frac{c_n}{\theta_n}\theta_k(u_{k-1}-u_k,u_k).
\end{aligned}$$

Summing both sides of this inequality from $n=1$ to m and dividing by $\sum_{n=1}^m \frac{c_n}{\theta_n}$, we get:

$$\begin{aligned}
&(\sum_{n=1}^m \frac{c_n}{\theta_n})^{-1}(u_0-u_m,c_ku_k)\\
&+\big(u_{k+1}-u_k,(\sum_{n=1}^m \frac{c_n}{\theta_n})^{-1}\sum_{n=1}^m \frac{c_n}{\theta_n}u_n\big)\\
&+\theta_k\big(u_{k-1}-u_k,(\sum_{n=1}^m \frac{c_n}{\theta_n})^{-1}\sum_{n=1}^m \frac{c_n}{\theta_n}u_n\big)\\
&\le c_k(\sum_{n=1}^m \frac{c_n}{\theta_n})^{-1}\sum_{n=1}^m \frac{1}{\theta_n}\big(u_{n+1}-u_n,u_n-u_k\big)\\
&+(\sum_{n=1}^m \frac{c_n}{\theta_n})^{-1}(\sum_{n=1}^m \frac{f_n}{\theta_n},c_ku_k)\\
&+\big(f_k,(\sum_{n=1}^m \frac{c_n}{\theta_n})^{-1}\sum_{n=1}^m \frac{c_n}{\theta_n}u_n\big)\\
&-c_k(\sum_{n=1}^m \frac{c_n}{\theta_n})^{-1}\sum_{n=1}^m(\frac{f_n}{\theta_n},u_n)-(f_k,u_k)\\
&+c_k(\sum_{n=1}^m \frac{c_n}{\theta_n})^{-1}\sum_{n=1}^m[\frac{1}{2}\|u_{n-1}\|^2-\frac{1}{2}\|u_n\|^2]\\
&+(u_{k+1}-u_k,u_k)+\theta_k(u_{k-1}-u_k,u_k).
\end{aligned}$$

Since $\{v_m\}$ is bounded, there exists a subsequence $\{v_{m_j}\}$ of $\{v_m\}$ such that $v_{m_j}\rightharpoonup p$; substituting m_j by m in the above inequality and letting $j\to+\infty$, we get:

$$\begin{aligned}
&(u_{k+1}-u_k,p)+\theta_k(u_{k-1}-u_k,p)\\
&\le \limsup_{j\to+\infty} c_k(\sum_{n=1}^{m_j}\frac{c_n}{\theta_n})^{-1}\sum_{n=1}^{m_j}\frac{1}{\theta_n}(u_{n+1}-u_n,u_n-u_k)
\end{aligned}$$

$$
\begin{aligned}
&+\lim_{j\to+\infty}\Big(\sum_{n=1}^{m_j}\frac{c_n}{\theta_n}\Big)^{-1}\Big(\sum_{n=1}^{m_j}\frac{\|f_n\|}{\theta_n}\Big)c_k\|u_k\|+(f_k,p)\\
&+(u_{k+1}-u_k,u_k)-(f_k,u_k)+\theta_k(u_{k-1}-u_k,u_k)\\
&=\limsup_{j\to+\infty}c_k\Big(\sum_{n=1}^{m_j}\frac{c_n}{\theta_n}\Big)^{-1}\sum_{n=1}^{m_j}\frac{1}{\theta_n}[(u_{n+1}-u_k,u_n-u_k)-\|u_n-u_k\|^2]\\
&+(f_k,p)+(u_{k+1}-u_k,u_k)-(f_k,u_k)+\theta_k(u_{k-1}-u_k,u_k)\\
&\leq\limsup_{j\to+\infty}c_k\Big(\sum_{n=1}^{m_j}\frac{c_n}{\theta_n}\Big)^{-1}\sum_{n=1}^{m_j}\frac{1}{2\theta_n}[\|u_{n+1}-u_k\|^2-\|u_n-u_k\|^2]\\
&+(f_k,p)+(u_{k+1}-u_k,u_k)-(f_k,u_k)+\theta_k(u_{k-1}-u_k,u_k)\\
&\leq\lim_{j\to+\infty}c_k\Big(\sum_{n=1}^{m_j}\frac{c_n}{\theta_n}\Big)^{-1}\sum_{n=1}^{m_j}\Big[\frac{1}{2\theta_{n+1}}\|u_{n+1}-u_k\|^2-\frac{1}{2\theta_n}\|u_n-u_k\|^2\Big]\\
&+(f_k,p)+(u_{k+1}-u_k,u_k)-(f_k,u_k)+\theta_k(u_{k-1}-u_k,u_k)\\
&=(f_k,p)+(u_{k+1}-u_k,u_k)-(f_k,u_k)+\theta_k(u_{k-1}-u_k,u_k)
\end{aligned}
$$

(where in the last inequality we used the assumption T_1). Therefore, we proved the following inequality:

$$
\begin{aligned}
(u_{k+1}-u_k,p)+\theta_k(u_{k-1}-u_k,p)\leq{}&(u_{k+1}-u_k,u_k)+\theta_k(u_{k-1}-u_k,u_k)\\
&+(f_k,p)-(f_k,u_k).
\end{aligned}
\tag{8.127}
$$

This implies that

$$
\begin{aligned}
\theta_k\|u_k-u_{k-1}\|^2\leq{}&(p,u_k-u_{k+1})+\theta_k(p,u_k-u_{k-1})+(f_k,p)+(u_{k+1}-u_k,u_k)\\
&-(f_k,u_k)+\theta_k(u_{k-1}-u_k,u_{k-1})\\
={}&(u_{k+1}-u_k,u_k-p)-\theta_k(u_k-u_{k-1},u_{k-1}-p)\\
&+(f_k,p)-(f_k,u_k).
\end{aligned}
$$

Dividing both sides of this inequality by θ_k and using the polarization identity, we get:

$$
\begin{aligned}
\|u_k-u_{k-1}\|^2\leq{}&\frac{1}{\theta_k}[(u_{k+1}-p,u_k-p)-\|u_k-p\|^2]\\
&-[(u_k-p,u_{k-1}-p)-\|u_{k-1}-p\|^2]\\
&-\frac{1}{\theta_k}(f_k,u_k-p)\\
\leq{}&\frac{1}{2\theta_k}\|u_{k+1}-p\|^2-\frac{1}{2\theta_k}\|u_k-p\|^2\\
&-\frac{1}{2}\|u_k-p\|^2-\frac{1}{2}\|u_{k-1}-p\|^2\\
&+\frac{1}{2}\|u_k-u_{k-1}\|^2+\|u_{k-1}-p\|^2+\|\frac{f_k}{\theta_k}\|\|u_k-p\|.
\end{aligned}
$$

Since $\{\theta_k\}_{k\geq 1}$ is nonincreasing, we deduce that:

$$\begin{aligned}\|u_k-u_{k-1}\|^2 \leq{}& \frac{1}{\theta_{k+1}}\|u_{k+1}-p\|^2-\frac{1}{\theta_k}\|u_k-p\|^2\\ &+\|u_{k-1}-p\|^2-\|u_k-p\|^2+M\|\frac{f_k}{\theta_k}\|,\end{aligned}$$

where $M:=2\sup_{k\geq 0}\|u_k-p\|$. Summing up this inequality from $k=1$ to $+\infty$, we get:

$$\sum_{k=1}^{+\infty}\|u_k-u_{k-1}\|^2<+\infty. \tag{8.128}$$

From (8.127), we have

$$\big(u_{n+1}-(1+\theta_n)u_n+\theta_n u_{n-1}-f_n,u_n-p\big)\geq 0, \tag{8.129}$$

for all $n\geq 1$. It follows that

$$\begin{aligned}\|u_n-p\|-\|u_{n-1}-p\| \leq{}& \frac{1}{\theta_n}(\|u_{n+1}-p\|-\|u_n-p\|)\\ &+\frac{\|f_n\|}{\theta_n}\\ \leq{}& \frac{1}{\theta_n\theta_{n+1}}(\|u_{n+2}-p\|-\|u_{n+1}-p\|)\\ &+\frac{\|f_{n+1}\|}{\theta_n\theta_{n+1}}+\frac{\|f_n\|}{\theta_n}\\ \leq{}& \cdots\\ \leq{}& \frac{1}{\theta_n\cdots\theta_{n+m}}(\|u_{n+m+1}-p\|-\|u_{n+m}-p\|)\\ &+\sum_{k=n}^{n+m}\frac{\|f_k\|}{\theta_n\cdots\theta_k}\\ \leq{}& \frac{1}{\theta_n\cdots\theta_{n+m}}\|u_{n+m+1}-u_{n+m}\|\\ &+\sum_{k=n}^{n+m}\frac{\|f_k\|}{\theta_k}\\ \leq{}& \frac{1}{\theta_n}\|u_{n+m+1}-u_{n+m}\|\\ &+\sum_{k=n}^{n+m}\frac{\|f_k\|}{\theta_k},\end{aligned}$$

since $\theta_n\geq 1$ for all $n\geq 1$. Letting $m\to+\infty$, and using (8.128), we obtain

$$\|u_n-p\|-\|u_{n-1}-p\|\leq\sum_{k=n}^{+\infty}\|\frac{f_k}{\theta_k}\|.$$

By Assumption (F_2), this implies that $\lim_{n\to+\infty}\|u_n-p\|$ exists. If q is another cluster point of v_n, then there exists $\lim_{n\to+\infty}\|u_n-q\|$. This implies that there exists $\lim_{n\to+\infty}(\|u_n-p\|^2-\|u_n-q\|^2)$ and therefore $\lim_{n\to+\infty}(u_n,q-p)$ exists. It follows that there exists $\lim_{n\to+\infty}(v_n,p-q)$, then $(p,p-q)=(q,p-q)$. Therefore $p=q$, and v_n converges weakly to p, as $n\to+\infty$. Now we prove that $p\in A^{-1}(0)$. Let $[x,y]\in A$. Then:

$$\begin{aligned}
(y,x-v_n) &= \Big(\sum_{k=1}^{n}\frac{c_k}{\theta_k}\Big)^{-1}\sum_{k=1}^{n}\frac{c_k}{\theta_k}(y,x-u_k)\\
&\geq \Big(\sum_{k=1}^{n}\frac{c_k}{\theta_k}\Big)^{-1}\sum_{k=1}^{n}(Au_k,x-u_k)\\
&= \Big(\sum_{k=1}^{n}\frac{c_k}{\theta_k}\Big)^{-1}\sum_{k=1}^{n}\Big[\frac{-1}{\theta_k}(u_{k+1}-x,u_k-x)\\
&\quad+\frac{1+\theta_k}{\theta_k}\|u_k-x\|^2-(u_{k-1}-x,u_k-x)+\Big(\frac{f_k}{\theta_k},u_k-x\Big)\Big]\\
&\geq \Big(\sum_{k=1}^{n}\frac{c_k}{\theta_k}\Big)^{-1}\sum_{k=1}^{n}\Big[\frac{1}{2\theta_k}\|u_k-x\|^2-\frac{1}{2\theta_{k+1}}\|u_{k+1}-x\|^2\\
&\quad+\frac{1}{2}\|u_k-x\|^2-\frac{1}{2}\|u_{k-1}-x\|^2-M\|\frac{f_k}{\theta_k}\|\Big]\\
&= \Big(\sum_{k=1}^{n}\frac{c_k}{\theta_k}\Big)^{-1}\Big[\frac{1}{2\theta_1}\|u_1-x\|^2-\frac{1}{2\theta_{n+1}}\|u_{n+1}-x\|^2\\
&\quad+\frac{1}{2}\|u_n-x\|^2-\frac{1}{2}\|u_0-x\|^2-M\sum_{k=1}^{n}\|\frac{f_k}{\theta_k}\|\Big]\to 0,
\end{aligned}$$

as $n\to+\infty$, so that we get $(y,x-p)\geq 0$ for all $[x,y]\in A$. Since A is maximal monotone, it follows that $p\in A^{-1}(0)$. Finally we prove that p is the asymptotic center of $\{u_n\}$. For $x\in H$, we have

$$\|u_n-p\|^2=\|u_n-x\|^2+\|x-p\|^2+2(u_n-x,x-p).$$

Multiplying both sides of this equality by $\frac{c_n}{\theta_n}$, summing up from $n=1$ to $n=m$, dividing by $\sum_{n=1}^{m}\frac{c_n}{\theta_n}$, and letting $m\to+\infty$, we get:

$$\begin{aligned}
\lim_{n\to+\infty}\|u_n-p\|^2 &\leq \limsup_{n\to+\infty}\|u_n-x\|^2-\|x-p\|^2\\
&< \limsup_{n\to+\infty}\|u_n-x\|^2,\ \text{if } x\neq p,
\end{aligned}$$

completing the proof of the theorem. □

Theorem 8.7.2 *Assume that u_n is a solution to* (8.4). *If* (T_2), (C_1) *and* (F_1) *are satisfied, then $v_n\rightharpoonup p\in A^{-1}(0)$, and p is the asymptotic center of $\{u_n\}$.*

Proof. First, assume that θ_n is nonincreasing. Let $v_{m_j} \rightharpoonup p$. By a similar computation as in Theorem 8.7.1, we get:

$$\begin{aligned}
&(u_{k+1}-u_k,p)+\theta_k(u_{k-1}-u_k,p) \\
&\le \limsup_{j\to+\infty} c_k\Big(\sum_{n=1}^{m_j}\frac{c_n}{\theta_n}\Big)^{-1}\sum_{n=1}^{m_j}\Big[\frac{1}{2\theta_{n+1}}\|u_{n+1}-u_k\|^2-\frac{1}{2\theta_n}\|u_n-u_k\|^2\Big] \\
&+(f_k,p)+(u_{k+1}-u_k,u_k)-(f_k,u_k)+\theta_k(u_{k-1}-u_k,u_k). \qquad (8.130)
\end{aligned}$$

(In the last inequality, we used the fact that $\{\theta_n\}$ is nonincreasing). Therefore we obtain (8.127). Again, by a similar computation as in Theorem 8.7.1, we get:

$$\begin{aligned}
\|u_k-u_{k-1}\|^2 &\le \frac{1}{2\theta_k}\|u_{k+1}-p\|^2-\frac{1}{2\theta_k}\|u_k-p\|^2-\frac{1}{2}\|u_k-p\|^2-\frac{1}{2}\|u_{k-1}-p\|^2 \\
&+\frac{1}{2}\|u_k-u_{k-1}\|^2+\|u_{k-1}-p\|^2+\|\frac{f_k}{\theta_k}\|\|u_k-p\|.
\end{aligned}$$

Since $\{\theta_k\}_{k\ge 1}$ is nonincreasing, we deduce that

$$\|u_k-u_{k-1}\|^2 \le \frac{1}{\theta_{k+1}}\|u_{k+1}-p\|^2-\frac{1}{\theta_k}\|u_k-p\|^2+\|u_{k-1}-p\|^2-\|u_k-p\|^2+2M\|\frac{f_k}{\theta_k}\|, \tag{8.131}$$

where $M := \sup_{k\ge 0}\|u_k-p\|$. Then we get (8.128) and (8.129). It follows that:

$$\begin{aligned}
M\|f_n\|+\|u_{n+1}-p\|^2-\|u_n-p\|^2 &\ge \theta_n(\|u_n-p\|^2-\|u_{n-1}-p\|^2) \\
&\ge \theta_n\|u_n-p\|^2-\theta_{n-1}\|u_{n-1}-p\|^2 \qquad (8.132)
\end{aligned}$$

Summing up (8.132) from $n=k$ to $n=m$, we get:

$$\theta_{k-1}\|u_{k-1}-p\|^2-\|u_k-p\|^2 \ge \theta_m\|u_m-p\|^2-\|u_{m+1}-p\|^2-M\sum_{n=k}^{m}\|f_n\|. \tag{8.133}$$

First, suppose that $\theta>1$. Taking lim sup when $m\to+\infty$ and then lim inf when $k\to+\infty$ in (8.133), and using (F_1), we get:

$$(\theta-1)\liminf_{n\to+\infty}\|u_n-p\|^2 \ge (\theta-1)\limsup_{n\to+\infty}\|u_n-p\|^2.$$

Hence there exists $\lim_{n\to+\infty}\|u_n-p\|$. If $\theta<1$, then from (8.133), we get:

$$\|u_{m+1}-p\|^2-\theta_m\|u_m-p\|^2 \ge \|u_k-p\|^2-\theta_{k-1}\|u_{k-1}-p\|^2-M\sum_{n=k}^{m}\|f_n\| \tag{8.134}$$

Now, first taking liminf when $m\to+\infty$ and then limsup when $k\to+\infty$ in (8.134), we get:

$$(1-\theta)\liminf_{n\to+\infty}\|u_n-p\|^2 \ge (1-\theta)\limsup_{n\to+\infty}\|u_n-p\|^2, \tag{8.135}$$

which implies again that there exists $\lim_{n\to+\infty}\|u_n - p\|$. If q is another cluster point of v_n, then there exists $\lim_{n\to+\infty}\|u_n - q\|$. This implies that there exists $\lim_{n\to+\infty}(\|u_n - p\|^2 - \|u_n - q\|^2)$ and therefore $\lim_{n\to+\infty}(u_n, q-p)$ exists. It follows that there exists $\lim_{n\to+\infty}(v_n, p-q)$, hence $(p, p-q) = (q, p-q)$, and thus $p = q$. Therefore $v_n \rightharpoonup p$ as $n \to +\infty$. Now we prove that $p \in A^{-1}(0)$. Let $[x,y] \in A$; then:

$$
\begin{aligned}
(y, x - v_n) &= (\sum_{k=1}^{n} \frac{c_k}{\theta_k})^{-1} \sum_{k=1}^{n} \frac{c_k}{\theta_k}(y, x - u_k) \\
&\geq (\sum_{k=1}^{n} \frac{c_k}{\theta_k})^{-1} \sum_{k=1}^{n} (Au_k, x - u_k) \\
&= (\sum_{k=1}^{n} \frac{c_k}{\theta_k})^{-1} \sum_{k=1}^{n} [\frac{-1}{\theta_k}(u_{k+1} - x, u_k - x) + \frac{1+\theta_k}{\theta_k}\|u_k - x\|^2 \\
&\quad - (u_{k-1} - x, u_k - x) + (\frac{f_k}{\theta_k}, u_k - x)] \\
&\geq (\sum_{k=1}^{n} \frac{c_k}{\theta_k})^{-1} \sum_{k=1}^{n} [\frac{1}{2\theta_k}\|u_k - x\|^2 - \frac{1}{2\theta_{k+1}}\|u_{k+1} - x\|^2 \\
&\quad + \frac{1}{2}\|u_k - x\|^2 - \frac{1}{2}\|u_{k-1} - x\|^2 - (M + \|x - p\|)\|\frac{f_k}{\theta_k}\|] \\
&\geq (\sum_{k=1}^{n} \frac{c_k}{\theta_k})^{-1} [\frac{1}{2\theta_1}\|u_1 - x\|^2 - \frac{1}{2\theta_{n+1}}\|u_{n+1} - x\|^2 \\
&\quad + \frac{1}{2}\|u_n - x\|^2 - \frac{1}{2}\|u_0 - x\|^2 - \frac{M + \|x - p\|}{\theta} \sum_{k=1}^{n} \|f_k\|] \to 0. \qquad (8.136)
\end{aligned}
$$

By letting $n \to +\infty$, we get $(y, x - p) \geq 0$ for all $[x,y] \in A$. Since A is maximal monotone, this completes the proof for the case $\{\theta_n\}$ nonincreasing. Now suppose that $\{\theta_n\}$ is nondecreasing. In the above proof, the assumption $\{\theta_n\}$ nonincreasing was used in (8.130), (8.131), (8.132) and (8.136). In the last inequality of (8.130), if $\{\theta_n\}$ is nondecreasing, then:

$$
\begin{aligned}
&\limsup_{j\to+\infty} c_k(\sum_{n=1}^{m_j} \frac{c_n}{\theta_n})^{-1} \sum_{n=1}^{m_j} \frac{1}{2\theta_n}[\|u_{n+1} - u_k\|^2 - \|u_n - u_k\|^2] \\
&+ (f_k, p) + (u_{k+1} - u_k, u_k) - (f_k, u_k) + \theta_k(u_{k-1} - u_k, u_k) \\
&\leq \limsup_{j\to+\infty} c_k(\sum_{n=1}^{m_j} \frac{c_n}{\theta_n})^{-1} \sum_{n=1}^{m_j} [\frac{1}{2\theta_{n+1}}\|u_{n+1} - u_k\|^2 - \frac{1}{2\theta_n}\|u_n - u_k\|^2] \\
&+ \limsup_{j\to+\infty} c_k(\sum_{n=1}^{m_j} \frac{c_n}{\theta_n})^{-1} \sum_{n=1}^{m_j} (\frac{1}{2\theta_n} - \frac{1}{2\theta_{n+1}})\|u_{n+1} - u_k\|^2 \\
&+ (f_k, p) + (u_{k+1} - u_k, u_k) - (f_k, u_k) + \theta_k(u_{k-1} - u_k, u_k).
\end{aligned}
$$

Since $\frac{1}{\theta_n} - \frac{1}{\theta_{n+1}} > 0$, and u_n is bounded, (8.127) follows. In (8.131), if $\{\theta_n\}$ is nondecreasing, we get:

$$\begin{aligned}\|u_k - u_{k-1}\|^2 \leq \frac{1}{\theta_{k+1}}\|u_{k+1} - p\|^2 - \frac{1}{\theta_k}\|u_k - p\|^2 + (\frac{1}{\theta_k} - \frac{1}{\theta_{k+1}})\|u_{k+1} - p\|^2 \\ + \|u_{k-1} - p\|^2 - \|u_k - p\|^2 + 2M\frac{1}{\theta_k}\|f_k\|\end{aligned}$$

Since $\{\theta_n\}$ is nondecreasing and u_n is bounded, we get (8.128) by summing up the above inequality from $n = 1$ to ∞. In (8.132) too $\{\theta_n\}$ nonincreasing was used. When $\{\theta_n\}$ is nondecreasing, from (8.129) we get:

$$\begin{aligned}M\|f_n\| + \|u_{n+1} - p\|^2 - \|u_n - p\|^2 &\geq \theta_n(\|u_n - p\|^2 - \|u_{n-1} - p\|^2) \\ &\geq \theta_n\|u_n - p\|^2 - \theta_{n-1}\|u_{n-1} - p\|^2 \\ &\quad + M^2(\theta_{n-1} - \theta_n)\end{aligned}$$

Now the rest of the proof for the weak convergence of v_n is similar to the case where θ_n is nonincreasing, because $\lim_{n\to\infty}\theta_n = \theta$. In the last two inequalities of (8.136) too, we used $\{\theta_n\}$ nonincreasing. If $\{\theta_n\}$ is nondecreasing, we get:

$$\begin{aligned}&(\sum_{k=1}^{n}\frac{c_k}{\theta_k})^{-1}\sum_{k=1}^{n}[\frac{-1}{\theta_k}(u_{k+1} - x, u_k - x) + \frac{1+\theta_k}{\theta_k}\|u_k - x\|^2 - (u_{k-1} - x, u_k - x) \\ &+ (\frac{f_k}{\theta_k}, u_k - x)] \\ &\geq (\sum_{k=1}^{n}\frac{c_k}{\theta_k})^{-1}\sum_{k=1}^{n}[\frac{1}{2\theta_k}\|u_k - x\|^2 - \frac{1}{2\theta_{k+1}}\|u_{k+1} - x\|^2 \\ &+ (\frac{1}{2\theta_{k+1}} - \frac{1}{2\theta_k})\|u_{k+1} - x\|^2 + \frac{1}{2}\|u_k - x\|^2 - \frac{1}{2}\|u_{k-1} - x\|^2 \\ &- (M + \|x - p\|)\|\frac{f_k}{\theta_k}\|] \to 0,\end{aligned}$$

since u_n is bounded and $\frac{1}{\theta_{k+1}} - \frac{1}{\theta_k} < 0$. Finally, a similar proof as in Theorem 8.7.1 shows that p is the asymptotic center of the sequence $\{u_n\}$. The proof is now complete. □

Theorem 8.7.3 *Assume that u_n is a solution to* (8.4). *If* (T_3), (C_1) *and* (F_2) *are satisfied, then $v_n \rightharpoonup p \in A^{-1}(0)$, and p is the asymptotic center of $\{u_n\}$.*

Proof. Again (8.129) holds, and therefore summing (8.132) from $n = k$ to $n = m$, and letting $m \to +\infty$, we get:

$$\begin{aligned}\|u_k - p\|^2 \leq (1-\theta)\liminf_{n\to+\infty}\|u_n - p\|^2 + \theta_{k-1}\|u_{k-1} - p\|^2 \\ + M\sum_{i=k}^{\infty}\|f_i\| + M^2(\theta - \theta_{k-1}).\end{aligned}$$

Since $\theta = 1$, we get:

$$
\begin{aligned}
\|u_n - p\|^2 &\le \theta_{n-1}\|u_{n-1} - p\|^2 + M\sum_{i=n}^{\infty}\|f_i\| + M^2(1-\theta_{n-1}) \\
&\le \|u_{n-1} - p\|^2 + \varepsilon_n,
\end{aligned}
$$

with $\sum_{n=1}^{\infty}\varepsilon_n < +\infty$, by (T_3) and (F_2). It follows that there exists $\lim_{n\to+\infty}\|u_n - p\|$. The rest of the proof is similar to that of Theorem 8.7.1. □

8.7.2 WEAK CONVERGENCE OF SOLUTIONS

Theorem 8.7.4 *Let $\{u_n\}_{n\ge 1}$ be a solution to* (8.4). *Assume that (T_2) and (F_1) are satisfied, and $c_n \ge c > 0$ for all $n \ge 1$. Then $u_n \rightharpoonup p \in A^{-1}(0)$ as $n \to +\infty$.*

Proof. Dividing both sides of (8.126) by c_n and using the Cauchy-Schwarz inequality, we get:

$$
\begin{aligned}
&(u_{k+1} - u_k, u_n) + \theta_k(u_{k-1} - u_k, u_n) \\
&\le \frac{c_k}{c_n}\|u_{n+1} - u_n\|\|u_k\| + \frac{\theta_n c_k}{c_n}\|u_{n-1} - u_n\|\|u_k\| \\
&\quad + \frac{c_k}{c_n}\|f_n\|\|u_k\| + (f_k, u_n) + \frac{c_k}{c_n}(u_{n+1} - u_n, u_n) \\
&\quad + \theta_n\frac{c_k}{c_n}(u_{n-1} - u_n, u_n) + \frac{c_k}{c_n}\|f_n\|\|u_n\| \\
&\quad + (u_{k+1} - u_k, u_k) + \theta_k(u_{k-1} - u_k, u_k) - (f_k, u_k).
\end{aligned}
$$

Since in this case also $\sum_{n=1}^{+\infty}\frac{c_n}{\theta_n} = +\infty$, by (8.128), we have $\|u_n - u_{n-1}\| \to 0$ as $n \to +\infty$; assume that $u_{n_j} \rightharpoonup p$. Replacing n by n_j in the above inequality and letting $j \to +\infty$, we get:

$$
\begin{aligned}
(u_{k+1} - u_k, p) + \theta_k(u_{k-1} - u_k, p) \le{}& (u_{k+1} - u_k, u_k) + \theta_k(u_{k-1} - u_k, u_k) \\
&+ (f_k, p) - (f_k, u_k)
\end{aligned}
$$

This implies (8.129). By the same proof as in Theorem 8.7.2, we deduce that $\lim_{n\to+\infty}\|u_n - p\|$ exists, and then $u_n \rightharpoonup p \in A^{-1}(0)$ as $n \to +\infty$. □

Theorem 8.7.5 *Assume that u_n is a solution to* (8.4). *If (T_2), (C_2) and (F_1) are satisfied, then $u_n \rightharpoonup p \in A^{-1}(0)$.*

Proof. Since in this case, we have $\sum_{n=1}^{+\infty}\frac{c_n}{\theta_n} = +\infty$, then (8.128) holds, and therefore, we have $\|u_n - u_{n-1}\| \to 0$ as $n \to +\infty$. Hence $\|Au_n\| \to 0$ as $n \to +\infty$. Assume that $u_{n_j} \rightharpoonup p$; then we get (8.129). Now the same proof as in Theorem 8.7.2 shows that $\lim_{n\to+\infty}\|u_n - p\|$ exists. The rest of the proof is now similar to that of Theorem 8.7.2. □

Theorem 8.7.6 *Assume that u_n is a solution to* (8.4). *If (T_3), (C_2) and (F_2) are satisfied, then $u_n \rightharpoonup p \in A^{-1}(0)$.*

Proof. The proof of Theorem 8.7.3 shows that (8.129) holds here. Now the rest of the proof is similar to that of Theorem 8.7.1. □

8.7.3 STRONG CONVERGENCE OF SOLUTIONS

Theorem 8.7.7 *Assume that the operator A in* (8.4) *is strongly monotone, and let $\{u_n\}_{n\geq 1}$ be a solution to* (8.4). *If (T_1), (C_1) and (F_2) are satisfied, then u_n converges strongly as $n \to +\infty$ to some $p \in A^{-1}(0)$.*

Proof. Assume that $(y_2 - y_1, x_2 - x_1) \geq \alpha \|x_2 - x_1\|^2$, for all $[x_i, y_i] \in A, i = 1,2$, and for some $\alpha > 0$. Then by using the strong monotonicity of A, a similar proof as in Theorem 8.7.1 gives the following inequalities:

$$
\begin{aligned}
(Au_n, u_n - v_k) &= (Au_n, u_n - (\sum_{i=1}^{k} \frac{c_i}{\theta_i})^{-1} (\sum_{i=1}^{k} \frac{c_i}{\theta_i} u_i)) \\
&= (\sum_{i=1}^{k} \frac{c_i}{\theta_i})^{-1} \sum_{i=1}^{k} \frac{c_i}{\theta_i} (Au_n, u_n - u_i) \\
&= (\sum_{i=1}^{k} \frac{c_i}{\theta_i})^{-1} \sum_{i=1}^{k} \frac{c_i}{\theta_i} [(Au_n - Au_i, u_n - u_i) + (Au_i, u_n - u_i)] \\
&\geq (\sum_{i=1}^{k} \frac{c_i}{\theta_i})^{-1} \sum_{i=1}^{k} \frac{c_i}{\theta_i} \alpha \|u_n - u_i\|^2 \\
&\quad + (\sum_{i=1}^{k} \frac{c_i}{\theta_i})^{-1} \sum_{i=1}^{k} \frac{1}{\theta_i} (u_{i+1} - (1+\theta_i) u_i + \theta_i u_{i-1} - f_i, u_n - u_i) \\
&= \alpha (\sum_{i=1}^{k} \frac{c_i}{\theta_i})^{-1} \sum_{i=1}^{k} \frac{c_i}{\theta_i} \|u_n - u_i\|^2 \\
&\quad + (\sum_{i=1}^{k} \frac{c_i}{\theta_i})^{-1} \sum_{i=1}^{k} \frac{1}{\theta_i} [(u_{i+1} - u_n, u_n - u_i) \\
&\quad + (1+\theta_i) \|u_n - u_i\|^2 + \theta_i (u_{i-1} - u_n, u_n - u_i) - (f_i, u_n - u_i)] \\
&\geq \alpha (\sum_{i=1}^{k} \frac{c_i}{\theta_i})^{-1} \sum_{i=1}^{k} \frac{c_i}{\theta_i} \|u_n - u_i\|^2 \\
&\quad + (\sum_{i=1}^{k} \frac{c_i}{\theta_i})^{-1} \sum_{i=1}^{k} [-\frac{1}{2\theta_i} \|u_{i+1} - u_n\|^2 - \frac{1}{2\theta_i} \|u_i - u_n\|^2 \\
&\quad + \frac{(1+\theta_i)}{\theta_i} \|u_n - u_i\|^2 - \frac{1}{2} \|u_{i-1} - u_n\|^2 - \frac{1}{2} \|u_i - u_n\|^2 \\
&\quad - \|u_i - u_n\| \frac{\|f_i\|}{\theta_i}] \\
&\geq \alpha (\sum_{i=1}^{k} \frac{c_i}{\theta_i})^{-1} \sum_{i=1}^{k} \frac{c_i}{\theta_i} \|u_n - u_i\|^2
\end{aligned}
$$

$$+(\sum_{i=1}^{k}\frac{c_i}{\theta_i})^{-1}\sum_{i=1}^{k}[\frac{1}{2\theta_i}\|u_n-u_i\|^2-\frac{1}{2\theta_{i+1}}\|u_n-u_{i+1}\|^2$$
$$+\frac{1}{2}\|u_i-u_n\|^2-\frac{1}{2}\|u_{i-1}-u_n\|^2-\|u_i-u_n\|\frac{\|f_i\|}{\theta_i}].$$

(where in the last inequality, we used the assumption that $\{\theta_n\}$ is nonincreasing). Letting $k\to+\infty$, by Theorem 8.7.2 and the Cauchy-Schwarz inequality, we get:

$$\begin{aligned}(Au_n,u_n-p)&\geq\alpha\liminf_{k\to+\infty}(\sum_{i=1}^{k}\frac{c_i}{\theta_i})^{-1}\sum_{i=1}^{k}\frac{c_i}{\theta_i}\|u_n-u_i\|^2\\&\geq\alpha[\liminf_{k\to+\infty}(\sum_{i=1}^{k}\frac{c_i}{\theta_i})^{-1}\sum_{i=1}^{k}\frac{c_i}{\theta_i}\|u_n-u_i\|]^2\\&\geq\alpha[\liminf_{k\to+\infty}\|u_n-v_k\|]^2\geq\alpha\|u_n-p\|^2.\end{aligned}\tag{8.137}$$

Multiplying both sides of (8.137) by $\frac{c_n}{\theta_n}$ and using (8.4), we get:

$$\frac{1}{\theta_n}(u_{n+1}-(1+\theta_n)u_n+\theta_nu_{n-1}-f_n,u_n-p)\geq\alpha\frac{c_n}{\theta_n}\|u_n-p\|^2.$$

It follows that

$$\begin{aligned}\alpha\frac{c_n}{\theta_n}\|u_n-p\|^2&\leq\frac{1}{\theta_n}(u_{n+1}-p,u_n-p)-\frac{(1+\theta_n)}{\theta_n}\|u_n-p\|^2\\&\quad+(u_{n-1}-p,u_n-p)-(\frac{f_n}{\theta_n},u_n-p)\\&\leq\frac{1}{2\theta_{n+1}}\|u_{n+1}-p\|^2-\frac{1}{2\theta_n}\|u_n-p\|^2\\&\quad+\frac{1}{2}\|u_{n-1}-p\|^2-\frac{1}{2}\|u_n-p\|^2+M\|\frac{f_n}{\theta_n}\|,\end{aligned}$$

where $M:=\sup_{n\geq0}\|u_n-p\|$. Summing up both sides of this inequality from $n=1$ to m and letting $m\to+\infty$, we get

$$\sum_{n=1}^{+\infty}\frac{c_n}{\theta_n}\|u_n-p\|^2<+\infty.$$

This implies that

$$\liminf_{n\to+\infty}\|u_n-p\|^2=0.$$

Since by Theorem 8.7.2, $\lim_{n\to+\infty}\|u_n-p\|$ exists, then $u_n\to p\in A^{-1}(0)$ as $n\to+\infty$. □

Theorem 8.7.8 *Assume that $\{u_n\}_{n\geq1}$ is a solution to (8.4), $(I+A)^{-1}$ is compact and $\{f_n\}_{n\geq1}$ satisfies (F_2). If $\sum_{n=1}^{+\infty}\frac{c_n^2}{(1+\theta_n)^2}=+\infty$, and (T_1) or (T_3) is satisfied, then u_n converges strongly as $n\to+\infty$ to some $p\in A^{-1}(0)$.*

Proof. Since in this case $\sum_{n=1}^{+\infty} \frac{c_n}{\theta_n} = +\infty$, by Theorems 8.7.1 and 8.7.3 we already know that v_n converges weakly as $n \to +\infty$ to some $p \in A^{-1}(0)$, and $\lim_{n\to+\infty} \|u_n - p\|$ exists. From (8.4), we have

$$u_{n+1} - p + \theta_n(u_{n-1} - p) \in c_n A u_n + f_n + (1+\theta_n)(u_n - p). \tag{8.138}$$

Squaring both sides of (8.138), we obtain

$$\begin{aligned}
&\|u_{n+1} - p\|^2 + \theta_n^2 \|u_{n-1} - p\|^2 + 2\theta_n(u_{n+1} - p, u_{n-1} - p) \\
&= c_n^2 \|Au_n\|^2 + \|f_n\|^2 + (1+\theta_n)^2 \|u_n - p\|^2 + 2c_n(Au_n, f_n) \\
&\quad + 2c_n(1+\theta_n)(Au_n, u_n - p) + 2(1+\theta_n)(f_n, u_n - p).
\end{aligned}$$

Dividing both sides of the above inequality by $(1+\theta_n)^2$ and using (8.4) and (8.129), we get:

$$\begin{aligned}
\frac{c_n^2}{(1+\theta_n)^2}\|Au_n\|^2 &\le \frac{1}{1+\theta_n}\|u_{n+1} - p\|^2 + \frac{\theta_n}{1+\theta_n}\|u_{n-1} - p\|^2 - \|u_n - p\|^2 \\
&\quad + \frac{2\|u_{n+1} - p\|}{(1+\theta_n)^2}\|f_n\| + \frac{2\theta_n\|u_{n-1} - p\|\|f_n\|}{(1+\theta_n)^2} + \|\frac{f_n}{1+\theta_n}\|^2 \\
&\le \frac{1}{1+\theta_n}(\|u_{n+1} - p\|^2 - \|u_n - p\|^2) \\
&\quad + \frac{\theta_n}{1+\theta_n}(\|u_{n-1} - p\|^2 - \|u_n - p\|^2) + 4M\|\frac{f_n}{\theta_n}\| + \|\frac{f_n}{\theta_n}\|^2 \\
&\le \frac{1}{1+\theta_{n+1}}\|u_{n+1} - p\|^2 - \frac{1}{1+\theta_n}\|u_n - p\|^2 + \frac{\theta_{n-1}}{1+\theta_{n-1}}\|u_{n-1} - p\|^2 \\
&\quad - \frac{\theta_n}{1+\theta_n}\|u_n - p\|^2 + 4M\|\frac{f_n}{\theta_n}\| + \|\frac{f_n}{\theta_n}\|^2.
\end{aligned}$$

Summing up both sides of the above inequality from $n = 1$ to m and letting $m \to +\infty$, we get:

$$\sum_{n=1}^{+\infty} \frac{c_n^2}{(1+\theta_n)^2}\|Au_n\|^2 < +\infty.$$

Therefore $\liminf_{n\to+\infty} \|Au_n\| = 0$. It follows that there exists a subsequence $\{u_{n_j}\}$ of $\{u_n\}$ such that $\lim_{j\to+\infty} \|Au_{n_j}\| = 0$. Then $\{u_{n_j} + Au_{n_j}\}$ is bounded. Since $(I+A)^{-1}$ is compact, there exists a subsequence of $\{u_{n_j}\}$, denoted again by $\{u_{n_j}\}$, such that $u_{n_j} \to q$. By the monotonicity of A, we have

$$(Au_k - Au_{n_j}, u_k - u_{n_j}) \ge 0.$$

Letting $j \to +\infty$, we get

$$(Au_k, u_k - q) \ge 0.$$

Multiplying both sides of this inequality by c_k, we get (8.129) with p replaced by q. Then we conclude that $\lim_{n\to+\infty} \|u_n - q\|$ exists. Therefore $u_n \to q \in A^{-1}(0)$. □

In the following theorem we assume that $A^{-1}(0)$ is nonempty.

Definition 8.7.9 *Let $P : H \to A^{-1}(0)$ be the metric projection onto the (closed and convex) zero set of A. The operator A is said to satisfy the "convergence condition" if $[x_n, y_n] \in A$, $\|x_n\| \le K$, $\|y_n\| \le K$ for a positive constant K, and $\lim_{n\to+\infty}(y_n, x_n - Px_n) = 0$ imply that $\lim_{n\to+\infty}(x_n - Px_n) = 0$.*

Theorem 8.7.10 *Suppose that A satisfies the "convergence condition" and $\{u_n\}_{n\ge1}$ is a solution to (8.4). If the conditions (T_1), (C_1) and (F_2) are satisfied, then u_n converges strongly to a zero of A.*

Proof. We have:

$$\begin{aligned}
0 &\le \frac{c_n}{\theta_n}(Au_n, u_n - Pu_n) \\
&= (\frac{u_{n+1}-u_n}{\theta_n} + u_{n-1} - u_n - \frac{f_n}{\theta_n}, u_n - Pu_n) \\
&= \frac{1}{\theta_n}(u_{n+1} - Pu_{n+1}, u_n - Pu_n) + \frac{1}{\theta_n}(Pu_{n+1} - Pu_n, u_n - Pu_n) \\
&\quad - \frac{1}{\theta_n}\|u_n - Pu_n\|^2 + (u_{n-1} - Pu_{n-1}, u_n - Pu_n) \\
&\quad + (Pu_{n-1} - Pu_n, u_n - Pu_n) - \|u_n - Pu_n\|^2 - (\frac{f_n}{\theta_n}, u_n - Pu_n) \\
&\le \frac{1}{2\theta_n}\|u_{n+1} - Pu_{n+1}\|^2 + \frac{1}{2\theta_n}\|u_n - Pu_n\|^2 - \frac{1}{\theta_n}\|u_n - Pu_n\|^2 \\
&\quad + \frac{1}{2}\|u_{n-1} - Pu_{n-1}\|^2 + \frac{1}{2}\|u_n - Pu_n\|^2 - \|u_n - Pu_n\|^2 + \|\frac{f_n}{\theta_n}\|\|u_n - Pu_n\| \\
&\le \frac{1}{2}(\frac{1}{\theta_{n+1}}\|u_{n+1} - Pu_{n+1}\|^2 - \frac{1}{\theta_n}\|u_n - Pu_n\|^2) + \frac{1}{2}(\|u_{n-1} - Pu_{n-1}\|^2 \\
&\quad - \|u_n - Pu_n\|^2) + M\|\frac{f_n}{\theta_n}\|
\end{aligned}$$

where $M := \sup_{n\ge0}\|u_n - Pu_n\|$. Summing up the above inequality from $n = 1$ to m and letting $m \to +\infty$, we get:

$$\sum_{n=1}^{+\infty} \frac{c_n}{\theta_n}(Au_n, u_n - Pu_n) < +\infty$$

It follows that $\liminf_{n\to+\infty}(Au_n, u_n - Pu_n) = 0$; hence by the convergence condition, there exists a subsequence $\{u_{n_k}\}$ of $\{u_n\}$ such that $\lim_{k\to+\infty}(u_{n_k} - Pu_{n_k}) = 0$. By a similar proof as in Theorem 8.7.1, we have:

$$\|u_{n+1} - Pu_{n+1}\| \le \|u_{n+1} - Pu_n\| \le \|u_n - Pu_n\| + \sum_{i=n+1}^{+\infty}\|\frac{f_i}{\theta_i}\|$$

Thus, $\lim_{n\to+\infty}\|u_n - Pu_n\|$ exists, and we have $u_n - Pu_n \to 0$ as $n \to +\infty$. Since

$$\begin{aligned}\|u_{n+m}-u_n\| &\le \|u_{n+m}-Pu_n\| + \|u_n - Pu_n\|\\ &\le 2\|u_n - Pu_n\| + \sum_{i=n+1}^{+\infty}(i-n)\|\frac{f_i}{\theta_i}\|\\ &\le 2\|u_n - Pu_n\| + \sum_{i=n+1}^{+\infty} i\|\frac{f_i}{\theta_i}\| \to 0\end{aligned}$$

as $n \to +\infty$, uniformly in $m \ge 0$, we conclude that $\{u_n\}$ converges strongly to a zero of A. □

Theorem 8.7.11 *Let u_n be a solution to* (8.4). *If* (T_2) *holds with* $\lim_{n\to+\infty}\theta_n := \theta <$ 1, $\limsup_{n\to+\infty} c_n > 0$ *and* $\sum_{n=1}^{\infty}(\sum_{i=1}^{n}\frac{c_i}{\theta^i})\frac{\|f_n\|}{c_n} < +\infty$, *then $u_n \to p$ as $n \to +\infty$, where* $p \in A^{-1}(0)$.

Proof. By the monotonicity of A, we have

$$(Au_{n+1} - Au_n, u_{n+1} - u_n) \ge 0$$

By (8.4), we get:

$$\begin{aligned}&\frac{1}{c_{n+1}}(u_{n+2}-u_{n+1}, u_{n+1}-u_n) - \frac{\theta_{n+1}}{c_{n+1}}\|u_{n+1}-u_n\|^2 - \frac{1}{c_n}\|u_{n+1}-u_n\|^2\\ &+\frac{\theta_n}{c_n}(u_n - u_{n-1}, u_{n+1}-u_n) - (\frac{f_{n+1}}{c_{n+1}} - \frac{f_n}{c_n}, u_{n+1}-u_n) \ge 0.\end{aligned}$$

This implies that

$$\begin{aligned}\frac{1}{c_n}\|u_{n+1}-u_n\| - \frac{\theta_n}{c_n}\|u_n-u_{n-1}\| \le{}& \frac{1}{c_{n+1}}\|u_{n+2}-u_{n+1}\| - \frac{\theta_{n+1}}{c_{n+1}}\|u_{n+1}-u_n\|\\ &+\|\frac{f_{n+1}}{c_{n+1}} - \frac{f_n}{c_n}\|.\end{aligned}$$

Summing up the above inequality from $n = k$ to $n = m-1$, we get:

$$\begin{aligned}\frac{1}{c_k}\|u_{k+1}-u_k\| - \frac{1}{c_m}\|u_{m+1}-u_m\| \le{}& \frac{\theta_k}{c_k}\|u_k - u_{k-1}\| - \frac{\theta_m}{c_m}\|u_m - u_{m-1}\|\\ &+2\sum_{n=k}^{m}\|\frac{f_n}{c_n}\|.\end{aligned}$$

Since $\limsup_{n\to+\infty} c_n > 0$, we have $\sum_{n=1}^{\infty}\frac{c_n}{\theta_n} = +\infty$; on the other hand, $\frac{\|f_n\|}{\theta_n} \le \frac{\|f_n\|}{\theta^n}$ for large n, and we have:

$$\sum_{n=1}^{\infty}\frac{\|f_n\|}{\theta^n} = \sum_{n=1}^{\infty}\frac{c_n}{\theta^n}\frac{\|f_n\|}{c_n} \le \sum_{n=1}^{\infty}(\sum_{i=1}^{n}\frac{c_i}{\theta^i})\frac{\|f_n\|}{c_n} < +\infty.$$

This implies that $\sum_{n=1}^{\infty} \frac{\|f_n\|}{\theta_n} < +\infty$, and hence (F_1) holds. In a similar way as in Theorem 8.7.4, (8.128) holds, and we have $\lim_{n\to+\infty} \|u_{n+1} - u_n\| = 0$. On the other hand, there exists a subsequence c_{n_j} of c_n and $\varepsilon > 0$ such that $c_{n_j} \geq \varepsilon$ for all $j \geq 1$. Substituting m by n_j in the above inequality, and letting $j \to +\infty$, we get:

$$\|u_{k+1} - u_k\| \leq \theta_k \|u_k - u_{k-1}\| + 2c_k \sum_{n=k}^{\infty} \|\frac{f_n}{c_n}\|.$$

This implies that

$$\|u_{k+1} - u_k\| \leq \lambda^k \|u_1 - u_0\| + \sum_{i=1}^{k} \lambda^{k-i} c_i a_i$$

where $\theta < \lambda < 1$ and $a_i = 2\sum_{n=i}^{\infty} \|\frac{f_n}{c_n}\|$. Then we get:

$$\begin{aligned}
\|u_n - u_m\| = \|\sum_{k=m}^{n-1} (u_{k+1} - u_k)\| &\leq \sum_{k=m}^{n-1} \|u_{k+1} - u_k\| \\
&\leq \sum_{k=m}^{n-1} (\lambda^k \|u_1 - u_0\| + \sum_{i=1}^{k} \lambda^{k-i} c_i a_i) \\
&= \frac{\lambda^m - \lambda^n}{1-\lambda} \|u_1 - u_0\| + \sum_{k=m}^{n-1} \lambda^k \sum_{i=1}^{k} \frac{c_i a_i}{\lambda^i} \\
&\leq \frac{\lambda^m - \lambda^n}{1-\lambda} \|u_1 - u_0\| + \sum_{k=m}^{n-1} \lambda^k \sum_{i=1}^{k} \frac{c_i a_i}{\theta^i}.
\end{aligned}$$

Since by assumption we have:

$$\sum_{i=1}^{k} \frac{c_i a_i}{\theta^i} \leq 2 \sum_{i=1}^{\infty} \frac{c_i}{\theta^i} \sum_{n=i}^{\infty} \|\frac{f_n}{c_n}\| = 2 \sum_{n=1}^{\infty} (\sum_{i=1}^{n} \frac{c_i}{\theta^i}) \|\frac{f_n}{c_n}\| < +\infty,$$

it follows that u_n is Cauchy, and therefore $u_n \to p \in H$. Now we prove that $p \in A^{-1}(0)$. Suppose that $[x,y] \in A$. By the monotonicity of A and (8.4), we have:

$$\begin{aligned}
(x - u_n, y) &= (x - u_n, y - Au_n) + (x - u_n, Au_n) \geq \frac{1}{c_n}(x - u_n, c_n A u_n) \\
&= \frac{1}{c_n}(x - u_n, u_{n+1} - (1+\theta_n)u_n + \theta_n u_{n-1} - f_n) \\
&= \frac{1}{c_n}(x - u_n, u_{n+1} - u_n) - \frac{\theta_n}{c_n}(x - u_n, u_{n-1} - u_n) - \frac{1}{c_n}(x - u_n, f_n) \\
&\geq \frac{-1}{c_n} \|x - u_n\| \|u_{n+1} - u_n\| - \frac{\theta_n}{c_n} \|x - u_n\| \|u_{n-1} - u_n\| \\
&\quad - \frac{\|f_n\|}{c_n} \|x - u_n\|.
\end{aligned}$$

Let's show that $\frac{\|f_n\|}{c_n} \to 0$ as $n \to +\infty$. Since $\frac{c_n}{\theta_n} < \frac{c_n}{\theta^n}$ for large n, we have $\sum_{n=1}^{\infty} \frac{c_n}{\theta^n} = +\infty$. Therefore since $(\sum_{i=1}^{n} \frac{c_i}{\theta^i})\frac{\|f_n\|}{c_n} \to 0$ as $n \to \infty$, and $\sum_{i=1}^{\infty} \frac{c_i}{\theta^i} = +\infty$, we must have $\frac{\|f_n\|}{c_n} \to 0$. Now substituting n_j for n in the above inequality, and letting $j \to +\infty$, then using the fact $\lim_{n\to+\infty} \|u_{n+1} - u_n\| = 0$, and since $\frac{\|f_n\|}{c_n} \to 0$ as $n \to +\infty$, we get $(x - p, y) \geq 0$. Since A is maximal monotone, then $p \in A^{-1}(0)$. The proof is now complete. □

8.8 APPLICATIONS TO OPTIMIZATION

(1) Assume that A is a maximal monotone operator. The sequence $\{u_n\}$ given by (8.4) provides an approximation scheme to the solutions of the stationary equation $0 \in A(x)$, which correspond to a minimum point of φ, when $A = \partial\varphi$ and φ is a proper, convex and lower semicontinuous function.
(2) We propose an application of (8.4) to a discrete variational problem.

Theorem 8.8.1 *If $A = \partial\varphi$, where $\varphi : H \to (-\infty, +\infty]$ is a proper, convex and lower semicontinuous function such that $\varphi \geq 0$ and φ has at least a minimum point, then problem (8.4) is equivalent to the following minimization problem:*
$F(\{u_n\}) = \inf\{F(\{v_n\});\ v_n \in \hat{H}\}$ *where*
$\hat{H} = \{(\{v_n\}) \subset H;\ v_0 = a,$ *and* $\sup_{n\geq 0} \|v_n\| < +\infty\}$,
and $\{f_n\}$, $\{c_n\}$, and $\{\theta_n\}$ satisfy the assumptions (F_2), (C_1), (T_1), and

$$F(\{v_n\}) = \sum_{n=1}^{\infty} \Big[\frac{c_n}{\theta_1 \cdots \theta_n}\varphi(v_n) + \frac{1}{2\theta_1 \cdots \theta_{n-1}}\|v_n - v_{n-1}\|^2 + \frac{1}{\theta_1 \cdots \theta_n}(f_n, v_n)\Big]$$

with $\theta_0 = 1$.

Proof. Suppose that $\{u_n\}$ is a solution of (8.4) and $\{v_n\} \in \hat{H}$. By the subdifferential inequality and (8.4), we have:

$$\begin{aligned}
c_n\varphi(u_n) - c_n\varphi(v_n) &\leq c_n(\partial\varphi(u_n), u_n - v_n) \\
&= (u_{n+1} - (1+\theta_n)u_n + \theta_n u_{n-1} - f_n, u_n - v_n) \\
&= (u_{n+1} - v_n, u_n - v_n) - \|u_n - v_n\|^2 - \theta_n\|u_n - u_{n-1}\|^2 \\
&\quad - \theta_n(u_n - u_{n-1}, u_{n-1} - v_{n-1}) + \theta_n(u_n - u_{n-1}, v_n - v_{n-1}) \\
&\quad - (f_n, u_n - v_n) \\
&\leq \frac{1}{2}\|u_{n+1} - v_n\|^2 - \frac{\theta_n}{2}\|u_n - v_{n-1}\|^2 - \frac{1}{2}\|u_n - v_n\|^2 \\
&\quad + \frac{\theta_n}{2}\|u_{n-1} - v_{n-1}\|^2 - \frac{1}{2}\|u_{n+1} - u_n\|^2 + \frac{\theta_n}{2}\|u_n - u_{n-1}\|^2 \\
&\quad - \frac{\theta_n}{2}\|u_n - u_{n-1}\|^2 + \frac{\theta_n}{2}\|v_n - v_{n-1}\|^2 - (f_n, u_n) + (f_n, v_n).
\end{aligned}$$

Multiplying both sides of the above inequality by $b_n = \frac{1}{\theta_1 \cdots \theta_n}$ (with $b_0 := 1$), and summing up from $n = 1$ to $n = N$, we obtain:

$$\begin{aligned}
\sum_{n=1}^{N} c_n b_n \varphi(u_n) - c_n b_n \varphi(v_n) \leq \frac{b_N}{2}\|u_{N+1} - v_N\|^2 - \frac{1}{2}\|u_1 - v_0\|^2 - \frac{b_N}{2}\|u_N - v_N\|^2 \\
+ \frac{1}{2}\|u_0 - v_0\|^2 - \frac{b_N}{2}\|u_{N+1} - u_N\|^2 + \frac{1}{2}\|u_1 - u_0\|^2 \\
- \sum_{n=1}^{N} \frac{b_{n-1}}{2}\|u_n - u_{n-1}\|^2 + \sum_{n=1}^{N} \frac{b_{n-1}}{2}\|v_n - v_{n-1}\|^2 \\
- \sum_{n=1}^{N} b_n(f_n, u_n) + \sum_{n=1}^{N} b_n(f_n, v_n) \\
\leq \frac{b_N}{2}\|u_{N+1} - u_N\|\|u_{N+1} + u_N - 2v_N\| \\
- \frac{b_N}{2}\|u_{N+1} - u_N\|^2 - \sum_{n=1}^{N} \frac{b_{n-1}}{2}\|u_n - u_{n-1}\|^2 \\
+ \sum_{n=1}^{N} \frac{b_{n-1}}{2}\|v_n - v_{n-1}\|^2 - \sum_{n=1}^{N} b_n(f_n, u_n) \\
+ \sum_{n=1}^{N} b_n(f_n, v_n).
\end{aligned}$$

Letting $N \to +\infty$, by (8.128) we get

$$F(\{u_n\}) \leq F(\{v_n\})$$

and this proves the theorem. □

REFERENCES

APR1. N. C. Apreutesei, Nonlinear second order evolution equations of monotone type and applications, Pushpa Publishing House, Allahabad, India, 2007.

APR2. N. C. Apreutesei, Existence and asymptotic behavior for a class of second order difference equations, J. Difference Equ. Appl. 9 (2003), 751–763.

APR3. N. C. Apreutesei, On a class of difference equations of monotone type, J. Math. Anal. Appl. 288 (2003), 833–851.

APR-APR1. G. Apreutesei and N. C. Apreutesei, Continuous dependence on data for bilocal difference equations, J. Difference Equ. Appl. 15 (2009), 511–527.

APR-APR2. G. Apreutesei and N. C. Apreutesei, A note on the continuous dependence on data for second order difference inclusions, J. Difference Equ. Appl. 17 (2011), 637–641.

BRU1. R. E. Bruck, On the strong convergence of an averaging iteration for the solution of operator equations involving monotone operators in Hilbert space, J. Math. Anal. Appl. 64 (1978), 319–327.

BRU2. R. E. Bruck, Periodic forcing of solutions of a boundary value problem for a second order differential equation in Hilbert space, J. Math. Anal. Appl. 76 (1980), 159–173.

DJA1. B. Djafari Rouhani, Ergodic theorems for nonexpansive sequences in Hilbert spaces and related problems, Ph.D. Thesis, Yale University, Part I, pp. 1–76 (1981).

DJA2. B. Djafari Rouhani, Asymptotic behaviour of quasi-autonomous dissipative systems in Hilbert spaces, J. Math. Anal. Appl. 147 (1990), 465–476.

DJA3. B. Djafari Rouhani, Asymptotic behaviour of almost nonexpansive sequences in a Hilbert space, J. Math. Anal. Appl. 151 (1990), 226–235.

ROU-KHA1. B. Djafari Rouhani and H. Khatibzadeh, A note on the asymptotic behavior of solutions to a second order difference equation, J. Difference Equ. Appl. 14 (2008), 429–432.

ROU-KHA2. B. Djafari Rouhani and H. Khatibzadeh, Asymptotic behavior of bounded solutions to a class of second order nonhomogeneous difference equations of monotone type, Nonlinear Anal. 72 (2010), 1570–1579.

ROU-KHA3. B. Djafari Rouhani and H. Khatibzadeh, New results on the asymptotic behavior of solutions to a class of second order nonhomogeneous difference equations, Nonlinear Anal. 74 (2011), 5727–5734.

ROU-KHA4. B. Djafari Rouhani and H. Khatibzadeh, Existence and asymptotic behaviour of solutions to first- and second-order difference equations with periodic forcing, J. Difference Equ. Appl. 18 (2012), 1593–1606.

ROU-KHA5. B. Djafari Rouhani and H. Khatibzadeh, Asymptotics of a difference equation and zeroes of monotone operators, Numer. Funct. Anal. Optim. 36 (2015), 350–363.

ROU-KHA6. B. Djafari Rouhani and H. Khatibzadeh, Asymptotic behavior for a general class of homogeneous second order evolution equations in a Hilbert space, Dynam. Systems Appl. 24 (2015), 1–15.

KHA1. H. Khatibzadeh, On the convergence of solutions to a second order difference equation with monotone operator, Numer. Funct. Anal. Optim. 32 (2011), 1271–1282.

KHA2. H. Khatibzadeh, Convergence of solutions to a second order difference inclusion, Nonlinear Anal. 75 (2012), 3503–3509.

MIT-MOR. E. Mitidieri and G. Morosanu, Asymptotic behavior of the solutions of second-order difference equations associated to monotone operators, Numer. Funct. Anal. Optim. 8 (1985–86), 419–434.

MOR1. G. Morosanu, 'Nonlinear Evolution Equations and Applications', Editura Academiei Romane (and D. Reidel publishing Company), Bucharest, 1988.

POF-REI. E. I. Poffald and S. Reich, A difference inclusion, in: Nonlinear semigroups, partial differential equations and attractors, in:Lecture notes in Mathematics, Vol. 1394, Springer, Berlin, 1989, pp. 122–130.

RAC-RAC1. L. Rachunek and I. Rachunkova, Strictly increasing solutions of nonautonomous difference equations arising in hydrodynamics, Adv. Difference Equ. 2010, Art. ID 714891, 11 pp.

RAC-RAC2. L. Rachunek and I. Rachunkova, Homoclinic solutions of non-autonomous difference equations arising in hydrodynamics, Nonlinear Anal. Real World Appl. 12 (2011), 14–23.

REI-SHA. S. Reich and I. Shafrir, An existence theorem for a difference inclusion in general Banach spaces, J. Math. Anal. Appl. 160 (1991), 406–412.

9 Discrete Nonlinear Oscillator Dynamical System and the Inertial Proximal Algorithm

9.1 INTRODUCTION

In this chapter we consider a discrete version of the heavy ball with friction dynamical system

$$\begin{cases} u''(t) + \gamma u'(t) + \nabla\varphi(u(t)) = 0 \\ u(0) = x_0, \ \ u'(0) = x_1 \end{cases} \tag{9.1}$$

that was studied in Chapter 6. By discretization and substituting $\nabla\varphi$ by a (possibly multi-valued) maximal monotone operator A, we get the following second order difference inclusion:

$$u_{k+1} - u_k - \alpha_k(u_k - u_{k-1}) + \lambda_k A(u_{k+1}) \ni 0 \tag{9.2}$$

where u_0, $u_1 \in H$ are given starting points, and α_k and λ_k are two non negative real sequences. This implicit iteration can be equivalently written as:

$$u_{k+1} = J^A_{\lambda_k}(u_k + \alpha_k(u_k - u_{k-1})) \tag{9.3}$$

where $J^A_{\lambda_k} = (I + \lambda_k A)^{-1}$ is the resolvent of A of order λ_k. Since J^A_λ is a single-valued operator defined on H, the sequence u_k is well defined. When $\alpha_k \equiv 0$, this iterative scheme is exactly the proximal point algorithm studied in Chapter 7. In fact (9.3) is obtained by adding an extrapolation term to the proximal point algorithm. In this brief chapter we want to study the asymptotic behavior of the sequence u_k satisfying

$$u_{k+1} = J^A_{\lambda_{k+1}}(u_k + \alpha_k(u_k - u_{k-1}) + e_k) \tag{9.4}$$

where $\{e_k\}$ is a sequence of error terms.

The first investigations for (9.3) were performed by Alvarez and Attouch [ALV-ATT] and by Jules and Maingé [JUL-MAI] and Maingé [MAI]. This chapter is essentially taken from [KHA-RAN].

Throughout the chapter φ is a proper, convex and lower semicontinuous function, A is a maximal monotone operator and

$$w_n := \Big(\sum_{i=1}^{n} \lambda_{i+1}\Big)^{-1} \sum_{i=1}^{n} \lambda_{i+1} u_{i+1}$$

is the weighted average of the sequence $\{u_n\}$. We denote by $\omega_w(x_n)$ the set of all weak cluster points of the sequence x_n, and by Au_{n+1} the element $\frac{u_n-u_{n+1}+\alpha_n(u_n-u_{n-1})+e_n}{\lambda_{n+1}}$ in H.

Lemma 9.1.1 *Suppose that $\{a_n\}$ and $\{b_n\}$ are positive sequences such that $\sum_{n=1}^{+\infty} b_n = +\infty$ and $\lim_{n\to+\infty} \frac{a_n}{b_n} = 0$, then $\lim_{m\to+\infty} \frac{\sum_{n=1}^{m} a_n}{\sum_{n=1}^{m} b_n} = 0$.*

Proof. See Chapter 7 Lemma 7.5.1 for the proof. □

Lemma 9.1.2 *[MAI] Assume that $\{\phi_n\} \subset [0,+\infty)$ and $\{\delta_n\} \subset [0,+\infty)$ satisfy*
$(a)\phi_{n+1} - \phi_n \leq \theta_n(\phi_n - \phi_{n-1}) + \delta_n$,
$(b)\sum_n \delta_n < +\infty$,
$(c)\{\theta_n\} \subset [0,\theta)$, *where* $\theta \in [0,1)$.
Then the sequence $\{\phi_n\}$ converges and $\sum_n[\phi_{n+1} - \phi_n]_+ < \infty$, where $[t]_+ := \max\{t,0\}$, $\forall t \in \mathbb{R}$.

Proof. Set $u_n = \phi_n - \phi_{n-1}$. By (a) and (c), we obtain $[u_{n+1}]_+ \leq \theta[u_n]_+ + \delta_n$, so that an easy computation entails $[u_{n+1}]_+ \leq \theta^n[u_1]_+ + \sum_{j=0}^{n-1}\theta^j\delta_{n-j}$. Hence by (b), and since $\theta \in [0,1)$, we obtain:

$$\sum_{n\geq 0}[u_{n+1}]_+ \leq \frac{1}{1-\theta}([u_1]_+ + \sum_{n\geq 1}\delta_n) < \infty.$$

Consequently, setting $w_n := \phi_n - \sum_{j=1}^{n}[u_j]_+$, we deduce that the sequence $\{w_n\}$ is bounded from below and there holds

$$w_{n+1} = \phi_{n+1} - [u_{n+1}]_+ - \sum_{j=1}^{n}[u_j]_+ \leq w_n.$$

Therefore the sequence $\{w_n\}$ is nonincreasing, hence $\{w_n\}$ is convergent, and so is $\{\phi_n\}$. □

9.2 BOUNDEDNESS OF THE SEQUENCE AND AN ERGODIC THEOREM

In this section, we first study the relation between the boundedness of the sequence $\{u_n\}$ and the assumption $A^{-1}(0) \neq \emptyset$, as well as the existence of $\lim\|u_n - p\|$ for each $p \in A^{-1}(0)$. Then we prove an ergodic theorem for the weak convergence of the weighted averages of the sequence $\{u_n\}$.

Proposition 9.2.1 *Suppose $A^{-1}(0) \neq \emptyset$ and $\{u_n\}$ is given by* (9.4).
(a) If $e_n \equiv 0$ and

$$\begin{cases} (\alpha_1)\ \exists\alpha \in [0,1[\ \ \textit{such that}\ \ \forall k \in \mathbb{N}, 0 \leq \alpha_k \leq \alpha, \\ (\alpha_2)\ \sum_{n=1}^{+\infty}\alpha_n\|u_n - u_{n-1}\|^2 < +\infty, \end{cases} \tag{9.5}$$

then $\lim_n \|u_n - p\|$ *exists and* $\sum_{n=1}^{\infty}[\|u_n - p\| - \|u_{n-1} - p\|]_+ < \infty$, $\forall p \in A^{-1}(0)$.
(*b*) *If* $\{u_n\}$ *is bounded,* $(E_1)\sum_{n=1}^{+\infty}\|e_n\| < +\infty$ *and* (α_1) *and* (α_2) *are satisfied, then for all* $p \in A^{-1}(0)$, $\lim_n \|u_n - p\|$ *exists and* $\sum_{n=1}^{\infty}[\|u_n - p\| - \|u_{n-1} - p\|]_+ < \infty$.
(*c*) *If* $(E_1)\sum_{n=1}^{+\infty}\|e_n\| < +\infty$ *and* (α_3) $\sum_{n=1}^{+\infty}\alpha_n\|u_n - u_{n-1}\| < +\infty$, *then for all* $p \in A^{-1}(0)$, $\lim_n \|u_n - p\|$ *exists.*

Proof. (a) follows by the same proof as (b) by taking $e_n \equiv 0$. So let us prove (b). Suppose that $p \in A^{-1}(0)$, then by (9.4) and the monotonicity of A, we have:

$$\begin{aligned}
0 &\geq (u_{n+1} - u_n - \alpha_n(u_n - u_{n-1}) - e_n, u_{n+1} - p) \\
&= (u_{n+1} - u_n, u_{n+1} - p) - \alpha_n(u_n - u_{n-1}, u_{n+1} - u_n) \\
&\quad - \alpha_n(u_n - u_{n-1}, u_n - p) - (e_n, u_{n+1} - p) \\
&= \frac{1}{2}\|u_{n+1} - p\|^2 - \frac{1}{2}\|u_n - p\|^2 - \frac{\alpha_n^2}{2}\|u_n - u_{n-1}\|^2 \\
&\quad + \frac{1}{2}\|u_{n+1} - u_n - \alpha_n(u_n - u_{n-1})\|^2 - \frac{\alpha_n}{2}\|u_n - u_{n-1}\|^2 \\
&\quad - \frac{\alpha_n}{2}\|u_n - p\|^2 + \frac{\alpha_n}{2}\|u_{n-1} - p\|^2 - (e_n, u_{n+1} - p)
\end{aligned}$$

Set $\phi_n = \frac{1}{2}\|u_n - p\|^2$. Then we have:

$$\phi_{n+1} - \phi_n - \alpha_n(\phi_n - \phi_{n-1}) \leq \alpha_n\|u_n - u_{n-1}\|^2 + \|e_n\|\|u_{n+1} - p\|.$$

Hence by Lemma 9.1.2, (α_2) and (E_1), there exists $\lim_n \|u_n - p\|$ and

$$\sum_{n=1}^{\infty}\left[\|u_n - p\| - \|u_{n-1} - p\|\right]_+ < \infty$$

Now we prove (c). Since the resolvent operator is nonexpansive, from (9.4) we have:

$$\|u_{n+1} - p\| \leq \|u_n - p\| + \alpha_n\|u_n - u_{n-1}\| + e_n.$$

Then it follows from (α_3) and (E_1) that $\lim_n \|u_n - p\|$ exists. □

In the following proposition, with suitable conditions, we prove that the boundedness of the sequence $\{u_n\}$ generated by (9.4) implies that $A^{-1}(0) \neq \emptyset$. Also, we show that the set of weak cluster points of the weighted averages w_n of the sequence u_n, is a subset of $A^{-1}(0)$.

Proposition 9.2.2 *Let* $\{u_n\}$ *be a bounded sequence generated by* (9.4). *If the following conditions hold:*

$$\begin{cases}
(\Lambda_1)\sum_{n=1}^{+\infty}\lambda_n = +\infty, \\
(\alpha_3)\sum_{n=1}^{+\infty}\alpha_n\|u_n - u_{n-1}\| < +\infty \ \text{or} \ (\alpha_4)\frac{\alpha_n\|u_n - u_{n-1}\|}{\lambda_{n+1}} \to 0, \\
(E_1)\sum_{n=1}^{+\infty}\|e_n\| < +\infty \ \text{or} \ (E_2)\frac{\|e_n\|}{\lambda_{n+1}} \to 0,
\end{cases} \tag{9.6}$$

then $A^{-1}(0) \neq \emptyset$ *and* $\omega_w(w_n) \subset A^{-1}(0)$, *where* $\omega_w(w_n)$ *is the set of weak cluster points of* w_n.

Proof. Suppose $[x,y] \in A$; since $\{u_n\}$ is bounded, there is a subsequence $\{w_{n_j}\}$ of $\{w_n\}$ such that $w_{n_j} \rightharpoonup p \in H$. On the other hand, by the monotonicity of A, we have:

$$
\begin{aligned}
(x-w_{n_j},y) &= (x-(\sum_{i=0}^{n_j-1}\lambda_{i+1})^{-1}\sum_{i=0}^{n_j-1}\lambda_{i+1}u_{i+1},y)\\
&= (\sum_{i=0}^{n_j-1}\lambda_{i+1})^{-1}\sum_{i=0}^{n_j-1}\lambda_{i+1}(x-u_{i+1},y)\\
&= (\sum_{i=0}^{n_j-1}\lambda_{i+1})^{-1}\sum_{i=0}^{n_j-1}\lambda_{i+1}\big((x-u_{i+1},y-Au_{i+1})+(x-u_{i+1},Au_{i+1})\big)\\
&\geq (\sum_{i=0}^{n_j-1}\lambda_{i+1})^{-1}\sum_{i=0}^{n_j-1}(x-u_{i+1},\lambda_{i+1}Au_{i+1})\\
&= (\sum_{i=0}^{n_j-1}\lambda_{i+1})^{-1}\sum_{i=0}^{n_j-1}(x-u_{i+1},u_i-u_{i+1}+\alpha_i(u_i-u_{i-1})+e_i)\\
&= (\sum_{i=0}^{n_j-1}\lambda_{i+1})^{-1}\sum_{i=0}^{n_j-1}\big[-(1+\alpha_i)(u_{i+1}-x,u_i-x)\\
&\quad+\alpha_i(u_{i+1}-x,u_{i-1}-x)+\|u_{i+1}-x\|^2+(x-u_{i+1},e_i)\big]\\
&= (\sum_{i=0}^{n_j-1}\lambda_{i+1})^{-1}\sum_{i=0}^{n_j-1}\big[-(u_{i+1}-x,u_i-x)+\alpha_i(u_{i+1}-x,u_{i-1}-u_i)\\
&\quad+\|u_{i+1}-x\|^2+(x-u_{i+1},e_i)\big]\\
&\geq (\sum_{i=0}^{n_j-1}\lambda_{i+1})^{-1}\sum_{i=0}^{n_j-1}\big[\frac{1}{2}\|u_{i+1}-x\|^2-\frac{1}{2}\|u_i-x\|^2+\frac{1}{2}\|u_{i+1}-u_i\|^2\\
&\quad-\alpha_i\|u_{i+1}-x\|\|u_i-u_{i-1}\|-\|e_i\|\|u_{i+1}-x\|\big]\\
&\geq (\sum_{i=0}^{n_j-1}\lambda_{i+1})^{-1}\sum_{i=0}^{n_j-1}\big[\frac{1}{2}(\|u_{i+1}-x\|^2-\|u_i-x\|^2)\\
&\quad-\alpha_i\|u_{i+1}-x\|\|u_i-u_{i-1}\|-\|e_i\|\|u_{i+1}-x\|\big]\\
&\geq (\sum_{i=0}^{n_j-1}\lambda_{i+1})^{-1}\big[-\frac{1}{2}\|u_0-x\|^2-\sum_{i=0}^{n_j-1}\alpha_i\|u_{i+1}-x\|\|u_i-u_{i-1}\|\\
&\quad-\sum_{i=0}^{n_j-1}\|e_i\|\|u_{i+1}-x\|\big].
\end{aligned}
$$

Letting $j\to\infty$, by (9.6) we get $(x-p,y)\geq 0$. Therefore by the maximality of A, we have $p\in A^{-1}(0)$, as desired. □

Remark 9.2.3 *If $e_n \equiv 0$ in (9.4) and the conditions $(\alpha_1), (\alpha_2)$, $[(\alpha_3)$ or $(\alpha_4)]$ and (Λ_1) are satisfied, then by Proposition 9.2.2 and part (a) of Proposition 9.2.1, $\{u_n\}$ is bounded if and only if $A^{-1}(0) \neq \emptyset$.*

Theorem 9.2.4 *Let $\{u_n\}$ be a bounded sequence generated by (9.4). Suppose that the conditions (α_1), (α_2), $[(\alpha_3)$ or $(\alpha_4)]$ and (Λ_1) are satisfied. If $(E_1)\sum_{n=1}^{+\infty}\|e_n\| < +\infty$, then $w_n \rightharpoonup p \in A^{-1}(0)$ as $n \to \infty$, which is also the asymptotic center of $\{u_n\}$.*

Proof. By Proposition 9.2.2, $A^{-1}(0) \neq \emptyset$ and $\omega_w(w_n) \subset A^{-1}(0)$, thus by part (b) of Proposition 9.2.1, there exists $\lim_n \|u_n - p\|$, $\forall p \in \omega_w(w_n)$. We show that $\omega_w(w_n)$ is singleton. Suppose $p, q \in \omega_w(w_n)$ and $p \neq q$, then there exists $\lim_n(\|u_n - p\|^2 - \|u_n - q\|^2)$, hence $\lim_{n\to+\infty}(u_n, p-q)$ exists. It follows that $\lim_{n\to+\infty}(w_n, p-q)$ exists. This implies that $(q, p-q) = (p, p-q)$ and hence $p = q$. So, $w_n \rightharpoonup p \in A^{-1}(0)$ as $n \to +\infty$. Now, we show that p is the asymptotic center of $\{u_n\}$. Suppose that $q \in H$ and $q \neq p$, then

$$\|u_n - p\|^2 = \|u_n - q\|^2 + 2(u_n, q-p) + \|p\|^2 - \|q\|^2.$$

Multiplying both sides of the above equality by λ_n, summing up from $n = 1$ to $n = m$ and dividing by $\sum_{n=1}^{m}\lambda_n$, then taking limsup as $m \to +\infty$, we get:

$$\begin{aligned}\lim_{n\to+\infty}\|u_n - p\|^2 &= \limsup_{m\to+\infty}\Big(\sum_{n=1}^{m}\lambda_n\Big)^{-1}\Big(\sum_{n=1}^{m}\lambda_n\|u_n - q\|^2\Big) - \|q-p\|^2 \\ &< \limsup_{n\to+\infty}\|u_n - q\|^2.\end{aligned}$$

This shows that p is the asymptotic center of the sequence $\{u_n\}$ as desired. □

Remark 9.2.5 *By Remark 9.2.3, if $e_n \equiv 0$, we can replace the boundedness of u_n by $A^{-1}(0) \neq \emptyset$ in Theorem 9.2.4.*

9.3 WEAK CONVERGENCE OF THE ALGORITHM WITH ERRORS

The main result of this section is to prove the weak convergence of the sequence $\{u_n\}$ generated by (9.4) to a zero of the maximal monotone operator A, provided that some suitable assumptions on the parameters λ_n and α_n, as well as appropriate assumptions on the error sequence $\{e_n\}$ hold.

Lemma 9.3.1 *Suppose that $\{u_n\}$ is a bounded sequence generated by (9.4). If the conditions*

$$(\alpha_5)\sum_{n=1}^{+\infty}\frac{\alpha_n^2\|u_n - u_{n-1}\|^2}{\lambda_{n+1}^2} < +\infty \text{ and } (E_3)\sum_{n=1}^{+\infty}\frac{\|e_n\|^2}{\lambda_{n+1}^2} < +\infty$$

are satisfied, then there exists $\lim_{n\to+\infty}\|Au_n\|$.

Proof. By the monotonicity of A, we have

$$(Au_{n+1} - Au_n, u_{n+1} - u_n) \geq 0.$$

The Equation (9.4) implies that

$$\big(Au_{n+1} - Au_n, \alpha_n(u_n - u_{n-1}) + e_n - \lambda_{n+1}Au_{n+1}\big) \geq 0.$$

Then

$$\begin{aligned}
\|Au_{n+1}\|^2 &\leq (Au_n, Au_{n+1}) + \big(Au_{n+1} - Au_n, \frac{\alpha_n}{\lambda_{n+1}}(u_n - u_{n-1}) + \frac{e_n}{\lambda_{n+1}}\big) \\
&\leq \frac{1}{2}\|Au_n\|^2 + \frac{1}{2}\|Au_{n+1}\|^2 - \frac{1}{2}\|Au_n - Au_{n+1}\|^2 + \frac{1}{2}\|Au_n - Au_{n+1}\|^2 \\
&\quad + \frac{1}{2}\|\frac{\alpha_n}{\lambda_{n+1}}(u_n - u_{n-1}) + \frac{e_n}{\lambda_{n+1}}\|^2 \\
&\leq \frac{1}{2}\|Au_n\|^2 + \frac{1}{2}\|Au_{n+1}\|^2 + \frac{\alpha_n^2}{\lambda_{n+1}^2}\|u_n - u_{n-1}\|^2 + \frac{\|e_n\|^2}{\lambda_{n+1}^2}.
\end{aligned}$$

Therefore

$$\|Au_{n+1}\|^2 \leq \|Au_n\|^2 + 2\frac{\alpha_n^2}{\lambda_{n+1}^2}\|u_n - u_{n-1}\|^2 + 2\frac{\|e_n\|^2}{\lambda_{n+1}^2}. \tag{9.7}$$

Now it follows from (α_5) and (E_3) that there exists $\lim_{n\to+\infty}\|Au_n\|$. □

Proposition 9.3.2 *Let $\{u_n\}$ be a bounded sequence generated by* (9.4). *If* (α_5), (E_3), *and the following conditions hold:*

$$\begin{cases}
(\Lambda_2)\sum_{n=1}^{+\infty}\lambda_n^2 = +\infty, \\
(\alpha_3)\sum_{n=1}^{+\infty}\alpha_n\|u_n - u_{n-1}\| < +\infty \ \text{or} \ (\alpha_6)\frac{\alpha_n\|u_n - u_{n-1}\|}{\lambda_{n+1}^2} \to 0, \\
(E_1)\sum_{n=1}^{+\infty}\|e_n\| < +\infty \ \text{or} \ (E_4)\frac{\|e_n\|}{\lambda_{n+1}^2} \to 0,
\end{cases} \tag{9.8}$$

then $\lim_{n\to+\infty}\|Au_n\| = 0$, $A^{-1}(0) \neq \varnothing$ *and* $\omega_w(u_n) \subset A^{-1}(0)$.

Proof. Set $L = \sup_{n\geq 1}\|Au_n\|$. By Proposition 9.2.2, $A^{-1}(0) \neq \varnothing$ (because $(\Lambda_2) \Rightarrow (\Lambda_1)$, $(\alpha_5) \Rightarrow (\alpha_4)$ and $(E_3) \Rightarrow (E_2)$). Assume that $p \in A^{-1}(0)$. By the monotonicity of A, we have:

$$(\lambda_{n+1}Au_{n+1}, u_{n+1} - p) \geq 0.$$

From (9.4), we get:

$$\big(u_n - u_{n+1} + \alpha_n(u_n - u_{n-1}) + e_n, u_{n+1} - p\big) \geq 0,$$

which implies that

$$(\alpha_n(u_n - u_{n-1}) + e_n, u_{n+1} - p) \geq (u_{n+1} - u_n, u_{n+1} - p).$$

Using (9.4), we get:

$$(\alpha_n(u_n-u_{n-1})+e_n,u_{n+1}-p)\geq(\alpha_n(u_n-u_{n-1})+e_n-\lambda_{n+1}Au_{n+1},u_{n+1}-p),$$

hence

$$\begin{aligned}
&2\alpha_n\|u_n-u_{n-1}\|\|u_{n+1}-p\|+2\|e_n\|\|u_{n+1}-p\|+\|u_n-p\|^2-\|u_{n+1}-p\|^2\\
&\geq\|\alpha_n(u_n-u_{n-1})+e_n-\lambda_{n+1}Au_{n+1}\|^2\\
&=\lambda_{n+1}^2\|\frac{\alpha_n}{\lambda_{n+1}}(u_n-u_{n-1})+\frac{e_n}{\lambda_{n+1}}-Au_{n+1}\|^2\\
&=\lambda_{n+1}^2[\|\frac{\alpha_n}{\lambda_{n+1}}(u_n-u_{n-1})+\frac{e_n}{\lambda_{n+1}}\|^2+\|Au_{n+1}\|^2\\
&\quad-2(\frac{\alpha_n}{\lambda_{n+1}}(u_n-u_{n-1})+\frac{e_n}{\lambda_{n+1}},Au_{n+1})]\\
&\geq\lambda_{n+1}^2[\|Au_{n+1}\|^2-2\|\frac{\alpha_n}{\lambda_{n+1}}(u_n-u_{n-1})+\frac{e_n}{\lambda_{n+1}}\|\|Au_{n+1}\|]\\
&\geq\lambda_{n+1}^2[\|Au_{n+1}\|^2-2\frac{\alpha_n}{\lambda_{n+1}}\|u_n-u_{n-1}\|\|Au_{n+1}\|-2\frac{\|e_n\|}{\lambda_{n+1}}\|Au_{n+1}\|].
\end{aligned}$$

So

$$\begin{aligned}
\lambda_{n+1}^2\|Au_{n+1}\|^2\leq\;&2L\lambda_{n+1}\alpha_n\|u_n-u_{n-1}\|+2L\lambda_{n+1}\|e_n\|\\
&+2\alpha_n\|u_n-u_{n-1}\|\|u_{n+1}-p\|+2\|e_n\|\|u_{n+1}-p\|\\
&+\|u_n-p\|^2-\|u_{n+1}-p\|^2.
\end{aligned}\tag{9.9}$$

Summing up both sides of (9.9) from $n=1$ to $n=k$ and then dividing by $\sum_{n=1}^k\lambda_{n+1}^2$, since $\|Au_n\|\to l$ as $n\to+\infty$ (by Lemma 9.3.1), it follows that $(\sum_{n=1}^k\lambda_{n+1}^2)^{-1}\sum_{n=1}^k\lambda_{n+1}^2\|Au_{n+1}\|^2\to l$ as $k\to+\infty$. Then by the assumptions and Lemma 9.1.1, the right hand side tends to zero, hence we get $l=0$. Now, if $u_{n_j}\rightharpoonup q$, then by the demiclosedness of A, we get $q\in A^{-1}(0)$, hence $\omega_w(u_n)\subset A^{-1}(0)$. □

Remark 9.3.3 *Proposition 9.3.2 shows that if $A^{-1}(0)$ is a singleton (which happens if, for example, A is strictly monotone), then $u_n\rightharpoonup p$, where p is the unique element of $A^{-1}(0)$.*

In the following theorem, we obtain the weak convergence of the sequence $\{u_n\}$ generated by (9.4). The result is an extension of the result of Theorem 2.1 of [ALV-ATT].

Theorem 9.3.4 *Let $\{u_n\}$ be a bounded sequence generated by (9.4). Suppose that the conditions (Λ_2), $(\alpha_1),(\alpha_2),(\alpha_5)$, (E_1) and (E_3) are satisfied. Then $u_n\rightharpoonup p\in A^{-1}(0)$ as $n\to\infty$.*

Proof. By Proposition 9.2.2, $A^{-1}(0)\neq\varnothing$. Assume that $p\in A^{-1}(0)$. By the monotonicity of A, we have

$$\big(u_{n+1}-u_n-\alpha_n(u_n-u_{n-1})-e_n,u_{n+1}-p\big)\leq 0.$$

Thus

$$
\begin{aligned}
\lambda_{n+1}^2\|Au_{n+1}\|^2+\|u_{n+1}-p\|^2 &\le \|u_n-p+\alpha_n(u_n-u_{n-1})+e_n\|^2\\
&=\|u_n-p\|^2+\alpha_n^2\|u_n-u_{n-1}\|^2+\|e_n\|^2\\
&\quad+2\alpha_n(u_n-p,u_n-u_{n-1})+2\alpha_n(e_n,u_n-u_{n-1})\\
&\quad+2(u_n-p,e_n)\\
&\le \|u_n-p\|^2+\alpha_n^2\|u_n-u_{n-1}\|^2+\|e_n\|^2\\
&\quad+\alpha_n\|u_n-p\|^2+\alpha_n\|u_n-u_{n-1}\|^2\\
&\quad-\alpha_n\|u_{n-1}-p\|^2+\alpha_n\|u_n-u_{n-1}\|^2\\
&\quad+\alpha_n\|e_n\|^2+2\|e_n\|\|u_n-p\|.
\end{aligned}
$$

Therefore

$$
\begin{aligned}
\lambda_{n+1}^2\|Au_{n+1}\|^2 &\le \|u_n-p\|^2-\|u_{n+1}-p\|^2+3\alpha_n\|u_n-u_{n-1}\|^2\\
&\quad+\alpha_n(\|u_n-p\|^2-\|u_{n-1}-p\|^2)+2\|e_n\|^2+2\|e_n\|\|u_n-p\|\\
&\le \|u_n-p\|^2-\|u_{n+1}-p\|^2+3\alpha_n\|u_n-u_{n-1}\|^2\\
&\quad+\alpha[\|u_n-p\|^2-\|u_{n-1}-p\|^2]_+ +2\|e_n\|^2+2\|e_n\|\|u_n-p\|.
\end{aligned}
$$

Summing up both sides of the above inequality from $n=1$ to k, we get:

$$
\begin{aligned}
\sum_{n=1}^{k}\lambda_{n+1}^2\|Au_{n+1}\|^2 &\le \sum_{n=1}^{k}(\|u_n-p\|^2-\|u_{n+1}-p\|^2)+3\sum_{n=1}^{k}\alpha_n\|u_n-u_{n-1}\|^2\\
&\quad+\alpha\sum_{n=1}^{k}[\|u_n-p\|^2-\|u_{n-1}-p\|^2]_+ +2\sum_{n=1}^{k}\|e_n\|^2\\
&\quad+2\sum_{n=1}^{k}\|e_n\|\|u_n-p\|.
\end{aligned}
$$

Letting $k\to\infty$, by part (b) of Proposition 9.2.1 and the assumptions, we obtain

$$
\sum_{n=1}^{\infty}\lambda_{n+1}^2\|Au_{n+1}\|^2<+\infty, \tag{9.10}
$$

which implies that $\liminf_n\|Au_n\|=0$. Since by Lemma 9.3.1, $\lim_n\|Au_n\|$ exists, then $\lim_n\|Au_n\|=0$. Now if $u_{n_j}\rightharpoonup q$, then by the demiclosedness of A, we have $q\in A^{-1}(0)$, hence $\omega_w(u_n)\subset A^{-1}(0)$. Now a similar proof as in Theorem 9.2.4 shows that $\omega_w(u_n)$ is a singleton. Therefore $u_n\rightharpoonup p\in A^{-1}(0)$. □

Remark 9.3.5 *Theorem 9.3.4 in exact form ($e_n\equiv 0$) extends Theorem 2.1 of [ALV-ATT], because if $e_n\equiv 0$, $A^{-1}(0)\neq\emptyset$ and $\lambda_n\ge\lambda>0$, then (Λ_2) is satisfied and*

$$
\sum_{n=1}^{\infty}\frac{\alpha_n^2\|u_n-u_{n-1}\|^2}{\lambda_{n+1}^2}\le\frac{\alpha}{\lambda^2}\sum_{n=1}^{\infty}\alpha_n\|u_n-u_{n-1}\|^2<+\infty,
$$

by (α_1) *and* (α_2)*. Then* (α_5) *is also satisfied.*

9.4 SUBDIFFERENTIAL CASE

In this section, we establish the weak convergence of the sequence $\{u_n\}$ generated by (9.4) to an element of $A^{-1}(0)$, when $A = \partial\varphi$.

Lemma 9.4.1 *Suppose that* $\{u_n\}$ *is a bounded sequence generated by (9.4) and* $A = \partial\varphi$*, where* $\varphi : H \to (-\infty, +\infty]$ *is a proper, convex and lower semicontinuous function. If the conditions*

$$(\alpha_7)\sum_{n=1}^{+\infty}\frac{\alpha_n^2\|u_n-u_{n-1}\|^2}{\lambda_{n+1}} < +\infty \text{ and } (E_5)\sum_{n=1}^{+\infty}\frac{\|e_n\|^2}{\lambda_{n+1}} < +\infty.$$

are satisfied, then there exists $\lim_{n\to+\infty}\varphi(u_n)$.

Proof. By the subdifferential inequality and (9.4), we get

$$\begin{aligned}
\lambda_{n+1}(\varphi(u_{n+1})-\varphi(u_n)) &\le (\lambda_{n+1}\partial\varphi(u_{n+1}), u_{n+1}-u_n)\\
&= (u_n-u_{n+1}+\alpha_n(u_n-u_{n-1})+e_n, u_{n+1}-u_n)\\
&= -\|u_{n+1}-u_n\|^2+(\alpha_n(u_n-u_{n-1})+e_n, u_{n+1}-u_n)\\
&\le \frac{1}{2}\|\alpha_n(u_n-u_{n-1})+e_n\|^2\\
&\le \alpha_n^2\|u_n-u_{n-1}\|^2+\|e_n\|^2.
\end{aligned}$$

Hence

$$\varphi(u_{n+1}) \le \varphi(u_n)+\frac{\alpha_n^2}{\lambda_{n+1}}\|u_n-u_{n-1}\|^2+\frac{\|e_n\|^2}{\lambda_{n+1}}. \tag{9.11}$$

It follows from the assumptions (α_7), (E_5) that there exists $\lim_{n\to+\infty}\varphi(u_n)$. □

The following proposition is the subdifferential version of Proposition 9.3.2.

Proposition 9.4.2 *Let* $\{u_n\}$ *be a bounded sequence generated by* (9.4) *and* $A = \partial\varphi$*. If* (α_7)*,* (E_5)*, and the following conditions hold:*

$$\begin{cases}
(\Lambda_1)\sum_{n=1}^{+\infty}\lambda_n = +\infty,\\
(\alpha_3)\sum_{n=1}^{+\infty}\alpha_n\|u_n-u_{n-1}\| < +\infty \text{ or } (\alpha_4)\frac{\alpha_n\|u_n-u_{n-1}\|}{\lambda_{n+1}} \to 0,\\
(E_1)\sum_{n=1}^{+\infty}\|e_n\| < +\infty \text{ or } (E_2)\frac{\|e_n\|}{\lambda_{n+1}} \to 0,
\end{cases} \tag{9.12}$$

then $\lim\varphi(u_n) = \inf_{x\in H}\varphi$*,* $(\partial\varphi)^{-1}(0) \neq \emptyset$ *and* $\omega_w(u_n) \subset (\partial\varphi)^{-1}(0)$.

Proof. By Proposition 9.2.2, $(\partial\varphi)^{-1}(0) \neq \emptyset$. Assume that $p \in (\partial\varphi)^{-1}(0)$; then we have:

$$\lambda_{n+1}(\varphi(u_{n+1})-\varphi(p)) \le (\lambda_{n+1}\partial\varphi(u_{n+1}), u_{n+1}-p)$$

$$= (u_n - u_{n+1} + \alpha_n(u_n - u_{n-1}) + e_n, u_{n+1} - p)$$
$$= (u_n - p, u_{n+1} - p) + (p - u_{n+1}, u_{n+1} - p)$$
$$+ (\alpha_n(u_n - u_{n-1}), u_{n+1} - p) + (e_n, u_{n+1} - p).$$

So,

$$\lambda_{n+1}(\varphi(u_{n+1}) - \varphi(p)) \leq \frac{1}{2}\|u_n - p\|^2 - \frac{1}{2}\|u_{n+1} - p\|^2 + \alpha_n\|u_n - u_{n-1}\|\|u_{n+1} - p\| + \|e_n\|\|u_{n+1} - p\|. \quad (9.13)$$

Summing up both sides of (9.13) from $n = 1$ to $n = k$ and dividing by $\sum_{n=1}^{k} \lambda_{n+1}$, then letting $k \to \infty$, since by Lemma 9.4.1, $\lim_{n\to+\infty} \varphi(u_n) - \varphi(p) = l$ exists, then

$$\lim_{k\to+\infty} (\sum_{n=1}^{k} \lambda_{n+1})^{-1} \sum_{n=1}^{k} \lambda_{n+1}(\varphi(u_{n+1}) - \varphi(p)) = \lim_{k\to+\infty} (\varphi(u_n) - \varphi(p)) = l.$$

It follows now from the assumptions that $l = 0$. Thus $\lim_k \varphi(u_k) = \varphi(p)$. Therefore, if $u_{n_j} \rightharpoonup q$, then $\varphi(q) \leq \liminf_j \varphi(u_{n_j}) = \varphi(p)$, which implies that $q \in (\partial\varphi)^{-1}(0)$. Hence $\omega_w(u_n) \subset (\partial\varphi)^{-1}(0)$. □

The following theorem shows that in the special case $A = \partial\varphi$, which is important from the optimization point of view, the conditions (Λ_2) and (α_5) in Theorem 9.3.4 can be replaced by the weaker conditions (Λ_1) and (α_7).

Theorem 9.4.3 *Let $\{u_n\}$ be a bounded sequence generated by (9.4) and $A = \partial\varphi$. Assume that the conditions $(\Lambda_1), (\alpha_1), (\alpha_2), (\alpha_7), (E_1)$ and $(E_5) \sum_{n=1}^{+\infty} \frac{\|e_n\|^2}{\lambda_{n+1}} < +\infty$ are satisfied. Then $u_n \rightharpoonup p \in (\partial\varphi)^{-1}(0)$ as $n \to \infty$.*

Proof. By Proposition 9.2.2, $A^{-1}(0) \neq \emptyset$. Assume that $p \in (\partial\varphi)^{-1}(0)$; then we have:

$$\lambda_{n+1}(\varphi(u_{n+1}) - \varphi(p)) \leq (\lambda_{n+1}\partial\varphi(u_{n+1}), u_{n+1} - p)$$
$$= (u_n - u_{n+1} + \alpha_n(u_n - u_{n-1}) + e_n, u_{n+1} - p)$$
$$= (p - u_{n+1}, u_{n+1} - p)$$
$$+ (u_n - p + \alpha_n(u_n - u_{n-1}) + e_n, u_{n+1} - p)$$
$$\leq -\frac{1}{2}\|u_{n+1} - p\|^2 + \frac{1}{2}\|u_n - p + \alpha_n(u_n - u_{n-1}) + e_n\|^2$$
$$= -\frac{1}{2}\|u_{n+1} - p\|^2 + \frac{1}{2}\|u_n - p\|^2 + \frac{\alpha_n^2}{2}\|u_n - u_{n-1}\|^2$$
$$+ \frac{1}{2}\|e_n\|^2$$
$$+ (u_n - p, e_n) + \alpha_n(u_n - u_{n-1}, u_n - p)$$
$$+ (\alpha_n(u_n - u_{n-1}), e_n)$$

$$\leq \frac{1}{2}\|u_n - p\|^2 - \frac{1}{2}\|u_{n+1} - p\|^2 + 3\frac{\alpha_n}{2}\|u_n - u_{n-1}\|^2$$
$$+ \frac{\alpha_n}{2}\|u_n - p\|^2 - \frac{\alpha_n}{2}\|u_{n-1} - p\|^2 + \|e_n\|^2$$
$$+ \|e_n\|\|u_n - p\|.$$

So

$$\lambda_{n+1}(\varphi(u_{n+1}) - \varphi(p)) \leq \frac{1}{2}\|u_n - p\|^2 - \frac{1}{2}\|u_{n+1} - p\|^2 + 3\frac{\alpha_n}{2}\|u_n - u_{n-1}\|^2$$
$$+ \frac{\alpha}{2}[\|u_n - p\|^2 - \|u_{n-1} - p\|^2]_+$$
$$+ \|e_n\|^2 + \|e_n\|\|u_n - p\|. \tag{9.14}$$

On the other hand, by Lemma 9.4.1, there exists $\lim_{n\to+\infty}(\phi(u_n) - \phi(p))$. Let $\lim_{n\to+\infty}(\phi(u_n) - \phi(p)) = l$. Summing up both sides of (9.14) from $n = 1$ to $n = k$, dividing by $\sum_{n=1}^{k} \lambda_{n+1}$, and letting $k \to +\infty$, we get:

$$\lim_{k\to+\infty} (\sum_{n=1}^{k} \lambda_{n+1})^{-1} \sum_{n=1}^{k} \lambda_{n+1}(\phi(u_n) - \phi(p)) = l.$$

It follows now from Proposition 9.2.1 (b) and the assumptions on $\{\lambda_n\}$, $\{\alpha_n\}$ and $\{e_n\}$ that $l = 0$. Therefore $\lim_{n\to+\infty} \phi(u_n) = \phi(p)$. Now if $u_{n_j} \rightharpoonup q$, then $\varphi(q) \leq \liminf_j \varphi(u_{n_j}) = \varphi(p)$, which implies that $q \in (\partial\varphi)^{-1}(0)$. Hence $\omega_w(u_n) \subset (\partial\varphi)^{-1}(0)$. The rest of the proof is now similar to that of Theorem 9.2.4. □

Remark 9.4.4 *Obviously* $(\Lambda_2) \Rightarrow (\Lambda_1)$*; on the other hand, by* (α_1), (α_2) *and* (α_5), *we have:*

$$\sum_{n=1}^{+\infty} \frac{\alpha_n^2\|u_n - u_{n-1}\|^2}{\lambda_{n+1}} \leq (\sum_{n=1}^{+\infty} \frac{\alpha_n^2\|u_n - u_{n-1}\|^2}{\lambda_{n+1}^2})^{\frac{1}{2}}(\sum_{n=1}^{+\infty} \alpha_n^2\|u_n - u_{n-1}\|^2)^{\frac{1}{2}}$$
$$\leq \alpha^{\frac{1}{2}}(\sum_{n=1}^{+\infty} \frac{\alpha_n^2\|u_n - u_{n-1}\|^2}{\lambda_{n+1}^2})^{\frac{1}{2}}(\sum_{n=1}^{+\infty} \alpha_n\|u_n - u_{n-1}\|^2)^{\frac{1}{2}} < +\infty.$$

Then (α_7) *is satisfied. This shows that, in the subdifferential case and when* $e_n \equiv 0$, *the weak convergence of* u_n *is proved with weaker assumptions on the parameters, than in Theorem 9.3.4.*

9.5 STRONG CONVERGENCE

In this final section, with additional assumptions on the maximal monotone operator A, we show the strong convergence of the bounded sequence $\{u_n\}$ generated by (9.4) to a zero of A.

Theorem 9.5.1 *Assume that* $\{u_n\}$ *is a bounded sequence generated by* (9.4). *If* $(I + A)^{-1}$ *is a compact operator and the conditions* $(\Lambda_2), (\alpha_1), (\alpha_3)$ *and* (E_1) *are satisfied, then* $u_n \to p \in A^{-1}(0)$.

Proof. By (9.10) and the assumptions, we get: $\liminf_n \|Au_n\| = 0$. Therefore there exists a subsequence $\{Au_{n_j}\}$ of $\{Au_n\}$ such that $\|Au_{n_j}\| \to 0$ and $\{u_{n_j} + Au_{n_j}\}$ is bounded. Since $(I+A)^{-1}$ is compact, $\{u_{n_j}\}$ has a strongly convergent subsequence (denoted again by $\{u_{n_j}\}$) to some $p \in H$. The maximality of A implies that $p \in A^{-1}(0)$. On the other hand, by part (b) of Proposition 9.2.1, $\lim_n \|u_n - p\|$ exists. Hence $u_n \to p \in A^{-1}(0)$. □

Theorem 9.5.2 *Let $\{u_n\}$ be a bounded sequence generated by* (9.4) *and A be a maximal monotone and strongly monotone operator. If (Λ_1), (α_3) and (E_1) are satisfied, then $u_n \to p$, where p is the unique element of $A^{-1}(0)$.*

Proof. By Proposition 9.2.2, $A^{-1}(0) \neq \varnothing$. Assume that p is the single element of $A^{-1}(0)$. By the strong monotonicity of A and (9.4), we get:

$$\beta \lambda_{n+1} \|u_{n+1} - p\|^2 \leq (u_n - u_{n+1} + \alpha_n(u_n - u_{n-1}) + e_n, u_{n+1} - p),$$

for some $\beta > 0$. It follows that

$$\begin{aligned} 2\beta\lambda_{n+1}\|u_{n+1} - p\|^2 \leq \|u_n - p\|^2 - \|u_{n+1} - p\|^2 \\ + 2\alpha_n \|u_n - u_{n-1}\| \|u_{n+1} - p\| + 2\|e_n\| \|u_{n+1} - p\|. \end{aligned}$$

Now summing up both sides of the above inequality from $n = 1$ to $n = k$, dividing by $\sum_{n=1}^{k} \lambda_{n+1}$, and letting $k \to \infty$, since by Proposition 9.2.1 (c), we know that $\lim_{n\to\infty} \|u_n - p\| = l$ exists, it follows that the left hand side is equal to l. Using the assumptions and Lemma 9.1.1, it follows that the right hand side is equal to zero. Therefore $l = 0$, and the proof is now complete. □

REFERENCES

ALV-ATT. F. Alvarez and H. Attouch, An inertial proximal method for maximal monotone operators via discretization of a nonlinear oscillator with damping, Set-Valued Anal. 9 (2001), 3–11.

JUL-MAI. F. Jules, P. E. Maingé, Numerical approach to a stationary solution of a second order dissipative dynamical system, Optimization 51 (2002), 235–255.

KHA-RAN. H. Khatibzadeh and S. Ranjbar, Inexact inertial proximal algorithm for maximal monotone operators. An. Ştiinţ. Univ. "Ovidius" Constanţa Ser. Mat. 23 (2015), 133–146.

MAI. P. E. Maingé, Convergence theorems for inertial KM-type algorithms, J. Comput. Appl. Math. 219 (2008), 223–236.

Part IV

Applications

10 Some Applications to Nonlinear Partial Differential Equations and Optimization

10.1 INTRODUCTION

In this final chapter of the book, we present some applications of the first and second order evolution equations of monotone type to minimization problems, variational problems and partial differential equations.

10.2 APPLICATIONS TO CONVEX MINIMIZATION AND MONOTONE OPERATORS

Consider the convex minimization problem:

$$\mathrm{Min}_{x\in H}\varphi(x), \tag{10.1}$$

where $\varphi : H \to (-\infty, +\infty]$ is a convex, proper and lower semicontinuous function. There are several methods for solving the above minimization problem. In this section, we consider some dynamic methods by using the solutions to first and second order evolution equations governed by the subdifferential of the convex function φ. Consider the first order system:

$$\begin{cases} -u'(t) \in A(u(t)), & t \in (0, +\infty), \\ u(0) = u_0 \in \overline{D(A)}. \end{cases} \tag{10.2}$$

As we saw in Chapter 4 of the book, solutions to (10.2) are not necessarily weakly convergent for a general monotone operator A. Bruck [BRU] proved the weak convergence of the solutions for a demipositive operator A, which includes the case when $A = \partial\varphi$, where φ is a proper, convex and lower semicontinuous function. In Chapter 5, we saw that the solutions to the following second order system:

$$\begin{cases} u''(t) \in Au(t), \quad t \in (0, +\infty), \\ u(0) = u_0 \in \overline{D(A)}, \quad \sup_{t\geq 0}\|u(t)\| < +\infty, \end{cases} \tag{10.3}$$

converge weakly for a general maximal monotone operator A. Then (10.2) and (10.3) provide dynamical approaches for finding a minimizer of the convex function φ.

From a numerical point of view, it is important to obtain strong convergence results, because weak open neighborhoods are unbounded in infinite dimensional Hilbert spaces. Güler [GUL1] for (10.2) and Véron [VER1] for (10.3) showed that the strong convergence does not hold in general, unless if we impose some additional assumptions on the convex function φ or the monotone operator A. (see Chapter 5 of the book). But as we showed in Theorem 5.8.11 of Chapter 5 of the book, the solutions to the second order evolution equation

$$\begin{cases} u''(t) + cu'(t) \in Au(t), & t \in (0, +\infty), \\ u(0) = u_0 \in \overline{D(A)}, & \sup_{t \geq 0} \|u(t)\| < +\infty, \end{cases} \tag{10.4}$$

with $c > 0$ converge strongly to a minimum point of φ. Moreover by Remark 5.8.12 of Chapter 5, the solutions to (10.4) converge to a zero of A exponentially. Therefore (10.4) seems more suitable for approximating a minimizer of φ.

This section contains three subsections. In the first subsection, we study the rate of convergence of the solutions to (10.2), (10.3) and (10.4) in the case when $A = \partial\varphi$, to a minimum point of the proper, convex and lsc function φ. In the second subsection, we study the rate of convergence for strongly maximal monotone operators. Finally in the third subsection, we show how one can apply (10.3) or (10.4) to approximate a zero of the monotone operator A.

10.2.1 RATE OF CONVERGENCE

This section is devoted to the study and comparison of the rate of convergence of the solutions to (10.2), (10.3) and (10.4).

Güler [GUL2] studied the gradient flow for (10.2) for the case that $A = \partial\varphi$. He proved that $\varphi(x(t)) - \varphi(p) = o(\frac{1}{t})$, provided that $x(t)$ converges strongly to p, where p is a minimum point of φ. Here, we give a simpler proof for Güler's result, without assuming the strong convergence of $x(t)$. First we prove some elementary lemmas.

Lemma 10.2.1 *Let $f : \mathbb{R}^+ \to \mathbb{R}^+$ be a bounded and differentiable function, then* $\liminf_{t\to+\infty} tf'(t) \leq 0$.

Proof. Suppose to the contrary that $\liminf_{t\to+\infty} tf'(t) \geq \lambda > 0$. Then, there exists $t_0 > 0$ such that for all $t \geq t_0 > 0$, $tf'(t) \geq \lambda$. Dividing by t and integrating from $t = t_0$ to T, we get

$$f(T) - f(t_0) \geq \lambda(\ln T - \ln t_0).$$

By letting $T \to +\infty$, we get a contradiction. □

Lemma 10.2.2 *Let $f, g : \mathbb{R}^+ \to \mathbb{R}^+$. Assume that f is nonincreasing and $\lim_{t\to+\infty} f(t) = 0$. If $\int_0^{+\infty} f(s)g(s)ds < +\infty$, then*

$$\lim_{t\to+\infty} f(t)\left(\int_0^t g(s)ds\right) = 0.$$

Proof. By assumption, for all $0 < t < T$, we have

$$f(T)(\int_0^T g(s)ds) \leq f(T)(\int_0^t g(s)ds) + \int_t^T f(s)g(s)ds.$$

The result follows by taking limsup as $T \to +\infty$, and then liminf as $t \to +\infty$. □

Theorem 10.2.3 *Suppose that $u(t)$ is a solution to* (10.2), *then $\varphi(u(t)) - \varphi(p) = o(\frac{1}{t})$, where p is a minimum point of φ.*

Proof. By the subdifferential inequality and (10.2), we get:

$$\varphi(u(t)) - \varphi(p) \leq (-u'(t), u(t) - p) = \frac{-1}{2}\frac{d}{dt}\|u(t) - p\|^2.$$

Integrating the above inequality from $t = 0$ to $t = T$, we get

$$\begin{aligned}\int_0^T (\varphi(u(t)) - \varphi(p))dt &\leq \frac{1}{2}(-\|u(T) - p\|^2 + \|u(0) - p\|^2) \\ &\leq \frac{1}{2}\|u(0) - p\|^2.\end{aligned}$$

This implies that

$$\int_0^{+\infty} (\varphi(u(t)) - \varphi(p))dt < +\infty. \tag{10.5}$$

Then $\liminf_{t \to +\infty} \varphi(u(t)) - \varphi(p) = 0$. On the other hand, by Lemma 2.2, pp. 57–58 of [MOR], we get

$$\frac{d}{dt}\varphi(u(t)) = (\partial\varphi(u(t)), u'(t)) = -\|u'(t)\|^2 \leq 0.$$

This shows that $\varphi(u(t)) - \varphi(p)$ is nonincreasing. Therefore

$$\lim_{t \to \infty} \varphi(u(t)) - \varphi(p) = 0.$$

Now the theorem follows from (10.5) and Lemma 10.2.2. □

The following theorem shows that the second order evolution Equation (10.3) provides a faster rate of convergence for $\varphi(u(t))$ to $\varphi(p)$ than (10.2).

Theorem 10.2.4 *Suppose that $u(t)$ is a solution to* (10.3), *then $\varphi(u(t)) - \varphi(p) = o(\frac{1}{t^2})$, where p is a minimum point of φ.*

Proof. By (10.3), for all $t, h > 0$, we get

$$u''(t+h) - u''(t) \in \partial\varphi(u(t+h)) - \partial\varphi(u(t)).$$

Multiplying both sides of the above inclusion by $u(t+h) - u(t)$, by the monotonicity of $\partial\varphi$, we get: $\frac{d^2}{dt^2}\|u(t+h) - u(t)\|^2 \geq 0$. Therefore $\|u(t+h) - u(t)\|^2$ is convex and bounded. Then, $\|u(t+h)-u(t)\|^2$ is nonincreasing i.e. for all $t>s>0$ and $h>0$, we have $\|u(t+h)-u(t)\|^2 \le \|u(s+h)-u(s)\|^2$. Dividing both sides of this inequality by h^2 and letting $h\to 0$, we get: $\|u'(t)\|^2$ is nonincreasing. Now, by Lemma 2.2, pp. 57–58 of [MOR], we get

$$\frac{d}{dt}\varphi(u(t)) = (\partial\varphi(u(t)),u'(t)) = (u''(t),u'(t)) = \frac{1}{2}\frac{d}{dt}\|u'(t)\|^2 \le 0. \tag{10.6}$$

By the subdifferential inequality and (10.3), we get:

$$\varphi(u(t))-\varphi(p) \le (u''(t),u(t)-p) = \frac{1}{2}\frac{d^2}{dt^2}\|u(t)-p\|^2 - \|u'(t)\|^2.$$

Multiplying both sides of the above inequality by t, integrating from $t=0$ to $t=T$, and integrating by parts, we get:

$$\begin{aligned}\int_0^T t(\varphi(u(t))-\varphi(p))dt &\le \frac{1}{2}T\frac{d}{dT}\|u(T)-p\|^2 - \frac{1}{2}\int_0^T \frac{d}{dt}\|u(t)-p\|^2 dt\\ &= \frac{1}{2}T\frac{d}{dT}\|u(T)-p\|^2 - \frac{1}{2}\|u(T)-p\|^2 + \frac{1}{2}\|u(0)-p\|^2.\end{aligned}$$

Taking liminf as $T\to+\infty$, by Lemma 10.2.1, we get

$$\int_0^\infty t(\varphi(u(t))-\varphi(p))dt < +\infty. \tag{10.7}$$

Therefore $\liminf_{t\to+\infty}\varphi(u(t))-\varphi(p)=0$. (10.6) implies that $\lim_{t\to+\infty}\varphi(u(t))-\varphi(p)=0$. Now the theorem follows from (10.7) and Lemma 10.2.2. □

Lemma 10.2.5 *If* $u(t)$ *is a solution to* (10.4), *then*

$$\frac{d}{dt}\|u'(t)\| + c\|u'(t)\| \le 0.$$

Proof. Let h be a small positive constant, and denote $y(t) := u(t+h)-u(t)$. By the monotonicity of A, we get

$$(y''(t),y(t)) + c(y'(t),y(t)) \ge 0.$$

Therefore

$$\frac{1}{2}\frac{d^2}{dt^2}\|y(t)\|^2 + \frac{c}{2}\frac{d}{dt}\|y(t)\|^2 \ge \|y'(t)\|^2 \ge 0. \tag{10.8}$$

By the proof of Theorem 5.8.11 of Chapter 5, $\|y(t)\|$ is nonincreasing. If there exists t_0 such that $\|y(t_0)\|=0$, then $\|y(t)\|=0, \forall t\ge t_0$. Therefore, eventually

$$c\int_r^t \|y(s)\|ds \le \|y(r)\| - \|y(t)\| \tag{10.9}$$

Otherwise $\|y(t)\|>0$ for all $t>0$. Then (10.8) implies that

$$\frac{1}{2}\frac{d}{dt}(2\|y(t)\|\frac{d}{dt}\|y(t)\|) + c\|y(t)\|\frac{d}{dt}\|y(t)\| \ge \|y'(t)\|^2.$$

Then

$$(\frac{d}{dt}\|y(t)\|)^2 + \|y(t)\|\frac{d^2}{dt^2}\|y(t)\| + c\|y(t)\|\frac{d}{dt}\|y(t)\| \geq \|y'(t)\|^2.$$

Since $\|y'(t)\|^2 \geq (\frac{(y'(t),y(t))}{\|y(t)\|})^2 = (\frac{d}{dt}\|y(t)\|)^2$, we get

$$\|y(t)\|(\frac{d^2}{dt^2}\|y(t)\| + c\frac{d}{dt}\|y(t)\|) \geq 0.$$

If $\|y(t)\| > 0, \forall t > 0$, then

$$\frac{d^2}{dt^2}\|y(t)\| + c\frac{d}{dt}\|y(t)\| \geq 0.$$

Hence $\|y(t)\| + c\int_0^t \|y(s)\|ds$ is a convex function. Since by Remark 5.8.12 of Chapter 5, for $p \in \text{Argmin}\varphi$, $\|u(t) - p\| \leq \frac{2}{c}\|u'(0)\|e^{\frac{-c}{2}t}$, we get

$$\begin{aligned}\|y(t)\| + c\int_0^t \|y(s)\|ds &\leq 2(M + \|p\|) + c\int_0^t \big[\|u(s+h) - p\| + \|u(s) - p\|\big]ds \\ &\leq 2(M + \|p\|) + \frac{8}{c}\|u'(0)\| < +\infty,\end{aligned}$$

where $M = \sup_{t\geq 0}\|u(t)\|$. Therefore, $\|y(t)\| + c\int_0^t \|y(s)\|ds$ is bounded from above. By a result in convex analysis, we have

$$\|y(t)\| + c\int_0^t \|y(s)\|ds \leq \|y(r)\| + c\int_0^r \|y(s)\|ds, \quad \forall t \geq r, \tag{10.10}$$

which implies (10.9). Dividing both sides of (10.9) by h and letting $h \to 0$, by Fatou's Lemma, we get:

$$\begin{aligned}c\int_r^t \|u'(s)\|ds &= c\int_r^t \liminf_{h\to 0}\frac{\|y(s)\|}{h}ds \\ &\leq c\liminf_{h\to 0}\int_r^t \frac{\|y(s)\|}{h}ds \\ &\leq \liminf_{h\to 0}(\frac{\|y(r)\|}{h} - \frac{\|y(t)\|}{h}) \\ &= \|u'(r)\| - \|u'(t)\|.\end{aligned}$$

Hence $\|u'(t)\| + c\int_0^t \|u'(s)\|ds$ is nonincreasing. Therefore

$$\frac{d}{dt}\|u'(t)\| + c\|u'(t)\| \leq 0,$$

as desired. □

Theorem 10.2.6 *Assume that $u(t)$ is a solution to (10.4); then $\|u'(t)\| = O(e^{-ct})$ and $\|u(t) - p\| = O(e^{-ct})$, where $p \in A^{-1}(0)$ is the strong limit of $u(t)$, as $t \to +\infty$.*

Proof. By Lemma 10.2.5, we get

$$\frac{d}{dt}(\|u'(t)\|e^{ct}) \leq 0 \Rightarrow \|u'(t)\| \leq e^{-ct}\|u'(0)\|.$$

Then $\|u'(t)\| = O(e^{-ct})$. On the other hand, by Theorem 5.8.11 of Chapter 5, we get

$$\begin{aligned}
\|u(t)-p\| &= \lim_{T\to+\infty}\|u(t)-u(T)\| \\
&= \lim_{T\to+\infty}\|\int_t^T u'(s)ds\| \\
&\leq \lim_{T\to+\infty}\int_t^T \|u'(s)\|ds \\
&\leq \lim_{T\to+\infty}\|u'(0)\|\int_t^T e^{-cs}ds \\
&= \frac{1}{c}e^{-ct}\|u'(0)\|.
\end{aligned}$$

Hence $\|u(t)-p\| = O(e^{-ct})$. □

Theorem 10.2.7 *Suppose that $u(t)$ is a solution to* (10.4)*; then*

$$\varphi(u(t)) - \varphi(p) = o(e^{-ct}),$$

where $p \in A^{-1}(0)$ is the strong limit of $u(t)$, as $t \to +\infty$.

Proof. By the subdifferential inequality and (10.4), we get:

$$\begin{aligned}
\varphi(u(t)) - \varphi(p) &\leq (u''(t)+cu'(t), u(t)-p) \\
&= \frac{1}{2}\frac{d^2}{dt^2}\|u(t)-p\|^2 - \|u'(t)\|^2 + \frac{1}{2}c\frac{d}{dt}\|u(t)-p\|^2.
\end{aligned}$$

Multiplying both sides of the above inequality by e^{ct}, we get

$$e^{ct}(\varphi(u(t)) - \varphi(p)) \leq \frac{1}{2}\frac{d}{dt}(e^{ct}\frac{d}{dt}\|u(t)-p\|^2).$$

Integrating the above inequality from $t=0$ to $t=T$, we get

$$\begin{aligned}
\int_0^T e^{ct}(\varphi(u(t)) - \varphi(p))dt &\leq \frac{1}{2}e^{cT}\frac{d}{dT}\|u(T)-p\|^2 - (u'(0), u(0)-p) \\
&\leq e^{\frac{c}{2}T}\|u'(T)\|e^{\frac{c}{2}T}\|u(T)-p\| - (u'(0), u(0)-p).
\end{aligned}$$

Letting $T \to +\infty$, by Theorem 10.2.6, we have

$$\int_0^{+\infty} e^{ct}(\varphi(u(t)) - \varphi(p))dt \leq \|u'(0)\|\|u(0)-p\| < +\infty. \tag{10.11}$$

Then $\liminf_{t\to+\infty}\varphi(u(t))-\varphi(p)=0$. On the other hand, by Lemma 2.2, pp. 57–58 of [MOR] and Lemma 10.2.5, we get

$$\begin{aligned}\frac{d}{dt}\varphi(u(t)) &= (\partial\varphi(u(t)),u'(t))\\ &= (u''(t)+cu'(t),u'(t))\\ &= \frac{1}{2}\frac{d}{dt}\|u'(t)\|^2+c\|u'(t)\|^2\\ &= (\frac{d}{dt}\|u'(t)\|+c\|u'(t)\|)\|u'(t)\|\le 0,\end{aligned}$$

if $\|u'(t)\|>0$, for all $t>0$. But if there exists t_0 such that $\|u'(t_0)\|=0$, since $\|u'(t)\|$ is nonincreasing (see the proof of Theorem 5.8.11 of Chapter 5), $\|u'(t)\|=0$ for all $t\ge t_0$. Then

$$\begin{aligned}\frac{d}{dt}\varphi(u(t)) &= (\partial\varphi(u(t)),u'(t))\\ &= (u''(t)+cu'(t),u'(t))\\ &= \frac{1}{2}\frac{d}{dt}\|u'(t)\|^2+c\|u'(t)\|^2=0\end{aligned}$$

This shows that $\varphi(u(t))-\varphi(p)$ is nonincreasing. Therefore

$$\lim_{t\to\infty}\varphi(u(t))-\varphi(p)=0. \tag{10.12}$$

The theorem is proved by (10.11), (10.12) and Lemma 10.2.2. □

10.2.2 STRONGLY MONOTONE CASE

One of the most important methods for solving (10.1), called the regularization method, is to consider a family of auxiliary problems:

$$\mathrm{Min}_{x\in H}\varphi_\alpha(x), \tag{10.13}$$

where $\varphi_\alpha(x):=\varphi(x)+\frac{\alpha}{2}\|x\|^2$, $\alpha>0$. Obviously if φ is convex with $\mathrm{Argmin}\varphi\neq\emptyset$, then φ_α is strongly convex and coercive, i.e. $\lim_{\|x\|\to+\infty}\varphi_\alpha(x)=+\infty$. By Theorem 2.4.3 of Chapter 2, φ_α has a unique minimum point that we denote by x_α. It is well known in the literature, that x_α converges strongly to a point x^* as $\alpha\to 0$, which is an element of $\mathrm{Argmin}(\varphi)$ with minimum norm (in fact, $x_\alpha=J_{\frac{1}{2\alpha}}0$, i.e. the resolvent of φ of order $\frac{1}{2\alpha}$ at 0, see [TIK, MOR]). Therefore the solutions to (10.13) approximate the least norm solution of (10.1). On the other hand, as we mentioned in the introduction, solutions to (10.2), (10.3) and (10.4) converge to a minimum point of φ. This convergence, in both cases of (10.2) and (10.3) is weak in general (except for (10.4)), and is strong when φ satisfies some additional conditions such as being strongly convex or even (see Theorems 4.10.1 and 4.10.2 of Chapter 4 and Theorem 5.7.24 of Chapter 5). Now, we plan to study and compare the convergence rate of the solutions to (10.2), (10.3) and (10.4), when φ is strongly convex, or even more generally, when A is α-strongly monotone.

Theorem 10.2.8 *Suppose that $u(t)$ is a solution to* (10.2)*, where A is α-strongly monotone, and p is the unique element of $A^{-1}(0)$, then $\|u(t)-p\| = O(e^{-\alpha t})$.*

Proof. Multiplying both sides of (10.2) by $u(t)-p$, and using the α-strong monotonicity of A, we get:

$$\alpha\|u(t)-p\|^2 \leq \frac{-1}{2}\frac{d}{dt}\|u(t)-p\|^2. \tag{10.14}$$

Since $\|u(t)-p\|$ is nonincreasing, if $\|u(t_0)-p\| = 0$ for $t_0 > 0$, then $\|u(t)-p\| = 0$, for all $t \geq t_0$ and the result holds. Otherwise, dividing both sides of (10.14) by $\|u(t)-p\|^2$, we get:

$$\frac{d}{dt}\ln\|u(t)-p\|^2 \leq -2\alpha.$$

Integrating from 0 to T, we get:

$$\ln\|u(T)-p\|^2 - \ln\|u(0)-p\|^2 \leq -2\alpha T. \tag{10.15}$$

Therefore

$$\|u(T)-p\| \leq \|u(0)-p\|e^{-\alpha T},$$

which yields the theorem. □

Theorem 10.2.9 *Suppose that $u(t)$ is a solution to* (10.4) *with $c \geq 0$, where A is α-strongly monotone, and p is the unique element of $A^{-1}(0)$, then $\|u(t)-p\| = o(e^{-\beta t})$, for each $c < \beta < \frac{c}{2} + \sqrt{\frac{c^2}{4} + \alpha}$, where c is the positive constant appearing in* (10.4).

Proof. Multiplying both sides of (10.4) by $u(t)-p$, and using the α-strong monotonicity of A, we get:

$$\alpha\|u(t)-p\|^2 + \|u'(t)\|^2 \leq \frac{1}{2}\frac{d^2}{dt^2}\|u(t)-p\|^2 + \frac{1}{2}c\frac{d}{dt}\|u(t)-p\|^2. \tag{10.16}$$

If there exists $t_0 > 0$ such that $\|u(t_0)-p\| = 0$, then $\|u(t)-p\| = 0$ for all $t \geq t_0$ and the result holds. Now suppose that $\|u(t)-p\| > 0$ for each $t > 0$. From (10.16), we have

$$\begin{aligned}
&\frac{d}{dt}(\|u(t)-p\|\frac{d}{dt}\|u(t)-p\|) + c\|u(t)-p\|\frac{d}{dt}\|u(t)-p\| \geq \|u'(t)\|^2 + \alpha\|u(t)-p\|^2 \\
&\Rightarrow (\frac{d}{dt}\|u(t)-p\|)^2 + \|u(t)-p\|\frac{d^2}{dt^2}\|u(t)-p\| + c\|u(t)-p\|\frac{d}{dt}\|u(t)-p\| \\
&\qquad \geq \|u'(t)\|^2 + \alpha\|u(t)-p\|^2.
\end{aligned}$$

Since $(\frac{d}{dt}\|u(t)-p\|)^2 \leq (\frac{(u'(t),u(t)-p)}{\|u(t)-p\|})^2 \leq \|u'(t)\|^2$, and $\|u(t)-p\| > 0$, we obtain

$$\frac{d^2}{dt^2}\|u(t)-p\| + c\frac{d}{dt}\|u(t)-p\| \geq \alpha\|u(t)-p\|. \tag{10.17}$$

Multiplying both sides of (10.17) by $e^{\beta t}$ for $c < \beta < \frac{c}{2} + \sqrt{\frac{c^2}{4} + \alpha}$ and integrating from 0 to T, we get:

$$\begin{aligned}\alpha \int_0^T e^{\beta t} \|u(t) - p\| dt &\leq e^{\beta T} \frac{d}{dT} \|u(T) - p\| - \frac{d}{dt} \|u(t) - p\|_{|t=0} \\ &\quad + (c - \beta) \int_0^T e^{\beta t} \frac{d}{dt} \|u(t) - p\| dt \\ &\leq C + (c - \beta) e^{\beta T} \|u(T) - p\| \\ &\quad - (c\beta - \beta^2) \int_0^T e^{\beta t} \|u(t) - p\| dt,\end{aligned}$$

for some constant C independent of T. Therefore $\int_0^{+\infty} (\alpha + c\beta - \beta^2) e^{\beta t} \|u(t) - p\| dt < +\infty$. Now the result follows from Lemma 10.2.2. □

Remark 10.2.10 *Theorems 10.2.8 and 10.2.9 show that the rate of convergence of solutions to* (10.4) *is much better than* (10.2)*, especially if c is chosen large enough. Even if $c = 0$, the rate of convergence of* (10.3) *is better than* (10.2)*, when* $0 < \alpha < 1$.

This advantage of (10.4) compared to (10.2), with respect to the rate of convergence, allows us to use (10.4) for the approximation of the unique minimum point of the strongly convex functions φ_α, which estimate a minimum point of φ, for sufficiently small α.

10.2.3 USING THE SECOND ORDER EVOLUTION EQUATION FOR APPROXIMATING A MINIMIZER

Even in $\mathbb{R}$, a simple example shows that the equation

$$\begin{cases} u''(t) + cu'(t) \in Au(t), \\ u(0) = u_0, \end{cases} \tag{10.18}$$

has infinitely many solutions, and only one of them is bounded. In general, we don't know of any method for finding this bounded solution. Each numerical method for solving (10.18) requires $u'(0)$ or some other additional condition, and for a given choice of $u'(0)$, the solution is not necessarily bounded. But by [APR2] (see also [APR1]), we know that if u_T is a solution to the two point boundary value problem below

$$\begin{cases} u_T''(t) + cu_T'(t) \in Au_T(t), \\ u_T(0) = u_0, \quad u_T(T) = v_T \end{cases} \tag{10.19}$$

on $[0, T]$, with $v_T = u_0$, then u_T converges uniformly to a solution u of (10.18) on each compact subinterval of $(0, +\infty)$, as $T \to +\infty$. Therefore we use (10.19) instead of (10.4) for approximating a zero of the monotone operator A, or a minimum point of φ_α. To this aim, we need to compute the solution u_T at an appropriate point in the interval $[0,T]$. To obtain a suitable point in the interval, let $L>0$ fixed and $L<S<T$. By relation (2.7) of [APR2], we have

$$\|u_T(t)-u_S(t)\|^2 \leq \frac{e^{cS}\|u_T(S)\|^2}{2\ln\frac{S}{L}}. \tag{10.20}$$

By relation (2.4) of [APR2], there exists $M>0$ independent of T and of $t\in[0,T]$ such that

$$e^{ct}\|u_T(t)\|^2 \leq M, \quad \forall t\in[0,T], \, \forall T>0. \tag{10.21}$$

Therefore $\lim_{T\to+\infty} u_T(t) = u(t)$, uniformly on every bounded interval $[0,L] \subset [0,+\infty)$. The function u is in fact the solution of (10.4). Now letting $T\to+\infty$ in (10.20), by (10.21), we get

$$\|u(t)-u_S(t)\|^2 \leq \frac{M}{2\ln\frac{S}{L}}. \tag{10.22}$$

In the above inequality, set $t=L=\sqrt{S}$, then

$$\|u(\sqrt{S})-u_S(\sqrt{S})\|^2 \leq \frac{M}{2\ln\sqrt{S}} = \frac{M}{\ln S}. \tag{10.23}$$

The last inequality shows that when S is sufficiently large, $u_S(\sqrt{S})$ is near $u(\sqrt{S})$. Since $u(\sqrt{S})$ converges to a zero of A (weakly when $c=0$ and strongly when $c>0$), as $S\to+\infty$, then $u_S(\sqrt{S})$ provides an approximation to a zero of A when S is sufficiently large. By Theorem 10.2.9 and Remark 10.2.10, although $u(\sqrt{S})$ approximates a zero of A with a rate of convergence better than (10.2) for large S, but we can compute only $u_S(\sqrt{S})$, which is an estimate to $u(\sqrt{S})$(by (10.23)). In addition, since $\|u_S(t)-p\|$, where p is a zero of A, is convex on $[0,S]$ (see Theorem 2.1.2 of [APR1]), we have

$$\|u_S(t)-p\| \leq \text{Max}\{\|u_S(S)-p\|, \|u_S(0)-p\|\} = \|u_0-p\|,$$

which shows the stability of the curve $u_S(t)$. To see an example and numerical implementation, the reader can consult [KHA-SHO].

10.3 APPLICATION TO VARIATIONAL PROBLEMS

Consider the second order evolution equation of the form

$$\begin{cases} u''(t) \in \partial\varphi(u(t)), \\ u(0)=x, \quad \sup_{t\geq 0}\|u(t)\| < +\infty \end{cases} \tag{10.24}$$

where $\varphi : H \to (-\infty,+\infty]$ is a proper, convex and lower semicontinuous function. In this brief section, we show that the solution to (10.24) is exactly the minimizer of the functional

$$F(v) = \int_0^{+\infty} (\frac{1}{2}\|v'(t)\|^2 + \varphi(v(t)))dt \tag{10.25}$$

on a suitable space.

Theorem 10.3.1 *Let $u(t)$, $t \geq 0$ be a solution to* (10.24). *Then u is the minimizer of the functional F defined in* (10.25) *on the space*

$$\mathscr{H} = \{v \in W^{2,2}_{\text{loc}}(\mathbb{R}^+;H); \frac{dv}{dt} \in L^2(\mathbb{R}^+;H) \text{ and } v(0) = x\}$$

Proof. By the subdifferential inequality for φ and the Equation (10.24), we get:

$$\begin{aligned}\varphi(u(t)) - \varphi(v(t)) &\leq (u''(t), u(t) - v(t)) \\ &= \frac{d}{dt}(u'(t), u(t) - v(t)) - \|u'(t)\|^2 + (u'(t), v'(t)) \\ &\leq \frac{d}{dt}(u'(t), u(t) - v(t)) - \frac{1}{2}\|u'(t)\|^2 + \frac{1}{2}\|v'(t)\|^2\end{aligned}$$

Integrating the above inequality from 0 to T, we get:

$$\begin{aligned}\int_0^T (\varphi(u(t)) + \frac{1}{2}\|u'(t)\|^2)dt &\leq \int_0^T (\varphi(v(t)) + \frac{1}{2}\|v'(t)\|^2)dt \\ &\quad + (u'(T), u(T) - v(T)) - (u'(0), u(0) - v(0))\end{aligned}$$

Now the result follows by letting T go to infinity, because $\|u'(T)\| \to 0$ as $T \to +\infty$. □

For more information, and the proof that the solution to a more general second order equation is the minimizer of a functional, we refer the interested reader to [APR1].

10.4 SOME APPLICATIONS TO PARTIAL DIFFERENTIAL EQUATIONS

The first and second order evolution equations have a lot of applications in partial differential equations. In this section we give only one concrete example for a first and a second order differential equation, as well as a difference equation that was discussed in this book. This section contains three subsections. In the first subsection, we give an example of a concrete partial differential equation that can be modeled by a first order nonlinear differential equation of monotone type. In the second subsection, we give an example of a second order differential inclusion associated to a maximal monotone operator. Finally in the last subsection, we give an application of the second order difference equation of monotone type discussed in Chapter 8. These examples were chosen from [MOR], [APR2, APR3](see also [APR1]).

10.4.1 A CONCRETE EXAMPLE OF THE FIRST ORDER EQUATION

Consider the nonlinear partial differential equation

$$\begin{cases} \frac{\partial^2 u}{\partial t^2} - \Delta u + \beta(\frac{\partial u}{\partial t}) \ni f(t,x), & (t,x) \in \mathbb{R}^+ \times \Omega \\ u(t,x) = 0, \quad (t,x) \in \mathbb{R}^+ \times \Gamma \\ u(0,x) = u_0, \quad \frac{\partial u}{\partial t}(0,x) = v_0, \quad x \in \Omega. \end{cases} \tag{10.26}$$

where $\Omega \subset \mathbb{R}^n$ is open and bounded with a sufficiently smooth boundary and $\beta : D(\beta) \subset \mathbb{R} \to \mathbb{R}$ is maximal monotone with $0 \in \beta(0)$. Equation (10.26) models the vibration of an elastic membrane with fixed boundary in the presence of a friction force $\beta(\frac{\partial u}{\partial t})$, where $f(t,x)$ on the right hand side represents an external force.
The Hilbert space $H = H_0^1(\Omega) \times L^2(\Omega)$ is equipped with the inner product

$$(U_1, U_2)_H = \int_\Omega (\nabla u_1, \nabla u_2)_{\mathbb{R}^n} dx + \int_\Omega v_1 v_2 dx$$

for every $U_1 = [u_1, v_1]$ and $U_2 = [u_2, v_2]$. By Proposition 1.3 of [MOR], $\beta = \partial j$ where $j : \mathbb{R} \to (-\infty, +\infty]$ is proper, convex and lower semicontinuous, and since $0 \in \beta(0)$ one can write the function j in the form

$$j(r) = \int_0^r \beta^0(s) ds.$$

Obviously, $\inf j = j(0) = 0$. Now consider the function $\varphi : H_0^1(\Omega) \to (-\infty, +\infty]$ defined by

$$\varphi(v) = \begin{cases} \int_\Omega j(v(x))dx, & \text{if } jov \in L^1(\Omega) \\ +\infty, & \text{otherwise} \end{cases}$$

Obviously φ is proper, convex and lower semicontinuous. By Proposition 3.5.1 of Chapter 3, the extension of φ to $L^2(\Omega)$ is also lower semicontinuous. The subdifferential $\tilde{\beta}$ of φ is defined by $\tilde{\beta} : D(\tilde{\beta}) \subset H_0^1(\Omega) \to H^{-1}(\Omega)$ with

$$z \in \tilde{\beta}(v) \Leftrightarrow \varphi(v) - \varphi(w) \leq (v - w, z)_{H_0^1(\Omega), H^{-1}(\Omega)}, \ \forall w \in H_0^1(\Omega).$$

By Theorem 3.4.2 of Chapter 3, $\tilde{\beta}$ is maximal monotone.

The Laplacian $-\Delta$ is an operator from $H_0^1(\Omega)$ to $H^{-1}(\Omega)$ defined by the bilinear form

$$a(v_1, v_2) = \int_\Omega (\nabla v_1, \nabla v_2)_{\mathbb{R}^n} dx, \quad \forall v_1, v_2 \in H_0^1(\Omega)$$

which is exactly the inner product of $H_0^1(\Omega)$. Therefore $-\Delta$ is precisely the canonical isomorphism from $H_0^1(\Omega)$ to $H^{-1}(\Omega)$, which is given by the Riesz representation Theorem (see Theorem 1.1.10 of Chapter 1). On the other hand, $L^2(\Omega)$ is isometrically isomorphic with its dual, and therefore algebrically and topologically we have $H_0^1(\Omega) \subset L^2(\Omega) \subset H^{-1}(\Omega)$. The first embedding is the canonical injection $i : H_0^1(\Omega) \to L^2(\Omega)$, and the second one is the adjoint map $i^* : L^2(\Omega) \to H^{-1}(\Omega)$. For simplicity, we denote $i^*(\omega)$ again by ω. By Rellich-Kondrachov theorem (see [ADA] p.143), the embedding $i : H_0^1(\Omega) \to L^2(\Omega)$ is compact, and by Schauder's theorem (see [YOS] p.282), $i^* : L^2(\Omega) \to H^{-1}(\Omega)$ is also compact. In other words, $H_0^1(\Omega)$ is compactly embedded in $H^{-1}(\Omega)$. Now define the operator $A : D(A) \subset H \to H$ by

$$D(A) = \{[p,q] \in H_0^1(\Omega) \times H_0^1(\Omega); \ (-\Delta p + \tilde{\beta}(q)) \cap L^2(\Omega) \neq \emptyset\}$$

$$A([p,q]) = \{-q\} \times (-\Delta p + \tilde{\beta}(q)) \cap L^2(\Omega), \ \forall [p,q] \in D(A).$$

It can be shown that A is maximal monotone (see [MOR]). Then the problem can be written as a first order evolution equation associated to A in the form:

$$\begin{cases} \frac{dU}{dt} + AU \ni F(t), \quad 0 < t < +\infty \\ U(0) = U_0 \end{cases}$$

where $F(t) = [0, f(t,\cdot)]$ with $f \in L^1(0,+\infty;\, L^2(\Omega))$, and $U_0 = [u_0, v_0] \in \overline{D(A)}$. To see more examples, we refer the interested reader to [MOR].

10.4.2 AN EXAMPLE OF A SECOND ORDER EVOLUTION EQUATION

Let $H = L^2(\Omega)$ where $\Omega \subseteq \mathbb{R}^n$ is a bounded domain with smooth boundary Γ. Let $j : \mathbb{R} \to (-\infty, +\infty]$ be a proper, convex and lower semicontinuous function, and $\beta = \partial j$. We assume for simplicity that $0 \in \beta(0)$. Define

$$Au = -\Delta u = -\sum_{i=1}^{n} \frac{\partial^2 u}{\partial x_i^2}$$

with

$$D(A) = \{u \in H^2(\Omega)\,,\ \frac{-\partial u}{\partial \eta}(x) \in \beta(u(x))\,,\ \text{a.e. on } \Gamma\}$$

where $(\frac{\partial u}{\partial \eta}(x))$ is the outward normal derivative to Γ at $x \in \Gamma$. It is known that $A = \partial\phi$, where $\phi : L^2(\Omega) \to (-\infty\,,\,+\infty]$, is the functional:

$$\phi(u) = \begin{cases} \frac{1}{2}\int_\Omega |\nabla u|^2 dx + \int_\Gamma \beta(u(x))d\sigma, & \text{if } u \in H^1(\Omega) \text{ and } \beta(u) \in L^1(\Gamma) \\ +\infty, & \text{otherwise.} \end{cases}$$

Consider the following equation

$$\begin{cases} \frac{\partial^2 u}{\partial t^2}(t,x) + \sum_i \frac{\partial^2 u}{\partial x_i^2}(t,x) = 0 & \text{a.e. on } \mathbb{R}^+ \times \Omega \\ -\frac{\partial u}{\partial \eta}(t,x) \in \beta u(t,x) & \text{a.e. on } \mathbb{R}^+ \times \Gamma \\ u(0,x) = u_0(x) & \text{a.e. on } \Omega \\ \sup_{t\geq 0} \int_\Omega u^2(t,x)dx < +\infty. \end{cases}$$

Then this is an example of a second order evolution equation.

10.4.3 AN EXAMPLE OF A SECOND ORDER DIFFERENCE EQUATION

Let $H = L^2(\Omega)$ where $\Omega \subseteq \mathbb{R}^n$ is a bounded domain with smooth boundary Γ. Denote by $\tilde{A}$ the operator from $W_0^{1,p}(\Omega)$ into $W^{-1,q}(\Omega)$(where $\frac{1}{p} + \frac{1}{q} = 1$, $p \geq 2$), given by

$$(\tilde{A}u, v) = \sum_{k=1}^{N} \int_\Omega |\frac{\partial u}{\partial x_k}|^{p-2} \frac{\partial u}{\partial x_k} \frac{\partial v}{\partial x_k} dx, \quad \forall u, v \in W_0^{1,p}(\Omega) \tag{10.27}$$

The operator $\tilde{A}$ coincides with the subdifferential of the convex and lower semicontinuous function

$$\varphi(u) = \frac{1}{p}\sum_{k=1}^{N}\int_{\Omega}|\frac{\partial u}{\partial x_k}|^p dx \tag{10.28}$$

Let A be the restriction of $\tilde{A}$ to $D(A) = W_0^{1,p}(\Omega)$. This operator is maximal monotone in $L^2(\Omega)$. Moreover, $A = \partial\psi$, where $\psi : L^2(\Omega) \to (-\infty, +\infty]$, is defined by

$$\psi(u) = \begin{cases} \varphi(u), & u \in W_0^{1,p}(\Omega), \\ +\infty, & \text{otherwise} \end{cases} \tag{10.29}$$

The sequence of the following boundary value problems

$$\begin{cases} u_{i+1} - (1+\theta_i)u_i + \theta_i u_{i-1} \in -c_i \sum_{k=1}^{N} \frac{\partial}{\partial x_k}(|\frac{\partial u_i}{\partial x_k}|^{p-2}\frac{\partial u_i}{\partial x_k}), & x \in \Omega \\ u_i(x) = 0, \quad x \in \partial\Omega, \\ u_0(x) = a(x), \quad \forall x \in \Omega, \quad \sup_{i\geq 1}\int_{\Omega} u_i^2(x)dx < +\infty \end{cases} \tag{10.30}$$

where $a \in L^2(\Omega)$, $c_i > 0$, $\theta_i > 0$, $\forall i \geq 1$, is an example of a second order difference equation of monotone type.

REFERENCES

ADA. R. Adams and J. J. F. Fournier, Sobolev spaces. First Edition, Academic Press, New York 1975. Second edition. Pure and Applied Mathematics (Amsterdam), 140. Elsevier/Academic Press, Amsterdam, 2003.

APR1. N. C. Apreutesei, Nonlinear second order evolution equations of monotone type and applications, Pushpa Publishing House, Allahabad, India, 2007.

APR2. N. C. Apreutesei, Second-order differential equations on half-line associated with monotone operators, J. Math. Anal. Appl. 223 (1998), 472–493.

APR3. N. C. Apreutesei, On a class of difference equations of monotone type, J. Math. Anal. Appl. 288 (2003), 833–851.

BAR. V. Barbu, Nonlinear semigroups and Differential Equations in Banach Spaces, Noordhoff, Leyden, 1976.

BRU. R. E. Bruck, Asymptotic convergence of nonlinear contraction semigroups in Hilbert space, J. Funct. Anal. 18 (1975), 15–26.

DJA-KHA1. B. Djafari Rouhani, H. Khatibzadeh, A strong convergence theorem for solutions to a nonhomogeneous second order evolution equation, J. Math. Anal. Appl. 363 (2010), 648–654.

GUL1. O. Güler, On the convergence of the proximal point algorithm for convex minimization, SIAM J. Control Optim. 29 (1991), 403–419.

GUL2. O. Güler, Convergence rate estimates for the gradient differential inclusion, Optim. Methods Softw. 20 (2005), 729–735.

KHA-SHO. H. Khatibzadeh and A. Shokri, On the first and second order strongly monotone dynamical systems and minimization problems, Optim. Methods and Soft. 30 (2015), 1303–1309.

MOR. G. Morosanu, Nonlinear Evolution Equations and Applications, Editura Academiei (and D. Reidel Publishing Company), Bucharest, 1988.

TIK. A. N. Tikhonov, V. Ya. Arsenin, Methods for the solution of ill-posed problems, Third edition, "Nauka", Moscow, 1986.

VER1. L. Véron, Un exemple concernant le comportement asymptotique de la solution bornée de l'équation $\frac{d^2u}{dt^2} \in \partial\varphi(u)$, Monatsh. Math. 89 (1980), 57–67.

YOS. K. Yosida, Functional analysis. Sixth edition. Grundlehren der Mathematischen Wissenschaften [Fundamental Principles of Mathematical Sciences], 123, Springer-Verlag, Berlin, New York, 1980.

Index

A

B

C

D

E

F

G

H

I

L

M

U

W

Y